LIVERPOOL INSTITUTE
OF HIGHER EDUCATION
LIBRARY
WOOLTON ROAD,
LIVERPOOL, L16 8ND

SHEETS OF MANY COLOURS

The mapping of Ireland's rocks
1750-1890

ROYAL DUBLIN SOCIETY SCIENTIFIC PUBLICATIONS

HISTORY OF SCIENCE
Historical Studies in Irish Science and Technology

Number 1
RICHARD GRIFFITH 1784-1878
Edited by G.L. Herries Davies & R.C. Mollan
1980, pp. vi + 221, ISBN 0 86027 005, IR£10.00*

Number 2
REPRINTED PAPERS IN BEE HUSBANDRY
Edited by R.C. Mollan
1980, pp. iv + 42, ISBN 0 86027 007 6, IR£3.00*

Number 3
JOHN TYNDALL — ESSAYS ON A NATURAL PHILOSOPHER
Edited by W.H. Brock, N.D. McMillan & R.C. Mollan
1981, pp. xii + 219, ISBN 0 86027 008 4, IR£10.00*

Number 4
SHEETS OF MANY COLOURS — THE MAPPING OF IRELAND'S ROCKS, 1750-1890
By. G.L. Herries Davies
1983, pp. xiv + 242, ISBN 0 86027 014 9, IR£15.00*

BIOLOGY/AGRICULTURE

PROCEEDINGS OF THE SYMPOSIUM ON GRASSLAND FAUNA
1978, pp. iv + 246, ISSN 0080 433 9, IR£12.00*

THE FLORA OF COUNTY CARLOW
By Evelyn Booth
1979, pp. viii + 172, ISBN 0 86027 003 3, Hardcover IR£6.50*, Softcover IR£3.50*

ENERGY MANAGEMENT AND AGRICULTURE
Edited by D.W. Robinson & R.C. Mollan
1982, pp. viii + 441, ISBN 0 86027 010 6, IR£25.00*

STUDIES ON IRISH VEGETATION
Edited by James White
1982, pp. iv + 368, ISBN 0 86027 009 2, IR£25.00*

FLORA OF CONNEMARA AND THE BURREN
By D.A. Webb & M.J.P. Scannell
Published by Cambridge University Press and the Royal Dublin Society
1983, pp. xlv + 322, ISBN 0 521 23395 X, IR£41.50*

A STUDENT'S GUIDE TO NORTH BULL ISLAND
By E. Walsh & D. Jeffrey
1983, pp. 44, ISBN 0 86027 012 2 IR£1.50*

JOURNAL OF LIFE SCIENCES
Vol. 1, 1979-80; Vol. 2, 1980-81; Vol. 3, 1982; Vol. 4, 1982-3; ISSN 0332 1800
Subscription per volume: Institutions IR£20.00*; Individuals IR£10.00*

*1983 prices, subject to change without notice.
Order from: Dr R. Charles Mollan, Royal Dublin Society, Science Section, Thomas Prior House, Ballsbridge, Dublin 4. **Add £1.50 for postage & package.*

ROYAL DUBLIN SOCIETY
HISTORICAL STUDIES IN IRISH SCIENCE AND TECHNOLOGY
NUMBER FOUR

Sheets of Many Colours

The Mapping of Ireland's rocks
1750-1890

GORDON L. HERRIES DAVIES

ROYAL DUBLIN SOCIETY
1983

© The Royal Dublin Society 1983
ISBN 0 86027 014 9

Maps are based on the Ordnance Survey, by permission
of the Government (Permit number 3.2.22).

Added E
910.509

Typeset by Print Prep Limited, Dublin.
Printed in the Republic of Ireland by Mount Salus Press Limited, Dublin.

FOREWORD

Sheets of Many Colours is the fourth volume in the Royal Dublin Society's series *Historical Studies in Irish Science and Technology*. The series was inaugurated in 1980 in order to draw attention to the achievements of Ireland's scientists and technologists of the past, to recognise their contributions to *scientia* and to knowledge internationally, and to help make known their part in the development of modern Ireland — a part which has not hitherto been given due credit in Irish historical studies.

This fourth volume marks a significant step in the evolution of the historical series, since it is the first one written by a single author. The first three books were, to an extent, 'manufactured'. *Richard Griffith, 1784-1878* was the edited proceedings of a symposium to mark the centenary of the death of this energetic Irishman. Indeed, the author of the present volume suggested the holding of the symposium in the first place and was senior editor of the book. *Reprinted Papers on Bee Husbandry* was a bound run-on of some interesting historical papers re-published in the Society's *Journal of Life Sciences* when that *Journal* was in its infancy and needed some extra papers to fill out a volume. In 1983, the problem is an oversupply of papers for the *Journal*: but the Bee book has been well received. One gratifying reviewer (admittedly a bee enthusiast) noted: 'Just very occasionally somebody does something... which is eminently worthwhile. This is just what the Royal Dublin Society has done'. *John Tyndall — Essays on a Natural Philosopher* resulted from the initiative of Dr Norman McMillan and his Tyndall Committee in Carlow, a group which has set the standard for other similar committees seeking to revive interest in their own particular distinguished 'sons of the parish'. (Apart from a few worthy naturalists, famous daughters of the parish seem, sadly, to be conspicuous by their absence.)

This volume, however, is different. The origin of the study owes nothing to the RDS Series: it would have been written anyway. It was in the author, and was not going to stay there. But it would seem to be important that there is now an Irish publishing house, using Irish setters and printers, which can publish such a work, based as it is firmly on Ireland, although with very significant if hitherto unrecognised international dimensions.

The importance of the Geological Survey of Ireland in promoting this publication cannot be overstated; in particular the personal support of the Director, Dr C.E. Williams, and of the past and present Assistant Directors, Dr D. Naylor and Dr R.R. Horne. The Society is pleased to acknowledge the substantial financial assistance of the Department of Industry and Energy towards the publication costs, obtained following the recommendation of the Survey Officers.

Professor Herries Davies is already an established figure in international science history, being author of *The Earth in Decay: a History of British Geomorphology 1578 to 1878* (Macdonald, London, 1969), and of numerous papers in various journals, including the Society's *Journal of Earth Sciences*, to which he contributed the first paper in Volume One (1978): 'The Earth Sciences in Irish Serial Publications 1787-1977'. He is also Chairman of the *National Committee for the History and Philosophy of Science* established by the Royal Irish Academy in 1980. The Society's historical series is fortunate to have the opportunity to include this major new work of Gordon Herries Davies in its listings.

But what of this new work? As an officer of the Royal Dublin Society, and a relative newcomer to its ranks, I found it particularly pleasing to note the enormous contributions made by the Society in encouraging the early flowering and subsequent development of Irish geological cartography. This is handsomely acknowledged, though realistically assessed, in the book. Indeed, to me the highlight of the volume is this discovery of the quality of early Irish work as it developed from geognostic to geological cartography. We can take the example of

William Tighe's map of County Kilkenny, published in one of the most successful of the Society's County surveys in 1802. The author notes:

> It is clear . . . that the map must occupy a position of honour among the earliest of all the geological maps produced within the British Isles.

And again, it:

> stands out as by far the most sophisticated piece of geological cartography to have been published within these islands down to 1802.

Perhaps my pleasure in reading this part of the story is too chauvanistic! It is important to recognise a more significant facet of the study, and one which makes this book of special historical value. This is the hitherto ignored importance to the development of British geology, and hence to the geology of the British Empire, of the power struggles that took place upon Irish territory. It is no overstatement, as demonstrated in Professor Herries Davies's meticulous research, to say that the key to the understanding of British geological mapping in the mid nineteenth century lies in the political battle for control of the Geological Survey in the 1843-5 period. Ireland had had an official Survey for more than a decade at this stage, and events in Ireland profoundly affected the structure of the British Survey as war was waged between the military Ordnance surveyors and the professional geologists.

Moving from historical significance to contemporary solace, it is a particular property of the story that it is immensely readable — and intelligible to the geological incognoscenti. Also, the choice of illustrations has been particularly carefully made in order to put a human face on the narrative. Read about the irascible George Henry Kinahan in Chapter Seven, then look at the expression on his wife's face in Figure 7.6, and add your own dimension to the story!

It now remains necessary only to acknowledge with thanks the help received in the preparation of this book for publication from Dr John Jackson of the RDS Committee of Science and its Industrial Applications, and from Eileen Byrne and Mary Butler in the Society's Science Secretariat.

R. Charles Mollan
Science Officer
Royal Dublin Society
June 1983

PREFACE

Around two hundred years ago geologists began first to take a serious interest in mapping the distribution of the rocks exposed at the surface of the earth. The story of the mapping of the rocks that form the British Isles has claimed the attention of many an historian, but virtually all of those historians have hailed from England, Scotland, or Wales. Hitherto scarcely so much as one Irish historian has arisen to add his own particular perspective to the telling of the tale. As a result the now internationally accepted account of the development of geological cartography within the British Isles is really a story marred by errors of interpretation, flawed by deficiencies of detail, and somewhat tainted with a pervasive British ethnocentricity. Unfamiliar with Irish geological literature, with Irish scientific institutions, or with Irish geologists, historians hailing from Cardiff, Edinburgh, and London have represented Ireland as a geological backwater of minimal concern to the historian of the geological map. In reality this is a travesty of the truth. Ireland actually nurtured some notable pioneers of geological cartography. She possessed an official geological survey some ten years before the establishment of a similar survey in England. In Griffith's quarter-inch map she gave birth to one of the world's finest geological maps produced as a private venture. During the early 1840s the entire future of geological mapping within the British Isles turned upon the outcome of events in Ireland, and later it was to Ireland that British geologists came to learn the art of six-inch field-mapping. Far from being a backwater where geologists merely applied skills that had been learned elsewhere, Ireland from 1750 onwards was in reality the scene of vigorous innovation within the realm of geological cartography. This book tells for the first time the story of those innovations and traces the fascinating history of Irish geological cartography down to the completion of the one-inch geological map of Ireland in 1890.

The writing of this book has taken me to libraries around the world from Galway in the west to Sydney in the east; from Edinburgh in the north to Melbourne in the south. The book nevertheless rests principally upon archival material preserved in Dublin libraries, and for assistance received over many years my thanks are due to the staffs of the libraries of Trinity College, the Royal Dublin Society, the Royal Irish Academy, and the National Library of Ireland. The later chapters of the book could never have been written in the absence of the comprehensive and enthralling series of guard-books and letter-books preserved in the archives of the Geological Survey of Ireland. There I am most grateful to Dr C.E. Williams, the Director of the Geological Survey of Ireland, for giving me access to the archives, and to Dr J.B. Archer, of the Survey's staff, in whose care the archives are now placed and who many times has drawn my attention to newly discovered material germane to my theme. I am also grateful for their interest in the project to Dr R.R. Horne, the Assistant Director of the Geological Survey of Ireland, and to his predecessor in that post, Dr D. Naylor. Sad acknowledgement must also be made to the late Commandant P.G. Madden of the Irish Ordnance Survey who gave me permission to consult and reproduce material preserved in the Survey's voluminous archives at Mountjoy House in Dublin's Phoenix Park. Among the many others who have rendered assistance and to whom my thanks are due there are the following individuals: Dr R.C. Cox and Dr G.D. Sevastopulo of Trinity College Dublin; Mr M. O'Meara formerly of the Geological Survey of Ireland; Commandant M.C. Walsh of the Ordnance Survey of Ireland; the Hon. A. Bonar Law, Mr J.F. Donnellan, and Dr J.S. Jackson, all of Dublin; Dr W.E. Nevill of University College Cork; Dr P.D. Ryan of University College Galway; Mr J. Griffith of Shannon Grove, County Limerick; Mr C.W.P. MacArthur of Marble Hill, County Donegal; Mr F.W. Dunning and Mr J.C. Thackray of the Geological Museum, the Institute of Geological Sciences, London; Mrs E.R. Nutt of the Geological Society of London; Mrs J. Pingree of Imperial College London;

Dr D.A. Bassett, Director of the National Museum of Wales; Mr N.E. Butcher of the Open University in Scotland; Miss C. Daniell; Colonel A.P. Daniell; Mrs J.M. Eyles; the late Dr V.A. Eyles; Mr P.J. McCartney; and Dr J. Parkin. The entire book was most efficiently typed by Eileen Russell, the diagrams were drawn by Martha Lyons, and the photography was executed by Terence Dunne, all three individuals being members of the staff of the Geography Department in Trinity College.

Two final debts of gratitude towards my Trinity colleagues remain to be mentioned. Firstly, Professor C.H. Holland took an interest in the project, read drafts of the concluding chapters, and made valued comments thereupon. Secondly, Professor J.H. Andrews read the entire work in typescript and allowed me to benefit from his own unrivalled knowledge of Irish cartographic history. Over the years Professor Andrews has regularly fed me with geological titbits unearthed during his own studies, it was he who guided me through the archives in Mountjoy House, and we were together when, many years ago, we found the then somewhat neglected guard-books and letter-books of the Geological Survey of Ireland lying in a top-floor room of 14 Hume Street. It is to Professor Andrews that my very special thanks must now be tendered.

In the story that follows two distinguished Irish institutions feature prominently: in their order of seniority, the Royal Dublin Society and the Geological Survey of Ireland. They both have proud records in the field of geological cartography and it gives me the greatest pleasure to see this book being published by the Royal Dublin Society with the financial assistance of the Geological Survey. To Dr Mollan of the Royal Dublin Society and to Dr Williams of the Geological Survey I must now express my heartfelt thanks.

Gordon L. Herries Davies

Trinity College
Dublin
18 June (Waterloo Day) 1983

CONTENTS

	Page
Foreword	v
Preface	vii
Contents	ix
List of Plates	xi
List of Figures	xi
Abbreviations	xii

1. **PIONEERS IN THE FIELD, 1750-1814** — 1
    - The Physico-Historical Society of Ireland — 1
    - William Hamilton and the Antrim basalts — 4
    - Geology and the Dublin Society — 5
    - Robert Fraser's County Wicklow — 8
    - William Tighe's County Kilkenny — 10
    - George Vaughan Sampson's County Londonderry — 13
    - Horatio Townsend's County Cork — 15
    - Richard Kirwan's proposed Mining Board — 17
    - Walter Stephens's central Leinster — 18
    - The Bog Commissioners and their maps — 20
    - The Dublin Society and the Leinster coalfield survey — 21
    - Ireland and the Geological Society of London — 27

2. **THE GENESIS OF GRIFFITH'S MAP, 1809-1830** — 35
    - Griffith's early years — 35
    - The map in vision — 38
    - New regional surveys — 40
    - Griffith and the base-map problem — 44
    - The lean years — 46
    - Griffith's earliest rivals — 51

3. **GRIFFITH'S UNOFFICIAL SURVEY, 1830-1857** — 57
    - The Valuation Survey — 57
    - The Railway Commission — 61
    - A cartographic masterpiece — 64
    - The map's later history — 72
    - Who made 'Griffith's Map'? — 77

4. **ORDNANCE SURVEY GEOLOGY, 1824-1845** — 85
    - Colby's strategy — 85
    - Colby's first defeat — 92
    - The Colonel's second sally — 95
    - The memoir in its heyday — 99
    - Portlock's Londonderry Report — 104
    - Captain James takes over — 106
    - Griffith creates a diversion — 115
    - Colby quits the field — 118

5.	A NEW SURVEY BREAKS GROUND, 1845-1850	126
	A team is chosen	127
	Mapping begins	132
	Exit James	137
	De La Beche's new broom	138
	Oldham sets to work	141
	The Survey's first publications	149
	Exit Oldham	152
6.	YEARS OF GREAT ACHIEVEMENT, 1850-1869	156
	Joseph Beete Jukes	156
	Conflict over scales	160
	Jukes's mapping programme	162
	Administrative changes	167
	The one-inch sheets	170
	The memoirs	172
	Geological problems	181
	Personal problems	184
	The death of Jukes	190
7.	THE TASK COMPLETED, 1869-1890	195
	The new Director	196
	The players in the final act	198
	Fresh scenery	205
	Old scenery revisited	208
	War in the wings	216
	The curtain falls	222
Epilogue		232
Index		233

LIST OF PLATES
Between pages 114-115 and 130-131

1. Tighe's map of County Kilkenny
2. Part of Griffith's quarter-inch map of 1855
3. The Geological Survey's half-inch map of County Carlow
4. The final one-inch Geological Survey sheet: Sheet 10

LIST OF FIGURES

		Page
1.1	Fraser's County Wicklow	9
1.2	Woodstock House	10
1.3	Sampson's County Londonderry	14
1.4	Townsend's County Cork	16
1.5	Griffith's Leinster Coal District	26
1.6	Northeastern Ireland by Berger, Buckland, and Conybeare	29
2.1	Griffith in 1840	37
2.2	Griffith's Connaught Coal District	42
2.3	Nimmo's geological map of Ireland	52
3.1	Griffith's map for the Railway Commissioners' *Atlas*	63
3.2	Part of Griffith's quarter-inch map of 1839	68
3.3	Ganly's field-stations	71
3.4	Ganly's contribution to Griffith's map	73
3.5	One of Ganly's letters to Griffith	74
3.6	'Phillipsia attacking Griffithides'	74
3.7	Subdivision of the Carboniferous Limestone 1839	76
3.8	Subdivision of the Carboniferous Limestone 1846	76
3.9	Griffith in old age	78
4.1	Thomas Colby	86
4.2	An Ordnance Survey panorama in County Londonderry	90
4.3	Part of an Ordnance Survey map of County Londonderry	92
4.4	Joseph E. Portlock	96
4.5	Part of Portlock's map from his Report of 1843	106
4.6	Sir Henry Thomas De La Beche	112
5.1	Henry James	128
5.2	James Flanagan	130
5.3	An early six-inch field-sheet from County Wexford	135
5.4	Thomas Oldham	139
5.5	An early six-inch field-sheet for Taghmon, County Wexford	144
5.6	The field-sheet for Taghmon in its final form	145
5.7	Number 51 St Stephen's Green East	146
6.1	J. Beete Jukes	157
6.2	One of G.V. Du Noyer's field-sheets	166
6.3	Sir Roderick I. Murchison	169
6.4	Progress of the Geological Survey's mapping	171
6.5	Staff of the Geological Survey about 1860	173
6.6	Production of the one-inch map	174
6.7	An early one-inch sheet: Sheet 153	176
6.8	Dates of publication of the one-inch map	180
7.1	Edward Hull	196
7.2	Number 14 Hume Street	199
7.3	Sir Andrew C. Ramsey	201
7.4	Sir Archibald Geikie	201
7.5	Changing interpretations of southwestern Ireland	211
7.6	Mr and Mrs G. Henry Kinahan	216

ABBREVIATIONS

ADB	*Australian Dictionary of Biography*
Advmt Sci.	*The Advancement of Science*
Ann. Ass. Am. Geogr.	*Annals of the Association of American Geographers*
Ann. Sci.	*Annals of Science*
Archs. Nat. Hist.	*Archives of Natural History*
Boulton Papers	In the Central Libraries, Birmingham
Br. J. Hist. Sci.	*The British Journal for the History of Science*
Brit. Mus.	British Museum
Bull. Geol. Soc. Am.	*Bulletin of the Geological Society of America*
Dawson Turner Collection	In the Department of Botany, British Museum (Natural History)
DLBP	*The De La Beche Papers*, in the Department of Geology, the National Museum of Wales, Cardiff
DNB	*Dictionary of National Biography*
DSB	*Dictionary of Scientific Biography*
Econ. Proc. R. Dubl. Soc.	*The Economic Proceedings of the Royal Dublin Society*
Ganly Papers	In the library of the Royal Irish Academy, Dublin
Geikie Papers	In the University Library, Edinburgh
Geogrl. J.	*The Geographical Journal*
Geol. Mag.	*The Geological Magazine*
Greenough Papers	In the University Library, Cambridge
GSI	Geological Survey of Ireland, Dublin
H.C.	Parliamentary Papers (House of Commons)
IGS	Institute of Geological Sciences, London
Ir. Geogr.	*Irish Geography*
Ir. Nat.	*The Irish Naturalist*
Ir. Nat. J.	*The Irish Naturalists' Journal*
J. Dep. Agric. Repub. Ire.	*Journal of the Department of Agriculture, Republic of Ireland*
J. Earth Sci. R. Dubl. Soc.	*Journal of Earth Sciences: Royal Dublin Society*
J. Geol. Soc. Dubl.	*Journal of the Geological Society of Dublin*
J. Life Sci. R. Dubl. Soc.	*Journal of Life Sciences: Royal Dublin Society*
J. R. Dubl. Soc.	*Journal of the Royal Dublin Society*
J. R. Geol. Soc. Irel.	*Journal of the Royal Geological Society of Ireland*
J. Soc. Biblphy. Nat. Hist.	*Journal of the Society for the Bibliography of Natural History*
Jl. Kerry Archaeol & Hist. Soc.	*Journal of the Kerry Archaeological and Historical Society*
Jl. R. Agric. Soc.	*Journal of the Royal Agricultural Society of England*
Lond. & Edinb. Phil. Mag.	*The London and Edinburgh Philosophical Magazine*
LP	*The Larcom Papers*, in the National Library of Ireland
McCoy Correspondence	In the Mitchell Library, Sydney
McCoy Papers	In the National Museum of Victoria, Melbourne
MMR	Manuscript Monthly Returns, in the Ordnance Survey, Phoenix Park, Dublin
Monteagle Papers	In the National Library of Ireland, Dublin
MP	*The Murchison Papers* in the Geological Society of London

NLI	National Library of Ireland
Notes Rec. R. Soc. Lond.	*Notes and Records of the Royal Society of London*
OSL	Typescript copies of two Ordnance Survey letter books. The copies were made in 1834 at the request of Dr F.J. North and they are now in the Department of Geology in the National Museum of Wales, Cardiff. The original letters were evidently destroyed by enemy action during World War II.
OSLR (I)	Ordnance Survey Letter Registers, Inwards. In the Ordnance Survey, Dublin.
OSLR (O)	Ordnance Survey Letter Registers, Outwards. In the Ordnance Survey, Dublin
PC	*The Phillips Correspondence*, in the Science Museum, Oxford
Phil. Mag.	*The Philosophical Magazine*
Phil. Trans. R. Soc.	*Philosophical Transactions of the Royal Society*
Pollock-Morris Papers	In the Geology Department, the Royal Scottish Museum, Edinburgh
Proc. Dubl. Soc.	*Proceedings of the Dublin Society*
Proc. Geol. Ass.	*Proceedings of the Geologists' Association*
Proc. Geol. Soc.	*Proceedings of the Geological Society of London*
Proc. R. Dubl. Soc.	*Proceedings of the Royal Dublin Society*
Proc. R. Instn Gt. Br.	*Proceedings of the Royal Institution of Great Britain*
Proc. R. Ir. Acad.	*Proceedings of the Royal Irish Academy*
Proc. R. Soc.	*Proceedings of the Royal Society*
Proc. Yorks. Geol. Soc.	*Proceedings of the Yorkshire Geological Society*
PROD	Public Record Office, Dublin
PROL	Public Record Office, London
Q. Jl. Geol. Soc. Lond.	*The Quarterly Journal of the Geological Society of London*
RDS	Royal Dublin Society
Rep. Br. Ass. Advmt. Sci.	*Report of the British Association for the Advancement of Science*
RGSIP	*The Royal Geological Society of Ireland Papers*, in the Department of Geology, Trinity College Dublin
RP	*The Ramsay Papers*, in the archives of Imperial College of Science and Technology, London
Scient. Proc. R. Dubl. Soc.	*The Scientific Proceedings of the Royal Dublin Society*
Scient. Trans. R. Dubl. Soc.	*The Scientific Transactions of the Royal Dublin Society*
SIN	Robert Lloyd Praeger, *Some Irish naturalists: a biographical note-book*, Dundalk, 1949
SP	*The Sedgwick Papers*, in the University Library, Cambridge
Trans. Cardiff Nat. Soc.	*Transactions of the Cardiff Naturalists' Society*
Trans. Dubl. Soc.	*Transactions of the Dublin Society*
Trans. Edinb. Geol. Soc.	*Transactions of the Edinburgh Geological Society*
Trans. Geol. Soc. Lond.	*Transactions of the Geological Society of London*
Trans. Geol. Soc. S. Afr.	*The Transactions of the Geological Society of South Africa*
Trans. Instn. Civ. Engrs. Ire.	*Transactions of the Institution of Civil Engineers of Ireland*
Trans. Manchr. Geol. Soc.	*Transactions of the Manchester Geological Society*
Trans. R. Ir. Acad.	*Transactions of the Royal Irish Academy*
Trinity College Board Minutes	In the Manuscript Department of the library of Trinity College Dublin
Victorian Nat.	*The Victorian Naturalist*
VOLB	Valuation Office Letter Books in the Public Record Office, Dublin

... it is good to be reminded that our present knowledge of the general geological structure of our island is by no means intuitive. When it is laid out before us on a map so that we can take it in almost at a glance, and it has become so familiar that we can hardly imagine it otherwise, we are apt to forget the toil that has been undergone, and the difficulties and obscurities that have been encountered by those who have prepared the way for us.

> Maxwell Henry Close in his Anniversary Address to the Royal Geological Society of Ireland delivered in the Royal Dublin Society's Lecture-Theatre at Leinster House, 17 February 1879

1

PIONEERS IN THE FIELD

1750-1814

Not much is known about Miss Susanna Drury. What we do know is that about 1740 she completed some splendid gouaches of the basaltic landscape at the Giant's Causeway in County Antrim — gouaches which the Dublin Society adjudged to be of such merit as to justify the award of a £25 premium to the artist.[1] A few years later — in February 1743/4 — those same gouaches became the basis of two magnificent and widely circulated engravings by François Vivarès. In the 1740s the Giant's Causeway was already an internationally famed phenomenon (the term 'a giant's causeway' was commonly employed for any area of columnar basalt wheresoever it might occur) but through this deployment of their skills Miss Drury and Mr Vivarès had served to focus attention yet more closely upon Ulster's basaltic coast. That attention rendered the Ulster basalts quite unique among Irish rocks; by the middle years of the eighteenth century they were without question the most observed, discussed, and illustrated of all Ireland's rocks. Other Irish rocks might have had far greater economic significance — the granites, the limestones, the mineralised slates, or the coal-bearing shales — but as yet no Irish rocks save the basalts had been deemed worthy of serious scientific investigation. As the eighteenth century moved into its second half all this was to change.

Throughout Europe from 1750 onwards there emerged a new and far more perceptive interest in the rocks of the earth's surface — not just in those rocks which displayed some local peculiarity but in *all* the rocks be they novel or commonplace, economically valuable or seemingly useless. Here two closely interconnected factors were at work. Firstly, there was a rapidly developing and purely scientific interest in rocks as being the key to a decipherment of the complex history of the earth's surface. Secondly, social and economic developments such as population growth and increasing industrialisation were for the first time riveting attention upon the earth's rocks as a natural resource whence there had to be obtained the basic essentials of the new order — essentials as varied as coal to fuel steam-engines, brick-clays for the construction of the spreading cities, and water-supplies for the burgeoning population. For both these reasons — the purely scientific and the severely practical — a knowledge of the distribution of the various rock types was a matter of prime importance, be that distribution either global in its purview or confined within some narrower compass. And how better to present this distributional information than incapsulated in cartographic form upon a geological map? Such maps were to become a feature of the new social and economic era. The concept of a geological map had made its earliest appearance upon the scene in a paper presented to the Royal Society of London on 12 March 1683 by Martin Lister[2] but the eighteenth century was well advanced before the earliest such maps were actually produced.[3] In Ireland the story of geological mapping begins among those of Susanna Drury's contemporaries who were members of the Physico-Historical Society of Ireland.

The Physico-Historical Society of Ireland

Ireland must have been one of the earliest nations in the world to display a concern for its international image. It was early in the 1740s that a group of Irish gentlemen began to feel disquiet at the regular presentation of their land as one populated by people who were primitive, barbarous, superstitious and unruly. In short they saw the image of Ireland being offered to the world through foreign publications as a grotesque caricature of reality.[4] Looking across the water they saw that the English county natural historians — and notably Robert Plot — had offered accurate accounts of their chosen territories; why, the Irishmen reasoned, should the same not be done for Ireland? The aforesaid group of Irish gentlemen therefore formed themselves into a loose association with the object of collecting materials for a book comparable with William Camden's *Britannia* and to be published under the title *Hibernia, or Ireland antient and modern*. To this end they despatched a circular 'to many curious and learned gentlemen in their several counties' soliciting their assistance in supplying information, and this move, we are told, elicited a good deal of material. The coverage of information nevertheless remained somewhat patchy and those involved in the project therefore prepared for circulation a list of specific questions seeking information upon a diversity of topics including epidemical diseases, meteors, holy wells and petrifying springs, tides and currents, tempests and hurricanes, thunder and lightning, echoes, rivers 'whether stony, gravelly, sandy, muddy?', waterfalls, the character of the lakes, the height and trend of mountains, whether there be any volcanoes, promontories 'whether hawks, eagles, &c. breed in them', soils, mines, woods, insects, birds, archaeology and manufactures. As still further indication of the type of information that was needed there seems also to have been circulated a skeleton account of County Down prepared, evidently, by one Charles Smith of whom more in a moment.

Again the results were disappointing, but the instigators of the project remained undeterred. They now concluded that their best hope of success lay in the formal establishment of a society founded expressly to give effect to their plan. On 14 April 1744, therefore, some twenty gentlemen gathered in the Lords' Committee Room of the Dublin Parliament House (today the Bank of Ireland in College Green) and there they founded the Physico-Historical Society of Ireland. The prime object of the society was now not to be the preparation of a single general volume covering the whole of Ireland, but rather to be the publication of detailed local surveys with one volume devoted to each of the thirty-two Irish counties. The hope was expressed that the appearance of such county volumes would encourage industrious foreigners to settle in Ireland, thus stimulating the national economy. Indeed, the members held that their activities would materially assist the work of national improvement that was already being conducted by the Dublin Society. The first president of the new body was Lord Southwell; there was a secretary for each of the four Irish provinces; and by April 1745 the society had 226 members. These members agreed each to contribute annually not less than one crown (some members gave much more) in order to finance the work of the society and they met in the Lords' Committee Room at noon on the first Monday of each month in order to transact their business. Among the membership there was a sprinkling of the nobility; Trinity College was represented by a number of the Fellows; from the Dublin Society there came Thomas Prior and John Rutty; and gentlemen of the cloth were well in evidence, there being three bishops at the inaugural meeting. Members were eager for clerical support because in the clergy they recognised a ready-made and nation-wide network of correspondents, and at the first meeting of the society there was passed the motion:

> That the Lords the Bishops be desired to recommend this scheme to the several clergymen & other gentlemen in the several dioceses.

The first year of the society's life seems to have been remarkably successful. At the close of that year — in April 1745 — the society issued a four-page pamphlet setting out its aims and detailing its earliest achievements, two thousand copies of the pamphlet being printed and distributed to the press and elsewhere. At that time it was believed that a period of not more than five years would see the publication of accounts of all the Irish counties, and from the pamphlet we learn that during the summer of 1744 a botanist — Isaac Butler — had been employed to travel in counties Dublin, Longford, Louth, Meath and Westmeath to collect both plants and geological specimens, the latter including 'many curious fossils'. Butler had also assembled a great deal of chorographical and antiquarian information. 'An ingenious gentleman' (again Charles Smith) had been enlisted to write an account of County Waterford and he had been travelling in that county at the society's expense. Two other gentlemen (Messrs Barton and Hacket) had undertaken to prepare a similar description of County Armagh, and it was likewise reported that surveys of both Dublin city and county (by Walter Harris and Lionel

Jenkins) were in an advanced state and should be ready for publication by the spring of 1746. Before the summer of 1745 surveys of counties Fermanagh and Monaghan seem also to have been started and by then a good deal of information had arrived from correspondents nationwide, the information ranging from notes and sketches to actual botanical, geological, and zoological specimens. The geological specimens seem to have been particularly numerous. In the minutes for 1 April 1745, for instance, we read:

> The sum of 8 shillings & 1 penny to Walter Harris to reimburse him for the cost of carriage of fossils from Co. Waterford to Dublin.

Most of the society's projected volumes failed to materialise, but surveys of four counties — Down, Waterford, Cork, and Kerry — were published between 1744 and 1756 under either the direct or indirect aegis of the society.[5] Those four surveys all have one significant feature in common: their authorship was either partly or entirely in the hands of the aforementioned Charles Smith. About Smith we can only wish that more was known.[6] He was evidently born around 1715, and he is described as being a native of Waterford. From some institution he obtained an M.D. degree, and he appears to have practised as an apothecary in Lismore from 1739 until 1744 and in Dungarvan from 1744 until 1760. Two years after that, in the early summer of 1762 (perhaps in July), Smith died in Bristol.

Each of Smith's county surveys contains a chapter with a title such as 'Of the most remarkable fossils discovered in this county'. In these chapters Smith offers accounts of various topics including 'soils, earths, and clays', 'marles', 'ochres or painting earths', 'coal', 'stones, slates and marbles', 'fossil shells', and 'spars, petrifactions and crystals', and all told the chapters display a surprisingly detailed knowledge of the regional geology, both solid and drift. Of the four published county surveys it is the two-volume survey of County Cork which holds the chief interest in the present context, the volumes being laid before the society on 6 November 1749 and published with the society's approbation in 1750. The volumes have an accompanying map at a scale of one inch to 3.5 statute miles (1:221,760) and entitled *A new and correct map of County of Cork*. The map in all its details was evidently constructed by Smith himself because in the society's minutes for 6 November 1749 we read:

> Mr Smith presented to the Board a large and accurate map of the County of Cork in two sheets, laid down by himself from an actual survey, assisted by astronomical observations.

Over the surface of the published version of the map there is a scattering of the symbol 'L.V.' which, from the key, we learn stands for 'Limestone Vales' — those long, narrow, synclinal, limestone-floored valleys which are such a distinctive feature of Cork's Ridge and Valley province. Smith has overlooked many places where the presence of limestone should have earned the locality an 'L.V.' but his map of 1750 is nevertheless of great interest as the earliest attempt to depict cartographically the solid geology of any part of Ireland.

The last dated meeting of the Physico-Historical Society to be recorded in the surviving minute book took place on 22 March 1752 and, for whatever reason, the society seems to have lapsed a year or so later. The doodles ornamenting the concluding pages of the minute book perhaps indicate that even the society's officers had by then lost their former zeal for the task at hand! The society certainly died leaving a great deal of material unpublished but Smith did give the society one posthumous offspring — his *The antient and present state of the county of Kerry* published in 1756. As usual, the volume contains a map, but this time it was a map without any attempt at the representation of solid geology. The text of the volume nevertheless contains the following remarkable observation:

> ... it appears, that the several substances beneath the surface of the earth, are ranged with more order and regularity than has been hitherto taken notice of, and that they are not scattered at random but are joined together in different ranges, so that they may be traced from one county to another: and if ever the whole of this island comes to be

accurately surveyed, it would not be very difficult to construct a map, whereon, by means of proper characters, its various soils might be expressed. This kind of inquiry might open an extensive field to geographers, and naturalists; and might form a connection between the two sciences, which are more dependant on each other in reality, than they have been hitherto supposed to be; and may give great light to the industrious searcher into minerals, and fossils, who, by knowing the true direction of any of these beds, may discover the same kind of matter as limestone, coal, slate, and several other metals at great distances from the place of their first appearance.[7]

Today this all seems so obvious, but in 1756 Smith was here writing a passage of remarkable prescience. His is a name to be respected by any student of the history of geological cartography. In County Cork in 1750 Smith had blazed a geological trail which was to reach its end in the wilds of County Donegal 140 years later. The trail that he inaugurated is the subject of this book.

William Hamilton and the Antrim basalts

Among those who made a careful study of the engravings of the Giant's Causeway derived from Susanna Drury's paintings there was a Frenchman who was destined never to set foot in Ireland — Nicolas Desmarest. In 1765 and 1771 he presented to the Académie Royale des Sciences a now famed memoir on the ancient volcanoes of central France in which he urged — correctly as we now know — that all basalts are volcanic in origin.[8] In this manner there was born what came to be known as the Volcanic Theory. From the evidence contained in Miss Drury's representations of the Antrim coast, Desmarest concluded that north-eastern Ireland must once have been held in Vulcan's grip, and thus from the outset Antrim featured prominently in the controversy that developed between Desmarest's Vulcanists and the opposing Neptunists who insisted that basalt was a chemical precipitate formed in some postulated chaotic fluid. Many were the visitors who came to Ireland in the decades around 1800 to inspect Ulster's Plateau Basalts, but of all the visitors to the northeast it was left to an Irishman — to the Rev. William Hamilton, a Fellow of Trinity College Dublin — to make the sole attempt at any kind of cartographic representation of the basalts.

William Hamilton (1755-1797) was born in the city of Londonderry, and he took his B.A. in Dublin in 1776. Three years later came election to his Trinity Fellowship and we are told that immediately thereafter he resolved to devote himself to the serious study of the natural sciences and especially to chemistry and mineralogy.[9] To this end he in 1784 toured north-eastern Ireland and the result was his fascinating *Letters concerning the northern coast of the County of Antrim*, first published in 1786.[10] The book offers a soundly reasoned exposition of the Volcanic Theory, to which Hamilton adhered, but of hardly less interest is the accompanying map depicting the coastline of northern County Antrim from Fair Head westwards to Portrush and drawn to a scale of one inch to one statute mile (1:63,360). This map is second only to Charles Smith's map of County Cork as an attempt to depict cartographically the rocks of any part of Ireland. The attempt is nevertheless rudimentary in the extreme, and Hamilton's map can hardly stand comparison with the fine map of Auvergne which had accompanied Desmarest's memoir published in 1771. So far as the solid geology is concerned, Hamilton merely employs clusters of tiny circles to represent the outcrop of columnar basalts, while along the coast the words 'Lime Stone' are inscribed wherever he had noted the presence of (Cretaceous) chalk in the cliffs. In the realm of drift geology two deposits are represented: a system of line-shading is employed to depict what are presumably peat-bogs, while stippling is used to represent the areas of blown-sand near Bushmills and Portrush. Perhaps Hamilton enjoyed making the map because he shortly embarked upon the more ambitious cartographic project which accompanied the 1790 edition of his book.[11] This new venture is a smaller scale map of the whole of north-eastern Ireland upon which Hamilton has plotted a line to represent the limits of the basaltic plateau, carefully differentiating between those localities where the boundary is clearly to be seen and those where its identification is less certain. His line is reasonably accurate, and while the modern geologist might protest that in many places the drawing of the line was a simple exercise in feature mapping, Hamilton must be accorded credit for being the first to attempt the cartographic delimitation of one of Ireland's major geological systems.

In his later years Hamilton extended his scientific interests to studies in climatology and terrestrial heat. His calling, too, was much in his mind, and in 1790 he resigned his Trinity Fellowship to become rector of Clondavaddog, a remote parish in northern County Donegal. There he performed a dual role: as well as ministering to the spiritual needs of the members of the established church, he also served as the local magistrate. Those were troubled times, and it seems that Hamilton's strict enforcement of the law brought him into collision with a section of the population inclined towards rebellion. In a manner all too characteristic of Ireland, his murder was arranged and, after one abortive attempt, the deed was finally executed by the shores of Lough Swilly one night in March 1797. Hamilton was buried in the churchyard of St Columb's Cathedral in his native Londonderry.

Geology and the Dublin Society

Charles Smith and William Hamilton represent the two different traditions evident in Irish geology during the period between 1750 and 1830. For Smith, geology was a useful and practical science to be enlisted in the search for new mineral lodes, ochres, or marl-pits — a science that held a vital key to future national prosperity. For Hamilton, on the other hand, geology was a fascinating intellectual pursuit — a science which allowed man to explore some of the deepest mysteries of the natural world, bringing him that much closer to the mind of his Creator. Although these two traditions — the applied and the pure — co-existed in Ireland for eighty years after 1750 it was always the applied tradition which was the stronger. Irish geologists such as William Hamilton, Richard Kirwan (1733-1812) and William Richardson (1740-1820) might participate in the international debate over the relative merits of the Volcanic and Neptunian theories — a debate to which Hutton's Plutonic Theory was added as a third ingredient in 1785 — but it was the severely practical aspects of geology which really concerned their Irish contemporaries. This is entirely understandable. Across the water in Britain an industrial revolution largely founded upon mineral wealth was beginning to transform the face of the nation and bring unprecedented prosperity. Might the same not happen in Ireland if only the appropriate mineral riches could be discovered? Few doubted but that Ireland must contain what was seen as its fair share of mineral wealth, and some foretaste of what it was felt must lie in store was witnessed in 1795 when the discovery of some large gold-nuggets in the drifts of the Gold Mines Valley of County Wicklow resulted in a minor gold-rush. What was needed, it seemed, was a national geological inventory and clearly such a document might very conveniently take cartographic form. This, for example, is the report of a speech made in the Irish House of Commons on 8 February 1786 by Charles O'Hara, one of the members for the county of Sligo:

> In his opinion it would be for the advantage of the kingdom if a proper and accurate chart of all the lands in Ireland was made, so as to point out the lands where there were mines and minerals, or where such were most likely to be found, which would be of infinite benefit to the nation to know the substrata of the kingdom.[12]

Ten years later, in an essay presented to the Royal Irish Academy on 29 September 1796, William Preston observed:

> It would be a measure of great national utility, were able mineralogists sent, at the public expence [sic], through the country, to examine its mineral productions; their quality, and the facility or difficulty of obtaining them; with other particulars, of that

kind, proper to guide the exertions of industry; and instructed, to combine their several discoveries, in something like a subterranean chart of the whole island.[13]

By around 1800 the Royal Irish Academy was evidently considering the practicability of conducting its own mineralogical survey of County Dublin.[14]

Whatever might be going on at the Royal Irish Academy, there can be no doubt but that throughout the decades down to 1830 the strong Irish interest in applied geology was firmly and squarely centred at the Dublin Society. The Dublin Society for Improving Husbandry, Manufactures and other useful Arts and Sciences was founded in 1731[15] (it assumed the title 'Royal' in June 1820) and from its earliest days the Society had taken an interest in what were then termed 'geological curiosities'. The members of the Society were drawn largely from the landlord class and from the establishment generally, and as the eighteenth century drew to its close it was only natural that such individuals should have displayed a concern for Ireland's geological exploration. If mineral wealth could be found, then the entire nation must benefit, but there is no escaping the fact that a sizeable proportion of the resultant profits would be going into the pockets of Dublin Society members. Self-interest was involved. Thus while travelling through Ireland in 1788 Richard Kirwan found 'the attention of many of the principal landholders awakened to researches after minerals'[16] and the following passage was written by the anonymous author of *An address to our countrymen, on the study of chemistry and mineralogy* published by the Dublin Society in 1802:

> Our countrymen are at last convinced, that the great riches of this island lie under ground; indeed, few countries can boast of so many valuable ores and fossils, as Ireland produces.... We therefore hope soon to see small mineralogical societies formed in every provincial city in Ireland.[17]

That ambitious objective was never realised but the Dublin Society itself did take its responsibilities in the earth-sciences very seriously. In 1792 the Society purchased for £1,350 the mineral collection – it contained 7,331 specimens – assembled by the German mineralogist Nathanael Gottfried Leske, one of the finest such collections then extant.[18] Other material flowed into the Society's museum from sites as varied as Vesuvius and the Harz Mountains,[19] and another interesting purchase was one of White Watson's 'tablets' of Derbyshire geology accompanied by numerous Derbyshire rock specimens.[20] To the museum there was ascribed a serious pedagogical role following the appointment in June 1795 of William Higgins (1763-1825)[21] as the Society's first professor of chemistry and mineralogy. Henceforth courses of mineralogical lectures were a regular feature of the Society's programme and not only were the lectures well attended, but they were clearly designed to present the subject in some depth. Those who enrolled for the course beginning in November 1801, for instance, were expected to continue in attendance until the following May and for this course a fee of five guineas was charged, reduced to two and a half guineas for Society members.[22] Those who felt confident of their mineralogical expertise could enter the Society's annual open examination in the subject with handsome cash prizes to be won: fifty guineas for the highest scoring candidate, thirty guineas for the second, and twenty guineas for the third.[23]

Many years later Isaac Weld (1775-1856), the Society's honorary secretary from 1828 until 1849, claimed that the purchasing of the Leskean collection and the establishment of the chair of chemistry and mineralogy were intended as initial steps towards the founding within the Society of a School of Mineralogy devoted expressly to the exploration of Ireland's geological resources.[24] Be that as it may, members of the Dublin Society were certainly keenly aware of a need for some kind of national geological survey. As far back as 1773 General Charles Vallancey (1726-1812), a leading member of the Society and its vice-president from 1799 until his death, proposed the organisation of a comprehensive survey of Ireland employing the questionnaire method and incorporating inquiries about the geology of respondents' home regions.[25] It looked very like the scheme which had been tried without success in the

1740s by the gentlemen who eventually founded the Physico-Historical Society, and it was perhaps the memory of that failure which discouraged interest in Vallancey's project both in 1773 and again in 1784 when he floated the idea for a second time. The Society, however, was soon to launch its own national survey devoted exclusively to Irish mineralogy.

It was on 15 June 1786 that a Scotsman named Donald Stewart was commissioned by the Society as its Itinerant Mineralogist at a wage of one guinea per week. Of Stewart's background we know little, but for the Society his first assignments were to investigate both the geology of County Wicklow and the exposures revealed along the line of the Grand Canal in the Irish midlands.[26] Other commissions soon followed and over the years Stewart's search for deposits of economic importance led him into most parts of Ireland. He is reputed to have made many discoveries of significance and his published report of 1800 describes in rudimentary fashion the economic geology of twenty-five of the thirty-two Irish counties.[27] Some idea of the kind of duty he was performing comes from the following entry in the Society's minutes for 17 November 1808:

> Resolved: That Donald Stewart be at liberty to go into the county of Tipperary to the estate of Sir Thomas J. Fitzgerald, Bart. and at his expence [sic] to inspect his estate, and point out where some Manganese is to be found thereon.[28]

Vallancey, himself a keen mineralogist, clearly held Stewart in low esteem; in a letter to Sir Joseph Banks on 23 April 1802 he referred to Stewart as 'a poor weak-minded Scotch man'[29] and many years later a Society report alluded to the former Itinerant Mineralogist as 'an industrious man, but uneducated, and ignorant of mineralogy'.[30] Not all members of the Society can have shared such harsh judgements because as from September 1808 Stewart's salary was raised to one hundred guineas per year.[31] But not for long did Stewart enjoy his new prosperity; he died in 1811 having served the Society for a quarter of a century. In the present context, however, his work is of little significance because his investigations always remained at the level of individual sites and specimens and he never gave to his studies that areal dimension necessary for the production of a geological map. When he died the post of Itinerant Mineralogist was left vacant because, as we will see shortly, by 1811 the Society had in mind more ambitious plans for the furtherance of a knowledge of Irish geology — plans involving the substitution of a modern stratigraphical approach for the mineralogical outlook of the earlier age.

Stewart may have produced no geological maps, but the early nineteenth-century Dublin Society is far from being a barren field for the student of geological cartography. In the 1790s there had occurred in Britain two publishing events which were to have their repercussion in Ireland: the publication of Sir John Sinclair's parochial survey of Scotland in 21 volumes between 1791 and 1799,[32] and the appearance of the county agricultural surveys of England and Wales which came from the Board of Agriculture between 1793 and 1817.[33] The Irish spirit of emulation which we noted earlier among the members of the Physico-Historical Society was still very much alive. Why should Ireland not have its own equivalent publications to match those just produced in Britain? So it came about that in 1800 the soon to be extinguished Irish parliament authorised the Dublin Society to expend a portion of its government grant 'in procuring agricultural examinations into all or any of the counties of this kingdom'.[34] Both General Vallancey and Richard Kirwan, Ireland's most distinguished scientist of the day, were active in the organisation of the project, and eventually surveys were published for twenty-three of the thirty-two Irish counties, the reports appearing between 1801 and 1832.[35] The surveys were designed to assist appraisal of the economic needs of each county and, although the volumes are primarily concerned with agriculture, they do broach many other issues. The Society's *Suggestions of enquiry for gentlemen who shall undertake the forming of agricultural surveys*[36] recommended that prospective authors should devote attention to 'soil and surface' and to 'minerals', and this advice was heeded. Indeed, some of the authors went so far as to offer a comprehensive survey of their county's geology together with a map depicting at least some aspects of the regional geology. In some cases the geology depicted is extremely rudimentary. On his 1802 map of County Down, for example, John Dubourdieu's plotting of geology consists merely of one boundary together with the explanation that 'the granite country lies south & west of this line'.[37] Rather more discriminating was Hely Dutton's map of County Clare published in 1808.[38] Here the author has with fair accuracy plotted the junction of the limestone both against the shale outliers in the west and against the Slieve Aughty and Slieve Bernagh inliers in the east. Interesting and significant though such maps are, they fall into a second rank as compared with the ambitious productions offered by four of the Dublin Society's authors: Robert Fraser, William Tighe, George Vaughan Sampson, and Horatio Townsend. In their reports these four gentlemen included comprehensive geological maps which deserve honourable mention in any international history of geological cartography. That their maps are not better known among historians of the earth-sciences is doubtless to be explained by the fact that many of the reports were issued in small editions of only 150 or 250 copies and in consequence the volumes are now rare even within Ireland itself.

Robert Fraser's County Wicklow

The earliest of those four maps accompanied Robert Fraser's survey of County Wicklow published in 1801.[39] Like Donald Stewart, Fraser was a Scotsman. He was a son of the manse, born at Redgorton near Perth about 1760, and he received his M.A. degree in the University of Glasgow in 1777.[40] Government service as a statistician and surveyor followed, and in the 1790s he prepared the accounts of both Cornwall and Devon for the Board of Agriculture.[41] He clearly had a special interest in geology and his report on the farming of Devonshire is among those Board of Agriculture surveys which contain a soil map. In 1802 he published a work incorporating a brief account of Ireland's economic geology[42] (it was not very accurate because it claimed that chalk was unknown in Ireland) and in 1807 he prepared the Dublin Society's statistical survey of County Wexford.[43] He lived until 1831, but his only Irish geological map is the one contained in the County Wicklow survey of 1801.

The scale of that map is approximately one inch to 2.5 statute miles (1:158,400) and, apart from bogs, which are shown by a marsh symbol, the region's geology is represented by coloured tints (Fig. 1.1). These tints are localised, however, and large areas of the map are left uncoloured because Fraser clearly felt justified in colouring only those places where he was convinced that a particular rock-type had been positively identified. Indeed, over the greater part of the sheet the tints have been applied only within the rather crudely drawn hachures which represent individual hills and mountains. The result is aesthetically unsatisfactory. The map gives the impression that a number of plump, fuzzy, and brightly coloured caterpillars are slowly crawling their way across the sheet!

The colours upon the map represent five different geological categories. Firstly, 'Metalliferous' (yellow) which has been applied to such mining localities as Croghan Mountain and the Vale of Avoca. Secondly, 'Argillaceous' (blue) employed for the Lower Palaeozoic slate uplands lying either side of the Leinster batholith. Thirdly, 'Hornstone' (orange-brown) representing the Bray Group (Cambrian?) rocks of Bray Head and the two Sugar Loafs in northeastern Wicklow. Fourthly, 'Granite' (red) applied widely (rather too widely upon some copies of the map) within the hachures representing the mountains developed upon the Leinster batholith. Finally, there is 'Calcareous' (green). Of all the colours used upon the map, this is the only one to have been washed over wide areas. Two portions of the county have been so treated: firstly the broad coastal plain lying between Bray Head and Wicklow town, and secondly the lowlands occupying the western part of the county around Blessington and Baltinglass. The rocks of both these areas are actually mostly slates of Ordovician or earlier age, and by no stretch of the imagination can they be classified as calcareous. Indeed, nowhere within County Wicklow do limestones exist *in situ*. It would seem that Fraser employed the green tint to represent the character of the soils rather than the nature of the underlying solid

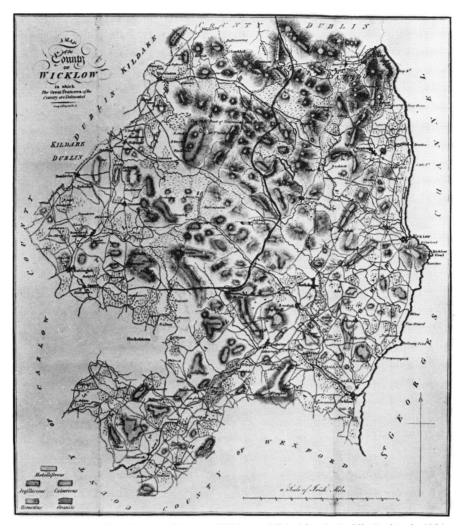

*Fig. 1.1. Robert Fraser's map of County Wicklow published by the Dublin Society in 1801.
(Reduced in size.)*

geology, and it is for this reason that he has allowed the green tint to lap all around the base of granite mountains. His mind was clearly harking back to the soil map of Devonshire which he had produced for the Board of Agriculture seven years before. In his text he refers to the calcareous region as containing 'an infinity of rounded and blunted limestones found in the strata of the earth'[44] and he recognised that this limestone debris must have been imported into the country by some process then unknown. His diagnosis was sound enough because, at least in western Wicklow, there are extensive spreads of calcareous till and outwash left by the Pleistocene glaciers, but another forty years were to elapse before there began to be recognised the enormously important role of former glaciation in the shaping of the modern Irish landscape.

Crude, incomplete, and inaccurate though Fraser's map may seem to modern eyes, it was a far more ambitious work than any previous Irish geological map. We are left to wonder what Fraser's conception of the geological structure of County Wicklow really was, and how he supposed his coloured 'caterpillars' to be related to each other and to the intervening country, but his map — the first coloured Irish geological map — is nevertheless a significant landmark in the history of Irish geological cartography. Its pre-eminence was nevertheless to be very short-lived.

William Tighe's County Kilkenny

We come now to a map which is one of the most remarkable examples of Irish geological cartography ever to be published – the map of County Kilkenny published in 1802 by William Tighe. Tighe's home was Woodstock House, County Kilkenny, standing two kilometres south of Inistioge. The house is now a gaunt ruin, a victim of Ireland's twentieth-century 'Troubles', but in Tighe's day it was a gracious residence overlooking the picturesquely incised valley of the River Nore. It was at Woodstock, in April 1802, that Tighe completed the dedication of his statistical survey of the county,[45] and on 8 July 1802 the newly published volume was placed upon the table of the Dublin Society by General Vallancey himself.[46] Today the work is recognised as one of the two most successful of the Dublin Society's county surveys, its sole rival being Isaac Weld's survey of County Roscommon published thirty years later.[47]

William Tighe was born in 1766, the eldest son of William Tighe of Rosanna, an estate near Rathnew in County Wicklow.[48] His father represented Athboy in the Irish parliament from 1762 until 1776, and his first cousin and sister-in-law was Mary Tighe, the author of the then much acclaimed poem *Psyche, or the legend of love*. William Tighe Junior, who had his own poetical aspirations,[49] was educated at Eton College and St John's College, Cambridge, and from 1790 until 1800 he was a member of the Irish parliament sitting first for the Borough of Wicklow and then for the Borough of Inistioge. He resolutely opposed the Act of Union but its passage was by no means the end of his political career because from 1806 until his death he sat for County Wicklow in the Westminster parliament and it was in London that he died on 19 March 1816.

Fig. 1.2. Woodstock House where William Tighe completed his statistical survey of County Kilkenny in April 1802. The house was burned in 1922 during the Irish Civil War. Reproduced from a pencil sketch by 'Donachie' which was originally published as a post-card.

County Kilkenny contains a wide variety of rock types ranging from slates of Lower Palaeozoic age, through the granites of the Caledonian Leinster batholith and the Old Red Sandstone, up to the limestone, shales, and coal measures of the Carboniferous. Tighe's geological map of the county is at a scale of one inch to 2.5 statute miles (1:158,400), its engraved area is 47.2 x 36.3 centimetres, and it was very attractively produced by J. Taylor of Donnybrook near Dublin (Plate 1). Such features as major rivers, projected canals, and towns are marked upon the map, and uplands are shown by very neat pictorial symbols, but Tighe clearly envisaged his map as primarily geological in character. He has, for example, omitted all roads, and the key contains nothing but an index to the map's geological colours. These colours occupy the entire map, and there are none of those blank, indeterminate areas which feature so prominently upon Fraser's map of Wicklow. Further, Tighe felt sufficiently confident of his mapping to allow the representation of the boundaries between the various geological formations by means of firm, engraved lines.

On the map he recognised the following eight geological categories:

1. 'Granite' (green). Tighe has applied this colour to that small portion of the Leinster batholith which lies within the county, and also to the batholith's associated granite apophyses.
2. 'Siliceous Schistus' (pink). This tint represents the schist produced by the metamorphism of the Cambro-Ordovician slates along the margins of the Leinster batholith.
3. 'Siliceous Breccia & Red Argillite' (brown). Later generations of geologists came to know the rocks thus coloured by Tighe as the Old Red Sandstone.
4. 'Secondary Siliceous Schistus & Ferruginous Argillite' (pale orange). Into these categories Tighe has placed the Upper Carboniferous (Namurian) shales of the Castlecomer and Slieve Ardagh plateaux.
5. 'Slaty Argillite' (lemon yellow). Here Tighe is representing both the Cambro-Ordovian slates lying beyond the batholith's metamorphic aureole and the Lower Palaeozoic rocks of the Slievenamon inlier in the southwest of the county.
6. 'Limestone & Calcareous Strata' (deep blue). This is the most extensive tint upon the map and it represents the Carboniferous Limestone.
7. 'Limestone gravel' (pale blue). Only a small area lying to the northeast of Thomastown is placed in this category and Tighe there seems to have identified a part of the morainic complex marking the front of the last (Midlandian) Irish ice-sheet.
8. 'Inundated or Wet Ground' (red-brown). A wide scattering of small areas has received this tint, most of them lying in valley bottoms.

In addition a few bogs are shown by the marsh symbol plus the red-brown tint, some collieries on the Castlecomer Plateau are indicated by groups of dots and the words 'C. Pitts', and terms such as 'Sand', 'Clay', 'Gravel', 'Siderocalcite', and 'Slate Quarry' are inscribed where appropriate.

The construction of any geological map involves three separate if closely related basic processes. Firstly, there is a physical task. The region has to be explored in search of rock exposures and at such sites a fresh, unweathered sample of the rock has to be removed for examination by the application to the rock-face of that most indispensable of all geological implements, the hammer. Secondly, there is an intellectual task. The rocks of the area have to be classified according to some method, the two methods normally employed depending either upon similarity of lithology (the older method) or upon contemporaneity of age as revealed by fossils (the newer method). Finally, there is a second physical task — a task which under Irish meteorological conditions can often be far from pleasant: the boundaries of the various rock-types involved in the classification have to be traced out laboriously in the field and plotted upon the map. Tighe has performed the first two of these tasks remarkably well and even a modern geologist would have little to add to Tighe's lithological classification.

In his third task — the tracing and plotting of the geological boundaries — Tighe was a trifle less successful. His delimitation of the Upper Carboniferous outliers in northern Kilkenny is sound enough, and he has with similar accuracy plotted the boundary of the Carboniferous strata against the older rocks both in central Kilkenny and further to the south in the Suir valley. His mapping of the various granite masses is also surprisingly good, and it was only among the Devonian, Silurian, and Ordovician rocks that he fell into error. Yet even there, amidst the relatively complex structures of southern Kilkenny, it must be conceded that he has done remarkably well, the more so since we have his admission that the entire survey was conducted with some haste and with limited objectives:

> In the map annexed to this work, the spaces occupied by the different species of stone, are marked out and coloured: the boundaries of each are delineated, not with that accuracy which a technical or tedious survey might have done, but in a manner sufficient for the objects of this enquiry.[50]

Slight errors though it may contain, Tighe's map was for its day a most remarkable achievement. In fact its excellence in both conception and execution poses a problem. How could Tighe possibly have come to possess the insight and expertise which allowed him to produce such a work? Had the map come as the culmination of a long period of personal geological study, then its outstanding merit might have been understandable, but there is no evidence that Tighe ever served such a geological apprenticeship. He did evidently have financial interests in some Cornish mines, but the Kilkenny map and its accompanying text are the only known evidence of his direct involvement with the earth-sciences. His survey appears upon the scene as a brilliant and quite unheralded flash in the Irish geological pan. But there is one further facet of Tighe's brilliance yet to be revealed. In his accompanying text we find the following passage:

> As the soil of this county is seldom much raised above the rock that forms its basis, it is not very difficult to trace the various stony substrata, which often appear near the surface; and by their position seem to shew the successive order in which they were deposited by the operations of nature.[51]

In short, Tighe recognised the existence of a chronological sequence among the rocks of Kilkenny. At the base of the sequence he placed the granite and then there followed the siliceous schistus (the granite's metamorphic aureole), the siliceous breccia (the Old Red Sandstone), varieties of argillite (Cambro-Ordovician, Silurian, and Carboniferous slates, sandstones and shales) and the limestone (Carboniferous). His sequence may have been wrong but the important point is that he saw his map as much more than a mere representation of the spatial distribution of the different lithologies. He quite clearly regarded the map as an essay in earth-history revealing how, through time, the various formations had been assembled one upon the other. Any modern geological map is a four dimensional document. It shows both the latitudinal and longitudinal extent of the various formations; it reveals as a third dimension the vertical arrangement of the rocks within the earth's crust; and, to the initiated, it indicates something of the fourth, temporal dimension, by offering an insight into the geological history of the region. As we have seen, Tighe coped with the representation of the first two dimensions well enough, but his grappling with the third, structural dimension is poor. He failed, for example, to distinguish between inliers (older rocks protruding through younger) and outliers (younger rocks surrounded by older), and this structural weakness made it impossible for him to achieve any valid understanding of the fourth, temporal dimension. Yet the attempt at a stratigraphical chronology had been made. Tighe was thus not only the first to compile a comprehensive geological map of any part of Ireland, but he was also the first Irishman to recognise that a geological map could offer an incapsulated geological history of the region that it depicted.

To consider Tighe merely within an Irish context is nevertheless to do him far less than justice. The complete story of geological cartography within the British Isles as a whole still remains to be written, and it is therefore impossible to place Tighe's map of Kilkenny in its full historical perspective. It is nevertheless clear that the map must occupy a position of honour among the earliest of all the geological maps produced within the British Isles. William Smith, for example, finished his manuscript map of the strata around Bath in 1799 and soon after he began to plot geological lines upon a map of England and Wales,[52] but it was 1815 before the first of his maps was actually published.[53] Among the published maps depicting the solid geology of the British Isles the present writer knows of only two which actually pre-date Tighe's Kilkenny. The first of these is John Boys's very rudimentary map of Kent published in 1794,[54] and the second is William George Maton's 'Mineralogical map of the western counties of England' published in 1797.[55] Neither of these maps, however, aspires to the detail and accuracy of Tighe's work and in the present state of our knowledge the Kilkenny sheet stands out as by far the most sophisticated piece of geological cartography to have been published within these islands down to 1802. It is sad that Tighe and his map should for so long have escaped the notice of historians of geology and it is to be hoped that further research will yield fresh insights into both the character of Tighe himself and the nature of his geological activities.

George Vaughan Sampson's County Londonderry

The third of the Dublin Society's county geological maps is contained in the Rev. George Vaughan Sampson's statistical survey of County Londonderry.[56] Sampson devoted more than three years to his project and the volume was finally published a few months after the appearance of Tighe's Kilkenny. Sampson's dedication of the work — a dedication to General Vallancey — is dated 16 August 1802 and ten days later the Dublin Society granted Sampson £80 for the completed survey.[57]

Sampson was born in the city of Londonderry in 1762 and he took his B.A. in Trinity College Dublin in 1784.[58] After a period as assistant chaplain to the British ambassador at the French court, he in 1790 became the headmaster of the Diocesan School in Londonderry. This post he held until 1794, although by then he had assumed the additional duties of curate of the local city parish of Templemore. In 1794 he left Londonderry to become the rector of the rural parish of Aghanloo, lying in the northern part of the county between the eastern shores of Lough Foyle and the towering pile of Binevenagh at the northwestern corner of the basaltic plateau. It was during his incumbency there that he completed his geological map of the county, but in 1807 he left Aghanloo to become the rector of Errigal, a parish up on the basaltic plateau itself and lying between the rivers Roe and Bann. It was there that in 1808 he must have heard of his election to Honorary Membership of the newly-established Geological Society of London,[59] and it was there that he died in March 1827.

Sampson's map of County Londonderry is drawn to the same scale as Fraser's Wicklow and Tighe's Kilkenny, namely one inch to 2.5 statute miles (1:158,400) but, unlike Tighe's map, Sampson's was intended to serve as a general, all-purpose map of the county (Fig. 1.3). It thus shows all the major rivers, and the relief is represented — though somewhat inadequately — by a system of hachures. It includes such 'cultural' features as roads, towns, forts, and old castles and, Sampson being of the cloth, he has added rectories and vicarages for good measure. But in addition to all these, he sought to represent both the solid and the drift geology of the county. In depicting the solid geology he employed a series of symbols (triangles, crosses, squares, etc.) and maps employing this method to show geology have been termed petrographic maps. Sampson devised symbols to represent the following seven types of rock: 'White Lime' (Cretaceous Chalk); 'Blue Lime' (metamorphic limestones amidst the Dalradian Schists); 'Basalt' (the Plateau Basalts); 'Schist' (the Dalradian Schists); 'Granite' (the granite near Slieve Gallion); 'Sand Stone' (Carboniferous sandstones and shales); and 'Indication of Coal' (marked

Fig. 1.3. George Vaughan Sampson's map of County Londonderry published by the Dublin Society in 1802. (Reduced in size.)

in a few places among the Carboniferous rocks). Upon the map the basalt symbol (a square) is fairly widespsread over the eastern portion of the sheet, but otherwise the symbols are not closely spaced and the reader is left to speculate about the character of the intervening areas. Sampson's map is thus comparable with Fraser's Wicklow in that it is only a partial representation of the geology and one suspects that Sampson has plotted a symbol only in those locations where he had actually identified a rock outcrop.

Sampson was a little more adventurous in his representation of the county's superficial geology. Here he was perhaps being influenced by the soil maps of some of the English counties published earlier by the Board of Agriculture because he attempts a generic classification of Londonderry's drift deposits. What he terms 'marle' he shows by a symbol, but otherwise the drift is represented by colour washes as follows: 'Heath & Bog (brown); 'Clay' (pale blue); 'Sand' (yellow); 'Gravel' (red); and 'Rich Loams' (green). Large areas of the map are nevertheless still left uncoloured and from the key we learn that 'The parts without Colour are unfertile'.

His representation of the solid geology by means of scattered symbols, while drift geology is indicated by continuous colour washes, suggests that in his thinking about drift geology

Sampson had a far greater spatial awareness than he displayed in his thinking about the underlying solid strata. It will be recollected that in the previous year Fraser had displayed precisely the same characteristics by mapping the solid geology of County Wicklow as a discontinuous phenomenon while at the same time he applied widespread colour washes to represent the county's calcareous drift. Perhaps it is easier to generalise one's observations over a visible drift surface than it is to extrapolate the lessons of isolated rock-exposure underground into the intervening drift-obscured strata. Such extrapolation must have seemed extremely hazardous in the days when geology was still lacking that secure theoretical basis which must guide the field-surveyor in his moments of doubt and difficulty. Certainly where the superficial deposits were concerned, Sampson felt confident enough to have firm lines engraved upon the map to indicate the limits of certain of his drift categories, although most of these engraved lines lie in the northwestern part of the map in the area which he presumably knew best as a result of his explorations based upon the Aghanloo rectory.

Significant though it is for the historian, Sampson's map cannot be rated very highly as a geological document, and it is certainly far inferior to Tighe's Kilkenny. Sampson himself nevertheless seems to have been reasonably happy about the map because in 1814 he published a very much larger scale map of the county for the Londonderry Grand Jury,[60] and this new map is basically a much enlarged version of the 1802 sheet. The scale of the Grand Jury map is 1.59 inches to the statute mile (1:39,849), a scale sufficient to carry a wealth of geological detail, but Sampson did nothing to avail himself of the opportunity presented. On the new map he continued to indicate the solid geology by means of symbols (he has changed some of the actual symbols and added the new category of 'gneiss') but the drift colour-washes of the 1802 map have now disappeared and instead the superficial deposits are represented by a second series of conventional signs. So far as the solid geology is concerned, the 1814 map shows no increase in the number of recorded exposures commensurate with the great increase in scale, and as a result the symbols are so widely scattered that the geological aspect of the map has become completely subordinate to the map's geographical function. Even the inclusion of some rather uninformative geological cross-sections of the county does little to redeem the 1814 map in the eyes of the geologist.

Horatio Townsend's County Cork

The final map in the Dublin Society's notable quartet is the Rev. Horatio Townsend's sheet of County Cork published in 1810[61] (Fig. 1.4). Townsend was born in 1750 and like Sampson he was a graduate of Trinity College Dublin where he took his B.A. in 1770.[62] He was ordained immediately after graduating, and from 1770 down until his death in March 1837 he held a variety of ecclesiastical appointments in County Cork where he himself owned an estate near Rosscarbery. He wrote to the Dublin Society in March 1808 agreeing to undertake their survey of County Cork,[63] and when it was published two years later the volume contained a fully coloured geological map of the county at a scale of one inch to 5.25 statute miles (1:332,640).

Townsend's map is a general, all-purpose representation of the county, but it lays considerable emphasis upon what its author regarded as geological structure. All his postulated geological boundaries have been engraved as pecked lines and, as in Tighe's map, the geological colours have been applied over the whole extent of the county. These colours serve to indicate the following four 'geological' categories:

1. 'Secondary Mountn. & high Moorland' (yellow). This tint has been applied to the mountainous Old Red Sandstone country in the west of the county, to the Old Red Sandstone of the Galty, Kilworth, and Knockmealdown mountains in the east, and to the Upper Carboniferous rocks forming the higher parts of the Munster Plateau in the county's northwestern corner.

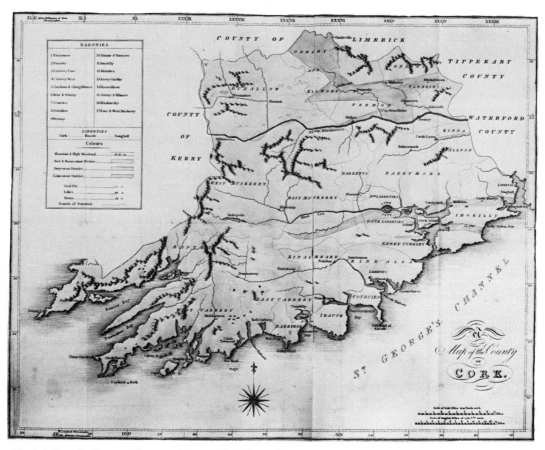

Fig. 1.4. Horatio Townsend's map of County Cork from the second edition of his A general and statistical survey of the County of Cork *published in 1815. (Reduced in size.)*

2. 'Red or Brownstone district' (red). In this category Townsend has included all the remaining, lower-lying portions of the Old Red Sandstone outcrop together with those Upper Carboniferous rocks which underlie the less elevated parts of the Munster Plateau.
3. 'Greystone district' (orange). This is the area of Carboniferous rocks lying in the southwest of the county and containing the strata which later geologists were to know as the Carboniferous Slate, the Culm facies, or the Cork Beds.
4. 'Limestone & calcareous Strata' (blue). Narrow bands of this tint represent the Carboniferous limestone lying on the floors of the county's synclinal valleys where sixty years earlier Smith had plotted his 'L.V.s'.

The geology of County Cork is relatively simple and the boundary between the Old Red Sandstone and the overlying Carboniferous rocks is easy to plot because in most places there is a sharp topographical break between the synclinal Carboniferous lowlands and the anticlinal Old Red Sandstone uplands. Townsend's mapping is fairly accurate in the eastern part of the county, where the ridge and valley character of the topography mirrors the geology most closely, but in the west and northwest his work is much less satisfactory. He made two really serious mistakes. Firstly, he divided the county's Old Red Sandstone into two categories on the basis of whether it formed subdued hills or rugged mountains, and secondly he included the Old Red Sandstone of west Cork in the same category as the Upper Carboniferous rocks of the Munster Plateau simply because they both formed high-standing areas. Indeed, there seems to have been some confusion in Townsend's mind as to whether his colours were intended to indicate rock-type or altitude above sea-level. Perhaps his experience in the eastern part of the county, with its neat correlation of limestone with lowlands and Old Red Sandstone with uplands, had misled him into supposing that geology could everywhere be correlated with topography. Maybe he imagined that in Cork all areas over an altitude of say 250 metres were *ipso facto* composed of one and the same rock!

Townsend published a second edition of his statistical survey of County Cork in 1815 [64] and for this he prepared a new version of the map. This second edition map comes much closer to being a specialised geological map, and all other details, including even place-names, have been reduced to a minimum. A number of geological changes have also been introduced, but, sadly, not all these represent improvements. In 1810, for example, the peninsulas of West Cork were correctly represented as consisting of Old Red Sandstone ('secondary mountn. & high moorland'), but in the new map the peninsulas are all re-classified as belonging to the Culm facies ('grey-stone district'). Neither is much improvement apparent up in the north-western corner of the county where in 1815 the Upper Carboniferous rocks of the Munster Plateau are still placed in the same category as the Old Red Sandstone of the remainder of the county. Perhaps Townsend appreciated that his knowledge of Carboniferous stratigraphy was somewhat shaky because in 1820, while on his way to Harrogate to take a cure at the age of seventy, he made a long geological detour through both England and Scotland with the express intention of improving his understanding of the coal-bearing strata.[65] Whatever lessons he may have learned seem never to have been applied to an Irish context and the map of 1815 was Townsend's final contribution to Irish geology. He died on 26 March 1837. In 1898 his great-granddaughter became Mrs George Bernard Shaw.

Richard Kirwan's proposed Mining Board

From the 1780s until his death in 1812 Richard Kirwan[66] was Ireland's most eminent man of science but he is a figure who has been sadly maligned by so-called historians of science. Among chemists he has been dismissed as one of the last defenders of the phlogiston theory; geologists have scoffed at him as a bibliolatrous Neptunist who in 1793 was foolish enough to cross swords with the great James Hutton; one biographer offered the dismissive assessment that Kirwan scarcely ever advocated any theory 'which was not almost immediately discovered to be unfounded'[67]; and his many eccentricities have all too often resulted in his being paraded merely as a figure of fun. Only recently have scholars begun to accord Kirwan a more sympathetic treatment and it still needs to be emphasised — and nowhere is such emphasis more necessary than in Ireland — that Kirwan was a remarkable polymath, a Copley medalist of the Royal Society (1782), and a man who in the fields of chemistry and mineralogy enjoyed the very highest of international reputations. In Dublin he was both the esteemed second president of the Royal Irish Academy (1799-1812) and the Dublin Society's *éminence grise* who was instrumental in securing for the Society the famed Leskean mineral collection (see p. 6). But, having made that point, it has to be admitted that as a geologist Kirwan made few lasting contributions to the science. He did in 1794 introduce the term 'Calp' for the dark, argillaceous, fine-grained Carboniferous limestones of the Dublin district[68] — a term familiar to Irish stratigraphers ever since — but he was essentially an arm-chair geologist. He is said to have made numerous geological excursions throughout Ireland, but his publications are devoid of geological maps and sections and they everywhere smack of the library and the specimen cabinet rather than of a first-hand acquaintance with nature in the field. Seemingly the only map to bear his name is a manuscript sheet dated December 1801 and depicting the gold-producing region at the foot of Croghan Mountain in County Wicklow,[69] but even here the presence of Kirwan's name was merely a courtesy gesture because he happened to hold the titular office of Inspector General of His Majesty's Mines in Ireland. The man really responsible for that map was a surveyor named Thomas Harding, and in any case the amount of geological information upon the sheet is so minimal as to render this mention almost superfluous.

Then why should Kirwan be dragged into the present story? Because about 1800, had the fates so decreed, he might easily have become the founder of a new geological institution – of a geological survey of Ireland. In the final days of the Irish parliament Kirwan produced for submission to the government a document entitled *A plan for the introduction and establishment of the most advantageous management of mines in the Kingdom of Ireland* and this was published by the Dublin Society in 1802.[70] In devising his recommendations Kirwan was clearly influenced by developments that had been taking place in the German states.[71] His scheme envisaged the establishment of an Irish Mining Board consisting of twelve expert and salaried members, all resident in Dublin, meeting twice weekly, and charged with the general oversight of all Irish mining enterprises. In order to be eligible for service upon the Board, candidates were to have studied for two years at the renowned Freiberg Mining Academy in Saxony and were then to have spent a further two years acquiring a practical knowledge of mining in both Germany and Britain. They were to be proficient in Latin, French, arithmetic, geometry, trigonometry, surveying, subterraneous geometry, drawing, mineralogy, chemistry, assaying, and the working of ores; their mathematical skills were to be tested by the Provost and Fellows of Trinity College Dublin and their chemical expertise by the Professors of Chemistry in Trinity College and the Dublin Society. These highly trained members of the Board, Kirwan proposed, should in addition be made responsible for a mineralogical survey of Ireland. His idea was that each summer two or more members of the Board should visit one of the Irish counties to conduct a mineralogical survey, the work proceeding year by year until the entire country had been covered.

Nothing ever came of Kirwan's proposal, partly because it was thought that the Mining Board might interfere with the constitutional rights of individual mine owners.[72] Had the scheme been implemented, then Ireland would seemingly have been the first nation in the world to possess the benefits of a geological survey. But the scheme was not quite stillborn. In April 1812 a group of Kirwan's friends and admirers founded in his honour the Kirwanian Chemical and Natural Historical Society of Dublin devoted to the study of mineralogy and its related sciences.[73] The members hoped that their investigations would further the knowledge of Ireland's geological resources, thus compensating to some extent for the non-existence of the Mining Board, but sadly their society was short-lived; by 1819 it was almost as dead as Kirwan himself.

Walter Stephens's central Leinster

The motives that impel the author of any particular geological map to undertake his cartographic task are doubtless highly complex. We are nevertheless justified in regarding almost all the maps discussed thus far as basically exercises in applied geology designed to assist Ireland's economic development. The only exceptions – the sole exercises in 'pure' geology – are the maps of William Hamilton emanating from within the ivory towers of Trinity College Dublin, but we come now to another map which in character was 'pure' rather than 'applied'. Appropriately, it too originated within Trinity College. The formal teaching of geology within the college seems to date from June 1806 when Whitley Stokes (1763-1845) received the permission of the College's Board to deliver lectures in natural history 'provided such lectures did not interfere with the other duties of the students'.[74] But Hamilton's activities show that some members of the university were already involved with geology well before 1806 and one who shared Hamilton's interest was the Rev. Walter Stephens (1772?-1808). Stephens took his B.A. in the university in 1791 and he later assumed responsibility for his college's geological museum[75] which had seemingly been founded shortly before by Hamilton.[76] Stephens evidently proved an admirable curator but he never allowed the museum to become his prison; he was an enthusiastic field-geologist and in some of his peregrinations he had the assistance of William Henry Fitton (1780-1861), a Trinity graduate who was destined to make a great name for himself amidst the English Mesozoic rocks. Fitton took his B.A. in 1799 and it is said that around then his geologising resulted in his being arrested on suspicion of involvement in the Irish insurrection of 1798.[77] In view of the primitive weapons carried by so many of the insurgents, the authorities can perhaps be forgiven for mistaking Fitton's hammer for a side-arm!

Stephens (with Fitton's assistance) explored the geology of a large area to the south and southwest of the city of Dublin, and after his companion's premature death in December 1808, Fitton edited Stephens's observations and printed them in 1812 as *Notes on the mineralogy of part of the vicinity of Dublin.*[78] Publication was assisted by a grant of £54 made by the board of Trinity College on 13 June 1812,[79] but the book seems never to have been placed upon commercial sale although complimentary copies were evidently distributed fairly widely.[80] The volume contains a folded map at a scale of one inch to about 3.8 statute miles (1:240,768) and entitled *Sketch of part of the mountainous country near Dublin*. All of County Wicklow is represented together with portions of counties Carlow, Dublin, Kildare, Leix (Queen's County), and Wexford, and the geology is indicated by colour washes. The geological categories represented are 'granite', 'slaty rocks', 'quartz', and 'limestone'. As on Fraser's map of 1801, large areas have been left devoid of colour; Stephens and Fitton were obviously willing to make a positive commitment only in those areas where they had seen a goodly number of exposures. Like Fraser — and like Sampson too — they perhaps felt uneasy about extrapolating geological conclusions from one outcrop to another and they were certainly at pains to emphasise those localities where they had actually seen a contact between two differing rock types. This they accomplished by means of a coloured line drawn at the geological junction, although here they have been poorly served by their colourist whose slap-dash methods have in many cases resulted in the line being drawn at some distance from the boundary which it was clearly intended to mark. The colourist cannot be held responsible for all of the map's shortcomings and the sheet is really a rather crude example of geological cartography. The various categories of rock are represented in essentially their correct positon but the map is lacking in all that precision and attention to detail which by 1812 was being increasingly expected of the geologist. Had the map by Stephens and Fitton appeared fifteen years earlier it must have been hailed as a notable achievement; by 1812 it was a somewhat dated production.

Before moving on from Stephens and Fitton there is one general point that may here be conveniently introduced. Any field-geologist records his observations upon some pre-existing base-map and he locates his outcrops in relation to the topographical detail already present upon the map in hand. Now if the base-map should be inaccurate, then the geologist's observations are displaced in relation to each other and any geological lines that may be drawn will perforce be distorted — the greater the error in the base-map, the greater the deformation of the geology. This difficulty is here termed 'the base-map problem' and it was a difficulty that plagued early geological cartographers the world over until accurate modern maps became available. In Ireland it was Stephens and Fitton who first highlighted the problem.[81] In compiling their map they had combined three earlier county maps: John Rocque's Dublin, Alexander Taylor's Kildare, and Arthur Richards Neville's Wicklow. But the two geologists found that although the best available, the three maps were really inadequate for geological purposes and in particular the accurate plotting of geological information near the county boundaries was impossible because of the relative displacement of topographical detail arising from internal errors within the county base-maps. Not until the appearance of the first Ordnance Survey maps in the 1830s did Irish geologists begin to feel themselves free of the constraints imposed by the base-map problem.

The Bog Commissioners and their maps

The exact size of the Irish population in the first decade of the nineteenth century may have been a matter for dispute, but on one fact there was general agreement — the population was growing rapidly. Population pressure upon the land was consequently increasing and very naturally attention came to focus upon the possibility of relieving that pressure by reclaiming the peat-bogs — peat-bogs which modern estimates regard as blanketing some 15 per cent of Ireland's surface. In 1809 the government therefore set up a commission 'to enquire into the nature and extent of the several bogs in Ireland, and the practicability of draining and cultivating them'.[82] Nowhere was there a greater awareness of the importance of land as a valuable resource than among those Anglo-Irish gentry who were members of the Dublin Society, and from the outset the Bog Commissioners and the Dublin Society operated in the closest association. General Vallancey, a pillar of the Society, was the chairman of the Commissioners until his death in August 1812; Bucknal McCarthy, the Society's assistant secretary, was secretary to the Commissioners; the Commissioners met in the Society's premises in Hawkins Street, Dublin; and all the chemical analyses of peat required by the Commissioners were conducted in the Society's laboratories. The Commissioners — there were nine of them — first met on 19 September 1809 and they proceeded to appoint a number of engineers among whom there was apportioned the task of conducting detailed investigations of the individual bogs. Each engineer was expected to produce a map of the bogs assigned to him and, by the time the Commissioners had completed their task in 1814, detailed surveys had been made of 1,013,358 acres (4,102 square kilometres) of bogland, while further reconnaissance surveys had been conducted of the peatlands over extensive areas of counties Galway, Mayo, and Wicklow. The fifty maps of the bogs accompanying the four published reports of the Commissioners[83] are mostly drawn to a scale of 1.5 inches to one statute mile (1:40,877) and they constitute a quite remarkable essay in the mapping of one facet of Ireland's drift geology. Under certain lighting conditions the reds and browns of the Irish peatlands can offer a surprisingly colourful landscape, but for most of the time the engineers and their assistants must have found themselves surveying a dank, rain-sodden landscape under a wind-swept drizzle falling from a sullen sky. The bog maps are a fine testimonial to their skill and tenacity of purpose.

Eventually a total of nine engineers was employed by the Bog Commissioners, among them being men such as David Aher, formerly a colliery engineer at Castlecomer, Richard Lovell Edgeworth, father of Maria, and Alexander Nimmo, a Scotsman of whom more anon (see pp. 52-53). In the present context, however, one of the engineers far exceeds all the others in importance — Richard Griffith (1784-1878), the man who has often been hailed as 'the father of Irish geology'.[84] Griffith, the son of a County Kildare landowner, was the first engineer to be appointed by the Commissioners (he was appointed on 28 September 1809) and it may be that nepotism secured him the post because his father happened to be one of the Commissioners. Should that be the case, then the passage of time was thoroughly to justify the parental action because Griffith was to enjoy a long career of the highest distinction in the public service.

For the Commissioners Griffith produced a total of ten maps, eight of them covering bogs in the Suck valley and parts of the Bog of Allen, but it is the other two sheets which must here detain our attention. On 2 July 1811 the Commissioners directed Griffith to go to north-western County Mayo to make a reconnaissance survey of the region's bogs, drainage, population, soil, and communications. There was also the following direction:

> Mr Griffith is to make such mineralogical or geological observations as may incidentally arise from his inspection.[85]

In addition Griffith was told to include in his report the most accurate map that he could achieve of northwestern Mayo, and this map he completed in 1812, basing some of it, he claimed, upon his own simple triangulation survey. Its title is *Map of the north western part of the County of Mayo*,⁸⁶ its scale is one inch to 1.7 statute miles (1:107,712), and it carries just one geological line, bearing the inscription 'Line of the junction of the Primary and Secondary countries'. The line is a somewhat crude plotting of the boundary along which the Carboniferous strata in the east overlap onto the Dalradian rocks to the west, the boundary running from the north Mayo coast near Belderg to the shores of Lough Conn.

During the following year – 1812 – Griffith conducted a similar reconnaissance survey in the Leinster Mountains. Again he employed his own triangulation to construct the map, this map being dated 1813 and entitled *Map of the mountainous & uncultivated parts of the counties of Dublin & Wicklow*.⁸⁷ Its scale is about one inch to 1.3 statute miles (1:84,480) and it, too, carries a single geological boundary, this time the junction selected being that of the granite of the Leinster batholith against the surrounding Ordovician slates. Griffith admitted that he had not been able to plot this line with any precision 'owing to the deep clay covering', but clearly he can have made only a feeble attempt to trace the line in the field because in some places his boundary is displaced up to eight kilometres from its true position. Neither his map of Mayo nor that of Wicklow can be hailed as a geological masterpiece, but they are both of interest as Griffith's earliest surviving attempts to represent aspects of Ireland's solid geology in cartographic form.

The Dublin Society and the Leinster coalfield survey

Here, momentarily, we leave Dublin for London. There the Society for the Encouragement of Arts, Manufactures and Commerce in Great Britain (later better known as the Royal Society of Arts) had been founded in 1754 to perform a role very similar to that fulfilled in Ireland by the Dublin Society. In 1802, inspired by the rising tide of interest in geology, the London society decided to further the knowledge of the science by offering either a gold medal or the sum of fifty guineas for the completion of the first mineralogical maps of England and Wales, of Scotland, and of Ireland. It was stipulated that any supplicants for the awards were to lodge their maps with the society before the first Tuesday in February 1804; that all the maps should be accompanied by certificates of accuracy; that the maps should be to a scale of not less than one inch to ten miles (1:633,600); and that they should indicate the location and character of all the mines in their area.⁸⁸

The closing date, 7 February 1804, came and went without any maps having been submitted, but the society continued to repeat its offer year by year until finally one claimant did appear. It was on 8 February 1815 that William Smith placed his splendid map of England and Wales before the society, and he was soon adjudged to be deserving of the society's fifty-guinea premium.⁸⁹ The scale of Smith's map is one inch to five miles (1:316,800) and in the light of this the society in 1815 revised the terms of its awards for any future maps of Scotland and Ireland by increasing the minimum demanded from one inch to ten miles to one inch to five miles.⁹⁰ Another development took place in the following year, 1816, when the society made the additional offer of a gold medal or a premium of fifty guineas for the first person to complete and publish a mineralogical map of any county within the United Kingdom on a scale of not less than one inch to the mile (1:63,360).⁹¹

No claims on behalf of either a geological map of Ireland or a geological map of an Irish county were ever lodged with the Royal Society of Arts despite the wide Irish publicity given to the London society's offers through the Dublin Society's own publications.⁹² The annually repeated offers of the Royal Society of Arts nevertheless did serve to keep the concept of a geological map of Ireland firmly in the public mind, and among the members of the Dublin Society there certainly persisted a strong conviction that such a map would be of the greatest value. In his *Statistical survey of the County of Wexford* of 1807 Robert Fraser observed:

> A complete mineralogical survey of Ireland would be of great national importance, and be a most desirable acquisition to that branch of science: and it is hoped that, with the return of quiet times, this important investigation will be taken up with spirit and determination.⁹³

Similarly on 18 February 1808 the Dublin Society itself resolved:

> That it is the opinion of this Society, that it is extremely desirable that a Mineralogical Survey of this Country should be made.[94]

At this juncture the Dublin Society decided to take matters into its own hands. Having passed this resolution on 18 February, the subject was referred to the Committee for Chemistry and Mineralogy for further consideration, and after due deliberation the committee reported back on 7 July 1808. Its report (Walter Stephens was one of the signatories) recommended that premiums should be offered for a geological map of one or more of the Irish counties, and this proposal the Society adopted immediately.[95] Not very surprisingly, County Dublin was selected as the area most urgently in need of mapping, and the committee was authorised to offer a premium of not more than £200 for the best geological and mineralogical map of the county. This plan seemed to offer much greater hope of success than the scheme devised earlier by the Royal Society of Arts because the financial inducement was now far greater and the geographical area to be mapped was far smaller. But, strangely, the scheme devised in Dublin was no more successful than that emanating from London, and a year elapsed without the Dublin Society receiving so much as a single claimant.

The Committee for Chemistry and Mineralogy met in the spring of 1809 to review the situation, and on 11 May it placed a series of fresh proposals before a general meeting of the Society.[96] Firstly, it recommended the withdrawal of the premium offered in the previous year, the committee pointing out that a county was in any case a poor and arbitrary unit upon which to base a geological survey. Secondly, the committee proposed that the £200 constituting the unclaimed premium should be put towards the cost of a commissioned geological survey of the Leinster coalfield located in parts of counties Carlow, Kilkenny, and Leix. Finally, it was suggested that further finance for such a survey should be drawn as needed from the £1,300 still remaining in the fund which had been established back in 1801 to finance the publication of the Society's series of county statistical surveys. At that time, in 1809, surveys still had to be published for thirteen counties, and the committee believed — wrongly as it transpired — that there was little hope of any of these outstanding surveys ever being published. In any case, the committee remarked, some of the surveys already in print were so mediocre that the abandonment of the series would be no great loss to the nation. The money, it was emphasised, would be far more profitably spent upon the proposed coalfield survey.

At this point Richard Griffith returns to the stage. He had been trained in Britain as a geologist and mining engineer between 1802 and 1808 and in the latter year, his apprenticeship over, he returned to his native land to seek a career for himself. Immediately, on St Patrick's Day 1808, he was elected to membership of the Dublin Society and he joined the Society's Committee for Chemistry and Mineralogy barely three months before it submitted its report to the Society on 11 May 1809.[97] It must surely have been his influence which caused the Society to change its policy by recommending a survey of the Leinster coalfield in substitution of the earlier project for the mapping of County Dublin. So much of Griffith's recent experience in Britain had been obtained in mining areas, and he doubtless recognised that the application of modern geological techniques to a region of known mineral potential was far more likely to yield beneficial economic results than was the somewhat academic exercise of mapping an arbitrarily chosen county. How far he was deliberately seeking to create a post for himself it is impossible to say, but if a coalfield survey was to be inaugurated, then clearly nobody in Ireland had better qualifications for the task than did Griffith. Certainly the committee was well aware of Griffith's credentials for the work, and it had broached with him the question of his assuming responsibility for the projected survey before even the general idea of a coalfield investigation had been considered by the Society at large. When on 11 May 1809 the proposal did come before the Society, the committee was able to report that Griffith had already intimated his willingness to undertake the survey on their behalf.

Further, the committee reported that if the Society authorised the project, Griffith would not be accepting the £200 fee proffered because, as a professional man, he set a higher value upon his services than this or any other sum which the Society was likely to be able to afford. He had therefore agreed to conduct the survey without any remuneration; he would merely seek reimbursement for his expenses and even that claim he would defer until after the survey had been completed. Griffith thus had no pecuniary motive in agreeing to undertake the work; outwardly it must have seemed that he was moved solely by an interest in geology and by a patriotic desire to see Ireland's mineral resources discovered and developed. But inwardly he did perhaps harbour an ulterior motive. Griffith's father had a good deal of capital invested in the Grand Canal Company, of which he was a director, and the company had recently acquired an interest in some of the coalmines near Castlecomer. Thus any enlightenment that might be shed upon the geological structure of the Leinster coalfield could be expected to rebound to the financial advantage of both the company and the Griffith family. Overtly it was nevertheless the Dublin Society which seemed to have itself a bargain and it was hardly surprising that the committee's proposals were accepted without hesitation.

Thus on 11 May 1809 there was commissioned the first detailed geological survey of any part of Ireland. Griffith's instructions were to prepare a large-scale map of the Leinster coalfield; to draw sections of the strata, and particularly of those in the vicinity of the mines; to elucidate the structure of the area in an accompanying memoir; and to assemble for the Society a collection of specimens illustrative of the region's geology. Limited though these immediate objectives might be, the Committee for Chemistry and Mineralogy clearly saw the Leinster coalfield survey as the first step towards a much more comprehensive, all-Ireland enterprise. In its report the committee observed:

> If such a beginning were once obtained, printed and circulated by the Society, it might serve as a useful pattern for further undertakings; and if executed with that degree of science, which the Committee flatter themselves with being able to obtain, might possibly appear of such national importance, as to obtain for the Society more ample funds for its further prosecution.[98]

The committee certainly entertained no doubts as to the desirability of a complete geological survey of Ireland:

> The Committee are of the opinion, that no measure would conduce more eminently to the advancement of the agriculture, manufactures, and general commerce of this country, than a complete and scientific Survey of its Mineral productions; but such a Survey as the Committee allude to, would require a degree of Geological science and practical knowledge, such as is possessed by very few, and if extended to the whole of Ireland, would demand an expense far beyond the means of the Society.[99]

Griffith started work on the coalfield survey immediately and he evidently had a strenuous field-season during the summer of 1809 because on 4 October he wrote to his close friend George Bellas Greenough:

> The constant field and hammer exercise my hands have undergone for the last four months, having rendered them nearly unfit for penmanship, will best excuse my long silence.[100]

As early as 28 March 1811 we find Griffith being granted permission to store his coalfield specimens on the Dublin Society's premises,[101] but it was 1814 before the survey was finally completed and published. Five years was an inordinately long time for the task in question, but the explanation for Griffith's lentor is to be found not so much in a lack of vigour upon his part as in his remarkable propensity — a propensity evident throughout his long life — for assuming a multiplicity of concurrent offices. Within five months of agreeing to undertake the

Leinster coalfield survey for the Dublin Society, Griffith was appointed an engineer to the Bog Commissioners and his work on their behalf must have made it difficult for him to escape from the dank brown boglands to the dripping black coalmines. Indeed, from the Commissioners he eventually claimed expenses for no less than 1300 days between 28 September 1809 and 31 December 1813[102] and, quite aside from the coalfield survey, such devotion to duty leaves us wondering how Griffith even found the time to go over to Scotland to get married in Kelso in September 1812. But his appetite for work was still by no means sated because in November 1812 he added yet a third string to his bow. On 4 June 1812 the Dublin Society had decided that, in view of the increasing importance of mining and the earth sciences generally, there should be created a new post bearing the title 'Mining Engineer to the Dublin Society'.[103] For the post there were only two candidates: Richard Griffith and Thomas Weaver (1773-1855), the latter a Freiberg graduate who since 1793 had been in charge of the copper mines at Cronbane and Tigroney in County Wicklow. The election was held on 12 November 1812 and the decision was overwhelmingly in Griffith's favour; he received 193 votes as compared with only 33 for Weaver.[104]

As the Society's Mining Engineer Griffith received a salary of £300 per annum to which in June 1815 the Society added an expense allowance of £1 for each day he was engaged in the field.[105] Initially, however, the Society was in some doubt as to precisely what his function should be. It was not until 18 February 1813 – three months after his appointment – that the Committee for Chemistry and Mineralogy reported to the Society outlining the duties which it was proposed that Griffith should perform.[106] The committee recommended that he should correspond with Irish mine proprietors in order to discover the state of the nation's mines; that he should inspect mines himself and report to the Society upon their potential with the intention of 'laying an extended foundation for a minute Mineralogical Survey of this country'; that he should examine and report upon the methods and machinery employed in Irish mines; and that he should suggest both improvements in existing mines, and places where mineral prospecting might be carried out with reasonable hope of success. Further, it was proposed that he should give an annual course of public lectures intended chiefly to be of use to landed proprietors, and treating both of the general geology of Ireland and of the basic techniques of mining and mineral exploration. Finally, the committee made the following recommendation which from our present standpoint was the most significant proposal:

> Your Committee further think it necessary, that the mining engineer should furnish accurate *maps*, on which are to be delineated the several objects of mineralogical interest which the country may afford, and also explanatory *sections* of the stratification, so as to enable a common observer to ascertain the situation of any particular bed or vein in which he may be interested.[107]

Griffith was expected to spend three months of each year in the field conducting surveys, and a further three months in Dublin completing his maps and sections, writing up his reports, and delivering his lectures. For the remainder of the year he was to be free of all responsibility to the Society. The Society itself was evidently happy with the committee's recommendations; they were accepted *in toto* on 4 March 1813.[108]

On 18 March 1813 the Dublin Society gave its Mining Engineer his first assignment; Griffith was to proceed to the Leinster coalfield to complete the survey which he had begun in 1809.[109] Perhaps in the issuing of this instruction we can detect a note of annoyance that after the passage of almost four years the coalfield survey – a project seen as one of national importance – was still incomplete. The further delay that ensued was not entirely Griffith's fault because no sooner had he been instructed to proceed to the coalfield, than he received a counter-directive to go off in search of coal around Kilbrew and Slane in County Meath. The search proved fruitless and the Society received his pessimistic report on 10 June 1813.[110] Following this diversion, he did manage to spend the greater part of the summer of 1813 down in the coalfield and he stayed there until October in which month he left for Scotland to visit his in-laws at Kelso. It was from there that he wrote to the Dublin Society on 1 November announcing the completion of his field operations and claiming, with elated exaggeration, that his investigations had revealed the presence of a coal-seam some four to six feet thick which would be sufficient to supply Ireland's fuel needs for centuries to come.[111] Now, he informed the Society, he was busy preparing his report in compliance with the instructions he had received in 1809, and by May 1814 the task was completed. On 2 June the Society authorised publication of the report[112] and agreed to pay Griffith the £200 which had been promised to him upon completion of the work, it being understood that his total expenses amounted to considerably more than the Society was now remitting.[113] Incidentally, that coal-seam mentioned by Griffith in his letter from Kelso proved a sad disappointment; the Leinster coalfield was never one of the major mining areas of the British Isles and all production finally ceased in the 1960s. Today Ireland has to import virtually all the coal she needs.

The Leinster coalfield lies between the rivers Nore and Barrow. It consists of an outlier of centripetally dipping, Upper Carboniferous strata so that the area forms a shallow geological basin. Topographically the coalfield is a plateau — the Castlecomer Plateau — because the Upper Carboniferous shales and sandstones are seemingly more resistant to denudation than are the surrounding Lower Carboniferous limestones flooring the Nore and Barrow valleys. At its highest points the surface of the upland rises some 250 metres above the adjacent lowlands, and the limits of the plateau are marked in most places by steep escarpments developed along the margins of the outlier.

Griffith's published report on this area consists of three elements. Firstly, there is the *Geological and mining report on the Leinster Coal District*, an octavo volume of some 160 pages published in Dublin in 1814 and printed by Graisberry and Campbell, the printers to the Dublin Society.[114] Secondly, there is a set of eight plates engraved by Wilson Lowry of London, bound in blue paper covers of size 27.4 x 42.9 centimetres without any title-page, author's name, or date, and simply bearing upon the cover a small printed label with the inscription 'Leinster Coal District Plates'. Of these plates numbers II, III, and IV are by far the most interesting because they are all large-scale geological cross-sections of the area. These sections are reminiscent of the profile of Derbyshire published by White Watson in 1811 at a scale of one inch to two miles (1:126,720),[115] but Griffith was working to far larger scales — a horizontal scale of up to eight inches to one mile (1:7,920) — and employing a wealth of detailed stratigraphical information collected from collieries lying astride his section lines. The result is three sections all aspiring to a degree of accuracy far higher than Watson had attempted to build into his Derbyshire transect. Admittedly, many of the minor flexures in Griffith's sections owe more to his imagination than to earthmovements, but nothing like these sections had been attempted before in Ireland and they are among the earliest geological sections produced for any part of the British Isles.

The third element in Griffith's report is his map of the coalfield. Occasionally the map is found bound in with the report, but more commonly it seems to have been included with the plates and both map and plates are now extremely rare. The map, engraved by Lowry and dated 1814, is entitled *Map of the Leinster Coal District and the surrounding country*. Its scale is one inch to 1.3 statute miles (1:84,480) — the same scale as Griffith's bog map of the Wicklow Mountains drawn the previous year — and its engraved area is a square with a side of 51.5 centimetres (Fig. 1.5). Relief is shown by hachures, and a wide range of 'cultural' features is included, but the map is basically a geological sheet representing the following four rock-categories:

1. 'Granite' — colour-washed in red.
2. 'Transition Rocks' — colour-washed in green and representing the Lower Palaeozoic slates to the east of Athy.

Fig. 1.5. Richard Griffith's map of the Leinster Coal District published by the Dublin Society in 1814. The original maps were coloured by Griffith's assistant John Kelly. (Reduced in size.)

3 'Limestone' – colour-washed in blue.
4 'Coal districts' – left uncoloured to represent the Upper Carboniferous strata, but with the outcrops of what Griffith regarded as the four major coal-seams indicated respectively by bands of yellow, purple, grey and orange.

The boundaries between the various rock categories are marked by engraved lines; continuous lines where Griffith had actually observed the junction and dotted lines where its existence could only be inferred because of the presence of a drift mantle. Dip arrows indicate the direction of inclination of coal-seams (although no degree values are given), and there are marked the sites of the boreholes for which logs are presented in the accompanying memoir.

Any critical assessment of the quality of Griffith's mapping in the coalfield is rendered difficult because of the inaccuracy of the base-map that perforce he had to employ. In so far as comment can be made upon the map, the following points are worthy of note. His mapping of the granite/limestone boundary is reasonably accurate except in the northern and southern extremities of the map, but his delimitation of the Transition rocks (the Lower Palaeozoic slates) is very poor. Again, his representation of the boundary between the limestone and the rocks of the Coal District is sound enough, although it may be observed that in most places in the field this particular junction follows the foot of the escarpment marking the edge of the Castlecomer Plateau. The plotting of the Limestone/Upper Carboniferous junction thus becomes a simple exercise in feature mapping. Griffith's chief failure was the result of his inability to cope with the many small faults which affect the region's strata. He himself was alive to the problem because in his memoir he observed that there were present in the region 'such frequent slips or down throws of the strata, as almost to baffle research'.[116] This difficulty has resulted in his joining up coal-seams in neighbouring fault-blocks and treating them as single expansive beds of coal when in reality they are fault-dislocated strata lying at very different stratigraphical horizons.[117] This must have diminished the value of his map for those mineral proprietors and mine managers whom the Dublin Society had in mind when it commissioned the survey. Griffith's work has nevertheless been adjudged superior to the survey of the same coalfield conducted forty years later by George Henry Kinahan on behalf of the Geological Survey of Ireland.[118] From the utilitarian standpoint the chief merit of Griffith's map was its clear demonstration that the margins of the plateau are developed in rocks lying well below the stratigraphical level of the coal-seams. It was therefore futile for landowners to search for coal in the strata lying beyond the limits of the lowest coal-seam to outcrop on the plateau-surface.

In 1802 William Tighe had regarded his map of Kilkenny as an essay in earth history in the sense that he believed he was depicting a sequence of rocks the chronological order of which he could discern. But in reality his misunderstanding of the county's structure (his failure, for instance, to differentiate inliers from outliers) led him to a false chronology. Now, in 1814, Griffith became the second Irish geologist to build a chronology into his map and, since he fully understood the structure of the Leinster coalfield, it follows that in his case the chronology inherent in the map is a valid one. Admittedly upon the map itself the only feature to reveal any concern for a stratigraphical sequence is Griffith's employment of the term 'Transition Rocks' with all its Wernerian chronological connotations, but the map cannot be considered apart from its accompanying sections. There we see displayed an implicit sequence from the granite of the Barrow valley, up through the limestone, to the coal-bearing formation that surfaces the Castlecomer Plateau, while in his memoir Griffith pauses briefly to consider the relative ages of the sedimentary members of this succession. He apologies for the intrusion of such theoretical matters into what was intended as a practical work, but he observes that the coal-bearing rocks are younger than the limestones, while for stratigraphical, lithological and palaeontological reasons he suggests that the local limestones, like those of Ireland in general, should be placed in the Floetz class rather than in the Transition.[119] It was only the second time that an Irish geological cartographer had concerned himself with such chronological problems, and Griffith's Leinster map must be hailed as the first successful modern geological map of any part of Ireland.

Ireland and the Geological Society of London

The Royal Society of Arts was not the only British institution to display an interest in Irish geology; there was also the Geological Society of London. Founded in 1807, the Geological Society of London was the world's first society devoted exclusively to the earth-sciences and, as part of their efforts to further a knowledge of the earth's structure, the members of the youthful society employed a geologist to conduct regional surveys upon their behalf. The man chosen for this task was Dr Jean François Berger (1779-1833). Berger, a Genevan by birth, was sent on his first assignment in 1809 when he investigated the geology of Devon and Cornwall.[120] The following year he worked in the Hampshire Basin[121] and in 1811 he was despatched to Ireland via the Isle of Man.[122] In Ireland, in 1811 and 1812, he spent most of his time examining the rocks of Ulster[123] and in connection with this visit there is told an amusing story revealing of the type of hazard that could confront the geologist in the superstitious Ireland of yore. It seems that one day upon a road in County Londonderry Berger encountered a cow being driven home for milking and he observed to its owner that it was a particularly fine beast. Shortly thereafter the animal sank to its knees with fatigue, but the simple peasant jumped to the conclusion that the mysterious stranger had put a spell upon the creature — a spell that only Berger could remove. He was therefore pursued and forcibly brought back to the recumbent cow which he was required to bless. A piece of Berger's coat-tail was then cut off and burned beneath the animal's nose, whereupon the cure supposedly effected, the cow was brought back to its feet by means of a few vigorous blows. We are told that one Murtaugh O'Grady, who had been enlisted to carry Berger's hammers and specimens, was so profoundly disturbed by his master's evident possession of the power of the evil eye that he jettisoned his load and declined to accompany Berger so much as one step further.[124] Despite such difficulties, Berger persevered and his work in northern Ireland was amplified by two distinguished English geologists, the Rev. W.D. Conybeare and William Buckland, during the course of a visit to the region during the summer of 1813. The result of their combined labours was a geological map of northeastern Ireland published by the Geological Society of London in 1816 and depicting the area to the northeast of a line drawn from Dundalk to Londonderry (Fig. 1.6).[125] The authors of the map expressed their pleasure at having encountered in Ireland representatives of many of the formations already familiar to them from their studies in England, and their map is the earliest representation of Irish geology to employ such imported stratigraphical terms as 'Old Red Sandstone' and 'Mountain Limestone', the latter being a somewhat inappropriate term in an island where the Carboniferous limestone rarely rises over the 100-metre contour. The use of such internationally recognised terms clearly implies for the map a built-in chronology thus making it the second true geological map of any part of Ireland, second only to Griffith's Leinster coalfield. The map is nevertheless only the result of a reconnaissance survey; large areas are left undifferentiated and in the south the 'Red Sandstone' (later to be identified as the Trias) of the Lagan valley is separated from the greywackes of the Newry Axis by a broad band of *terra incognita*.

In a sense this has of necessity been a somewhat disjointed chapter. It is impossible to give cohesion to a story involving a multiplicity of geologists associated with various institutions and mapping in different parts of the country. Yet the chapter really does possess a basic unity; all these geologists were convinced that, for intellectual or economic reasons, a fuller understanding of the Irish rocks was desirable and, further, that such an understanding would be facilitated by the delineation of Ireland's rocks in cartographic form. In no other part of the British Isles does there seem around 1800 to have been so much interest in geological cartography as there was in Ireland, and in this respect great credit is due to the Dublin Society. Under the Society's aegis there had been published by 1815 partial geological maps of counties Wicklow and Londonderry, a complete if not very successful map of County Cork, a detailed map of the Leinster coalfield, and an admirable map of County Kilkenny which in 1802 had rasied geological cartography within these islands onto quite a new plane of achievement. To ensure that this fine tradition was continued the Society had in 1812 created the post of Mining Engineer, the incumbent of that office being specifically charged with the duty of preparing further regional geological maps.

Many of the maps discussed in this chapter — those of Fraser and Sampson for instance — represent rocks as classified according to lithology rather than according to age. Such maps, devoid of an inbuilt chronology, may be termed 'geognostical maps' and, since there was little understanding of the principles of biostratigraphy before 1815, most early maps depicting the rocks of the earth's surface perforce fall into this category. So far as Ireland is concerned Griffith's Leinster coalfield and Berger and Conybeare's northeastern Ireland are the only two exceptions. They both possess inbuilt chronologies and they are both therefore true modern geological maps. Intermediate between the geognostical map and the geological there stands Tighe's Kilkenny where the rocks are classified according to lithology but where Tighe makes a

Sheets of Many Colours

Fig. 1.6. The map of northeastern Ireland by Jean François Berger, William Buckland, and William Daniel Conybeare. The original is 35.3 x 57.7 centimetres in size. From Transactions of the Geological Society of London, *3, 1816.*

valiant attempt to introduce some idea of chronological sequence. All these maps, be they geognostical or geological, represent a remarkable early flowering of Irish geological cartography. It is a flowering of which the Royal Dublin Society and Irish geologists as a whole may feel justly proud. When, in the 1760s, Nicolas Desmarest had sought information about the Irish basalts it was to the artistry of Susanna Drury that he was forced to turn. Now, by 1815, the geologist seeking a knowledge of Ireland's rocks could turn to sources which, if less artistic, were far more satisfactory — to the earliest of our geognostical and geological maps.

REFERENCES AND NOTES FOR CHAPTER ONE

1. M. Anglesea and J. Preston, 'A philosophical landscape': Susanna Drury and the Giant's Causeway, *Art History* 3 (3), 1980, 252-273.
2. M. Lister, An ingenious proposal for a new sort of maps of countrys, together with tables of sands and clays, such chiefly as are found in the north parts of England, *Phil. Trans. R. Soc.* 14 (no. 164), 1684, 739-746.
3. Some sources for the history of geological cartography: R.C. Boud, The early development of British geological maps, *Imago Mundi* 27, 1975, 73-96; V.A. Eyles, Mineralogical maps as forerunners of modern geological maps, *Cartographic Journal* 9 (1), 1972, 133-135; Archibald Geikie, *The founders of geology*, London, 1905, pp. xii + 486; H.A. Ireland, History of the development of geologic maps, *Bull. Geol. Soc. Am.* 54, 1943, 1227-1280; Frederick John North, *Geological maps: their history and development with special reference to Wales*, Cardiff, 1928, pp. vi + 134.
4. Some sources for the history of the Physico-Historical Society: the MS minute-book of the society 1744-1752 in the Library of the Royal Irish Academy, MS 24 E 28; *An account of the rise and progress of the Physico-Historical Society*, [Dublin, 1745], pp. 4 (a copy is in the Royal Irish Academy Tracts, box 200, tract 10); G.L. Herries Davies, The making of Irish geography, IV: the Physico-Historical Society of Ireland, 1744-1752, *Ir. Geogr.* 12, 1979, 92-98.
5. Charles Smith and Walter Harris, *The antient and present state of the County of Down*, Dublin, 1744, pp. 293; Charles Smith, *The antient and present state of the County and City of Waterford*, Dublin, 1746, pp. 386; Charles Smith, *The antient and present state of the County and City of Cork*, Dublin, 1750, in two volumes, pp. 434 and 436; Charles Smith, *The antient and present state of the County of Kerry*, Dublin, 1756, pp. 424.
6. Some sources for the life of Smith: DNB; Edmund Downey, *The story of Waterford*, Waterford, 1914, pp. x + 398; *Dublin Magazine*, July 1762, 448 for a notice of his death; M.J. Hurley, Charles Smith, our county historian, and his works, *Journal of the Waterford & southeast of Ireland Archaeological Society* 1, 1894-95, 44-47; Alfred Webb, *A compendium of Irish biography*, Dublin, 1878.
7. Smith, op. cit., 1756, 395, (ref. 5).
8. N. Desmarest, Mémoire sur l'origine & la nature du basalte à grandes colonnes polygones, déterminées par l'histoire naturelle de cette pierre, observée en Auvergne, *Histoire de l'Académie Royale des Sciences*, 1771 part 2, 705-775, and 1773 part 2, 599-670.
9. Some sources for the life of Hamilton: DNB; Elizabeth Andrews, *Ulster folklore*, London, 1913, 105-118; Thomas Frederick Colby, *Ordnance Survey of the County of Londonderry. Vol. 1. Memoir of the City and north western Liberties of Londonderry. Parish of Templemore*, Dublin, 1837, 94-96; William Hamilton, *Letters concerning the northern coast of the County of Antrim*, Belfast, 1822, xiv.
10. William Hamilton, *Letters concerning the northern coast of the County of Antrim. Containing a natural history of its basaltes: with an account of such circumstances as are worthy of notice representing the antiquities, manners and customs of that country*, Dublin, 1786, pp. viii + 195.
11. William Hamilton, *Letters concerning the northern coast of the County of Antrim. Containing such circumstances as appear worthy of notice respecting the antiquities, manners and customs of that country. Together with the natural history of the Basaltes and its attendant fossils, in the northern counties of Ireland*, Dublin, 1790, pp. xi +190.
12. *The Parliamentary Register: or, history of the proceedings and debates of the House of Commons of Ireland*, 6, Dublin, 1786, 94.
13. W. Preston, Essay on the natural advantages of Ireland, the manufactures to which they are adapted and the best means of improving those manufactures, *Trans. R. Ir. Acad.* 9, 1803, part 1, 287.
14. G. Knox, Observations on calp, *Trans. R. Ir. Acad.* 8, 1802, part 1, 207.
15. Henry Fitzpatrick Berry, *A history of the*

Royal Dublin Society, London, 1915, pp. xv + 460; Terence De Vere White, *The story of the Royal Dublin Society*, Tralee, N.D. [1955], pp. viii + 228; James Meenan and Desmond Clarke (editors), *R.D.S.: the Royal Dublin Society 1731-1981*, Dublin, 1981, pp. x + 288.

16. R. Kirwan, Observations on coal-mines, *Trans. R. Ir. Acad.* 2, 1788, part 1, 157.
17. An address to our countrymen, on the study of chemistry and mineralogy, *Trans. Dubl. Soc.* 2 (2), 1801 (dated 1802), 227-234.
18. Dietrich Ludwig Gustavus Karsten, *A description of the minerals in the Leskean Museum*, (translated by George Mitchell), Dublin, 1798, in two volumes, pp. x + 369 and 297. For an account of Leske and his collection see J.M. Sweet, Robert Jameson's Irish journal, 1797, *Ann. Sci.* 23, 1967, 97-126. See also *Proc. Dubl. Soc.* 28, 1791-92, 62, 73, 102, 133; Berry, *op. cit.*, 1915, 156-157, (ref. 15).
19. Georg Sigismund Otto Lasius, *Catalogue of a collections of fossils, from the Hartz Mountains*, Dublin, 1805, pp. 11. See also *Trans. Dubl. Soc.* 5, 1806, 257-271.
20. White Watson, *Catalogue of a collection of fossils, the production of Derbyshire; arranged according to the respective strata, in which they are found; accompanied with a tablet, representing a section of the strata*, Dublin, 1805, pp. 15. See also T.D. Ford, White Watson (1760-1835) and his geological sections, *Proc. Geol. Ass.* 71, 1960, 349-363.
21. *Proc. Dubl. Soc.* 31, 1794-95, 126-127. See also T.S. Wheeler, William Higgins, chemist (1763-1825), *Studies* 43, 1954, 78-91, 207-218, 327-388; Thomas Sherlock Wheeler and James Riddick Partington, *The life and work of William Higgins, chemist (1763-1825)*, Oxford and London, 1960, pp. viii + 696.
22. *Trans. Dubl. Soc.* 2 (1), 1800, 40.
23. *Proc. Dubl. Soc.* 33, 1796-97, 78; *Trans. Dubl. Soc.* 2 (1), 1800, 24; 3 (1), 1802 (published 1803), 39-40.
24. Isaac Weld, *Observations on the Royal Dublin Society, and its existing institutions, in the year 1831*, Dublin, 1831, 9-11.
25. J.H. Andrews, Charles Vallancey and the map of Ireland, *Geogrl. Journ.* 132, 1966, 48-61.
26. *Proc. Dubl. Soc.* 22, 1785-86, 171; 23, 1786-87, 10, 83. Berry, *op. cit.*, 1915, 154-156, (ref. 15).
27. *The report of Donald Stewart, Itinerant Mineralogist to the Dublin Society*, Dublin, 1800, pp. 142. It was also included in *Trans. Dubl. Soc.* 1 (2), 1799 (dated 1800). The Royal Dublin Society has the following MSS relating to Stewart's work: *Museum Hibernicum, regnum minerale, or a catalogue of Irish minerals in the museum of the Dublin Society*, circa 1800-1803; *A description of fossil substances discovered by Donald Stewart Mineralogist to the Dublin Society in the years 1803, 1804, & 1805 in the countys of Kilkenny, Waterford & Tipperary*; five volumes entitled *Catalogue of minerals*, circa 1797-1800, and listing 3095 specimens collected in 15 Irish counties. Some of his work also appeared in Joseph Archer's *Statistical survey of the County Dublin*, Dublin, 1801, 248-259. See also K. Dillon, Donal [*sic*] Stewart and the mineral survey of Co. Clare, *North Munster Antiquarian Journal* 7, 1953-57, 187-189, and 'Catalogue of Irish minerals in the museum of the Dublin Society', *ibid.*, 190-196.
28. *Proc. Dubl. Soc.* 45, 1808-9, 14.
29. Charles Vallancey to Sir Joseph Banks 23 April 1802. *Dawson Turner Collection* 13, 73-75.
30. *Proc. Dubl. Soc.* 61, 1824-25, 109.
31. *Ibid.*, 44, 1807-8, 166.
32. John Sinclair, *The Statistical Account of Scotland. Drawn up from the communications of the Ministers of the different parishes*, Edinburgh.
33. Ernest Clarke, The Board of Agriculture, 1793-1822, *Jl. R. Agric. Soc.*, Series 3, 9, 1898, 1-41.
34. *The statutes at large passed in the Parliaments held in Ireland*, 20, Dublin, 1801, 40 George 3, chap. 31, 361-363.
35. D. Clarke, Dublin Society's statistical surveys, *An Leabharlann* 15(2), 1957, 47-54.
36. These suggestions were reprinted in each of the Society's county surveys.
37. John Dubourdieu, *Statistical survey of the County of Down, and observations on the means of improvement*, Dublin, 1802, pp. xvi + 319.
38. Hely Dutton, *Statistical survey of the County of Clare, with observations on the means of improvement*, Dublin, 1808, pp. xxiv + 369 + 13.
39. Robert Fraser, *General view of the agriculture and mineralogy, present state and circumstances of the County Wicklow, with observations on the means of their improvement*, Dublin, 1801, pp. xvi + 289.
40. William Innes Addison, *A roll of the graduates of the University of Glasgow from 31st December, 1727 to 31st December, 1897*, Glasgow, 1898, 205.
41. Robert Fraser, *General view of the County of Cornwall with observations on the means of its improvement*, London, 1794, pp. 70; *General view of the County of Devon. With observations on the means of its improvement*, London, 1794, pp. 75.
42. Robert Fraser, *Gleanings in Ireland; particularly respecting its agriculture, mines, and fisheries*, London, 1802, pp. viii + 88.
43. Robert Fraser, *Statistical survey of the County of Wexford*, Dublin, 1807, pp. vi + 156.
44. Fraser, *op. cit.*, 1801, 9, (ref. 39).

45. William Tighe, *Statistical observations relative to the County of Kilkenny, made in the years 1800 & 1801*, Dublin, 1802, pp. xvi + 644 + 199.
46. *Proc. Dubl. Soc.* 38, 1801-2, 146.
47. Isaac Weld, *Statistical survey of the County of Roscommon*, Dublin, 1832, pp. xx + 710 + lxxii.
48. Bernard Burke, *A genealogical and heraldic history of the landed gentry of Ireland*; George Dames Burtchaell, *Genealogical memoirs of the Members of Parliament for the County and City of Kilkenny*, Dublin and London, 1888, 192; John Archibald Venn, *Alumni Cantabrigienses*, part 2, vol. 6, Cambridge, 1954, 190.
49. William Tighe, *The plants: a poem, cantos the first and second, with notes; and occasional poems*, London, 1808, pp. viii + 159.
50. Tighe, *op. cit.*, 1802, 90, (ref. 45).
51. *Ibid.*, 23.
52. T. Sheppard, William Smith: his maps and memoirs, *Proc. Yorks. Geol. Soc.*, N.S. 19, 1914-1922, 75-253; L.R. Cox, New light on William Smith and his work, *ibid.*, N.S. 25, 1942-45, 1-99.
53. V.A. Eyles and J.M. Eyles, On the different issues of the first geological map of England and Wales, *Ann. Sci.* 3, 1938, 190-212.
54. John Boys, *General view of the agriculture of the County of Kent, with observations on the means of its improvement*, Brentford, 1794, pp. 107.
55. William George Maton, *Observations relative chiefly to the natural history, picturesque scenery, and antiquities, of the western counties of England, made in the years 1794 and 1796*, Salisbury, 1797, in two volumes, pp. xi + 336 and pp. 232. The map is in vol. 2.
56. George Vaughan Sampson, *Statistical survey of the County of Londonderry, with observations on the means of improvement*, Dublin, 1802, pp. xxvi + 510 + 42. See also G.A.J. Cole, Early geological mapping in Ireland, *Ir. Nat.* 10, 1901, 10-11.
57. *Proc. Dubl. Soc.* 38, 1801-2, 155.
58. George Dames Burtchaell and Thomas Ulick Sadleir, *Alumni Dublinenses*, London, 1924, 731; James Blennerhassett Leslie, *Derry clergy and parishes*, Enniskillen, 1937, 219; Colby, *op. cit.*, 1837, 97-98, (ref. 9).
59. Horace Bolingbroke Woodward, *The history of the Geological Society of London*, London, 1907, 270.
60. George Vaughan Sampson, *County of Londonderry*. The map is dated 1813 but it was not published until 14 June 1814. Sampson prepared a memoir to accompany the map. See G.V. Sampson, *A memoir, explanatory of the chart and survey of the County of London-Derry, Ireland*, London, 1814, pp. xviii + 359.
61. Horatio Townsend, *Statistical survey of the County of Cork, with observations on the means of improvement*, Dublin, 1810, pp. xx + 749 + 96.
62. William Maziere Brady, *Clerical and parochial records of Cork, Cloyne, and Ross*, vol. 1, Dublin, 1863, 63-64.
63. *Proc. Dubl. Soc.* 44, 1807-8, 90.
64. Horatio Townsend, *A general and statistical survey of the County of Cork*, Cork, 1815, in two volumes, pp. xiv + 493 and vi + 238 + 248.
65. Horatio Townsend, *A tour through Ireland and the northern parts of Great Britain. With remarks on the geological structure of the places visited, made for the purpose of forming some judgment respecting the nature and extent of the coal formation in Ireland*, Cork, 1821, pp. 80.
66. Some sources for the life of Kirwan: J. O'Reardon, The life and works of Richard Kirwan, *The National Magazine* 1, 1830, 330-342; M. Donovan, Biographical account of the late Richard Kirwan, Esq., *Proc. R. Ir. Acad.* 4, 1847-1850, lxxxi-cxviii; J. Reilly and N. O'Flynn, Richard Kirwan, an Irish chemist of the eighteenth century, *Isis* 13, 1930, 298-319; P.J. McLaughlin, Richard Kirwan: 1733-1812, *Studies* 28, 1939, 461-474, 593-605 and 29, 1940, 71-83, 281-300.
67. Richard Ryan, *Biographia Hibernica*, London, 1821, vol. 2, 358.
68. T.R. Marchant and G.D. Sevastopulo, The Calp of the Dublin district, *J. Earth Sci. R. Dubl. Soc.* 3(2), 1980, 195-203.
69. 'A mineralogical map comprising the base and vicinage of Croaghan Mountain in the County of Wicklow, Ireland', Brit. Mus. Add. MS. 32, 451G.
70. R. Kirwan, A plan, for the introduction and establishment of the most advantageous management of mines in the Kingdom of Ireland, *Trans. Dubl. Soc.* 1(1), 1799 (dated 1800), 277-284; 2(2), 1801 (published 1802), 245-251. See also Richard Kirwan to Sir Joseph Banks 15 March 1802, *Dawson Turner Collection* 13, 53-54.
71. T.M. Porter, The promotion of mining and the advancement of science: the chemical revolution of mineralogy, *Ann. Sci.* 38(5), 1981, 543-570.
72. Richard Griffith, *Geological and mining report on the Leinster Coal District*, Dublin, 1814, ii.
73. The Kirwanian Society of Dublin, *Phil. Mag.* 39, 1812, 319-320; P.J. McLaughlin, The Kirwanian Society (1812-1818?), *Studies* 43, 1954, 441-450; the MS. minute book of the society from 1815-1818 in the Library of the Royal Irish Academy at 12 B2 16.
74. Trinity College Board Minutes 21 June 1806.
75. Whitley Stokes, *A catalogue of the minerals in*

75. *[continued] ...the museum of Trinity College, Dublin. Part I. Containing the systematic collection*, Dublin, 1807, iii.
76. *Ibid.*; James Apjohn, *A desciptive catalogue of the simple minerals in the systematic collection of Trinity College, Dublin*, Dublin, 1850, v-vii.
77. *Q. Jl. Geol. Soc. Lond.* 18, 1862, xxx-xxxiv; P.I. & J.M.E. Manning, The centenary of a pioneer Irish geologist: W.H. Fitton, M.D., F.R.S., 1780-1861, *Ir. Nat. J.* 13(11), 1961, 241-244.
78. Walter Stephens, *Notes on the mineralogy of part of the vicinity of Dublin*, edited by W.H. Fitton, London, 1812, pp. 57. See also W.H. Fitton, Notice respecting the geological structure of the vicinity of Dublin; with an Account of some rare minerals found in Ireland, *Trans. Geol. Soc. Lond.* 1, 1811, 269-280, and *Nicholson's Journal* 31, 1812, 280-290.
79. Trinity College Board Minutes 13 June 1812.
80. W.H. Fitton to G.B. Greenough 17 November 1812, *Greenough Papers*.
81. Stephens, *op. cit.*, 1812, 6f-7f, (ref. 78).
82. 'Commissioners for improvement of bogs in Ireland minute book', 19 September 1809 to 31 December 1813, a MS. in private hands in Ireland.
83. H.C. 1810 (365), X, pp. 55; 1810-11 (96), VI, pp. 206; 1813-14 (131), VI second part, pp. 218.
84. See, for example, George Henry Kinahan, *Manual of the geology of Ireland*, London, 1878, dedication, and John Kaye Charlesworth, *Historical geology of Ireland*, Edinburgh and London, 1963, 3.
85. MS. minute book, (ref. 82).
86. *The fourth report of the Commissioners appointed to enquire into the nature and extent of the several bogs in Ireland, and the practicability of draining and cultivating them*, H.C. 1813-14 (131), VI second part, plate XV.
87. *Ibid.*, plate XVI.
88. *Transactions of the society instituted at London, for the encouragement of arts, manufactures, and commerce* 20, 1802, 41-42; Henry Trueman Wood, *A history of the Royal Society of Arts*, London, 1913, 301-302; Derek Hudson and Kenneth William Luckhurst, *The Royal Society of Arts 1754-1954*, London, 1954, 98-99.
89. *Transactions...*, *op. cit.*, 33, 1815, vii, 22, 51-60, (ref. 88). See also Eyles and Eyles, *op. cit.*, 1938, (ref. 53).
90. *Transactions...*, *op. cit.*, 33, 1815, 12, (ref. 88).
91. *Ibid.*, 34, 1816, 16.
92. *Premiums offered by the Dublin Society, for agriculture, planting, and fine arts. Also, premiums of the Board of Agriculture, and London Society, extending to Ireland*, Dublin, 1802, pp. 53. A similar list was published by the Dublin Society each year. See *Trans. Dubl. Soc.* 3, 1802 (published 1803), 102-103; 6, 1810, 17.
93. Fraser, *op. cit.*, 1807, 19, (ref. 43).
94. *Proc. Dubl. Soc.* 44, 1807-8, 70.
95. *Ibid.*, 166-167.
96. *Ibid.*, 45, 1808-9, 135-139. See also *Nicholson's Journal* 23, 1809, 237-239.
97. *Ibid.*, 45, 1808-9, 54.
98. *Ibid.*, 137.
99. *Ibid.*, 137.
100. *Greenough Papers*.
101. *Proc. Dubl. Soc.* 47, 1810-11, 114.
102. Griffith's work returns are in the reports of the Bog Commissioners. See ref. 83.
103. *Proc. Dubl. Soc.* 48, 1811-12, 138-140, 147. In 1812 the Dublin Society also created the new post of Professor of Mineralogy, a post to which in 1813 there was appointed the German Karl Ludwig Metzler Von Giesecke (1761-1833). He held the post until his death and he made many tours in Ireland, but his work was purely mineralogical and he never depicted his findings in cartographic form.
104. *Proc. Dubl. Soc.* 49, 1812-13, 21-22.
105. *Ibid.*, 48, 1811-12, 147; 51, 1814-15, 195-196, 203.
106. *Ibid.*, 49, 1812-13, 84-87.
107. *Ibid.*, 85.
108. *Ibid.*, 108.
109. *Ibid.*, 112.
110. R. Griffith, Mineralogical observations on the country, between Kilbrew and Slane, in the County of Meath, *ibid.*, 191-195.
111. *Ibid.*, 50, 1813-14, 16-18.
112. *Ibid.*, 166.
113. *Ibid.*, 188-189, 195.
114. Richard Griffith, *Geological and mining report on the Leinster Coal District*, Dublin, 1814, pp. xxiv + 135.
115. White Watson, *A delineation of the strata of Derbyshire, forming the surface from Bolsover in the east to Buxton in the west*, Sheffield, 1811, pp. xiii + 72. Reprinted in 1973.
116. Griffith, *op. cit.*, 1814, 16, (ref. 114).
117. W.E. Nevill, The Millstone Grit and Lower Coal Measures of the Leinster Coalfield, *Proc. R. Ir. Acad.* 58B (1), 1956, 1-16.
118. T.R. Marchant and G.D. Sevastopulo, Richard Griffith and the development of stratigraphy in Ireland, pp. 173-196 in Gordon Leslie Herries Davies and Robert Charles Mollan (editors), *Richard Griffith 1784-1878*, Royal Dublin Society, 1980, pp. vi + 221.
119. Griffith, *op. cit.*, 1814, xiv-xvi, (ref. 114).
120. J.F. Berger, Observations on the physical structure of Devonshire and Cornwall, *Trans. Geol. Soc. Lond.* 1, 1811, 93-184. See also Woodward, *op. cit.*, 1907, 13, 44, 49-51, (ref. 59).

121. J.F. Berger, A sketch of the geology of some parts of Hampshire and Dorsetshire, *Trans. Geol. Soc. Lond.* 1, 1811, 249-268.
122. J.F. Berger, Mineralogical account of the Isle of Man, *ibid.*, 2, 1814, 29-65.
123. W. Conybeare and J.F. Berger, On the geological features of the northeastern counties of Ireland, *ibid.*, 3, 1816, 121-222. On 19 November 1813 Berger presented a map of County Wicklow to the Geological Society of London but it is not clear whether the map carried any geological information. See *ibid.*, 2, 1814, 540.
124. J.B., Popular Irish superstitions, *The Newry Magazine* 1, 1815, 214-218.
125. Conybeare and Berger, *op. cit.*, 1816, (ref. 123).

2

THE GENESIS OF GRIFFITH'S MAP

1809-1830

Within the city of Dublin the small area lying between Trinity College and the murky waters of the River Liffey is hardly one noted for the quality of its urban landscapes. Hawkins Street shares in the character of the district. Extending from Trinity College to the Liffey at Burgh Quay, the street commemorates Alderman William Hawkins who died in 1680, but if he could see his street today he might be less than happy with the intended honour. What would greet his gaze is the side-wall of a cinema, the rear-end of a chapel, a quick-food bar, a turf-accountant's, and a forest of bus-stops, the *tout ensemble* dominated by a high-rise office block which is entirely devoid of any architectural pretensions. But things were rather different in Dublin's gracious Georgian era when, from 1796 to 1815, a three-storey, pilastered building in Hawkins Street served as the headquarters of the Dublin Society. Among the facilities contained within that building there were exhibition galleries, a large library, two spacious rooms for the Leskean collection, a chemical laboratory, and a lecture-theatre said to have been capable of accommodating 800 people. Such a large theatre was evidently a necessity because the Society's programme of lectures seems to have attracted large audiences. In 1811, for instance, when the Society brought over Humphry Davy to deliver two lecture courses, one in chemistry and the other in geology, the tickets for the courses cost two guineas apiece but all the 550 tickets available were sold within a week. Disappointed applicants, it seems, were willing to pay up to twenty guineas in order to secure a seat at Davy's prelections.[1] One small portion of the Dublin Society's Hawkins Street headquarters perhaps still survives, but the lecture-theatre, where Davy once held his audiences in rapt attention, has long since disappeared. That theatre must nevertheless hold some interest for us because it was probably there, in the spring of 1814, that a geological map of the whole of Ireland was first placed upon public exhibition. The author of that pioneer map was Richard Griffith, the Dublin Society's Mining Engineer. This chapter explains how that map came to be compiled and traces the story of Griffith's early efforts to refine his work and to secure its publication on behalf of the Royal Dublin Society. Those efforts bore no fruit. When Griffith resigned his post with the Society in November 1829 publication of the map was still almost ten years distant and consideration of the later stages in the map's protracted gestation will have to be reserved until a later chapter.

Griffith's early years

Born in the city of Dublin on 20 September 1784, Richard Griffith was the son of the owner of the Millicent estate which is picturesquely located upon the banks of the Liffey between Sallins and Clane in County Kildare.[2] Educated at schools in Portarlington, Rathangan, and Dublin, the young Griffith in 1800 took a commission in the Royal Irish Regiment of Artillery, but when his regiment was merged into the British Royal Regiment of Artillery at the Act of Union in 1801 he elected to retire from the army, although for the next half century his name continued to feature in the Army List as an officer retired upon full ensign's pay. Now, perhaps inspired by his father, Griffith resolved to become a civil and mining engineer. He took himself off to Hawkins Street to examine the rocks and minerals in the Dublin Society's collections[3], and about 1802 he went to London to study chemistry, geology, and mineralogy at the Scientific Establishment for Pupils run in Soho Square by William Nicholson. Next, in 1804, Griffith travelled to Cornwall to acquaint himself with practical geology in what was then a major British mining region, and while there he made a profound impression upon one of the wealthiest of the local mining magnates, Lord de Dunstanville. When Griffith proposed to visit Birmingham to inspect Matthew Boulton's famed foundry and manufactory it was de

Dunstanville who gave him the necessary letter of introduction, the baron observing that:

> ... the young man has a great turn for scientific pursuits.... He is not more than 20 years of age. I am sure therefore you will be surprised at his natural talents & acquired knowledge considering his extreme youth.[4]

The visit to Birmingham seemingly took place in 1805 and it was evidently part of a major excursion through the mining districts of Bristol, Glamorgan, Lancashire, and Yorkshire, but Griffith's absence from Ireland can hardly have been protracted because he tells us that in the same year he attended the lectures given in Trinity College Dublin by Robert Perceval,[5] the University's first Professor of Chemistry.

In 1806, Griffith departed for Edinburgh where the engineering works connected with the development of the New Town doubtless provided him with many a precept. But Auld Reekie had another and more important claim upon his attention. He was becoming increasingly involved with geology and the Scottish capital was then an internationally renowned centre of geological debate as two schools of thought within the city vied with each other for dominance. On the one hand there were the Huttonian Plutonists — the advocates of an igneous origin for granite — while on the other hand, and led by the Freiberg-trained Robert Jameson, there were the Wernerian Neptunists — the advocates of the theory that granite and various other rocks were chemical precipitates formed in a mysterious primordial fluid. Edinburgh was the place to be, and many were the students who flocked thither to meet the leading protagonists and to examine the field evidence adduced by the two schools in support of their respective systems.

Although he never matriculated in the University of Edinburgh, Griffith tells us that he did attend Professor Jameson's lectures[6], and it was perhaps in these classes that he acquired that familiarity with the Wernerian terminology which is so evident in his earlier geological writings (see p. 27). This is not to say that Jameson had found an easy convert in his young Irish student; Griffith was always at pains to explain to his readers that his use of such terms as Formation, Primitive, Transition, and Floetz implied upon his part no general acceptance of the Wernerian system.[7] His Edinburgh contacts were certainly by no means exclusively within the Neptunian camp; he mixed freely with such leading Plutonists as John Playfair and Lord Webb Seymour, and he tells us that he actually assisted Sir James Hall in some of his famed experiments on the heating and cooling of rocks.[8] Griffith's claim that Jameson once privately confessed to him that as a mineralogist he was no match for his young Irish pupil hardly has about it the ring of truth (modesty was never one of Griffith's virtues), but he certainly does seem to have made some impact upon his Edinburgh confrères. How else can we explain the fact that in 1807, at the early age of 22, he was elected a Fellow of the Royal Society of Edinburgh?

Among Griffith's fellow students in Edinburgh there is one who deserves especial mention — the Genevan Louis Albert Necker (1786-1861). With all the abounding ambition of youth, Necker was compiling nothing less than a geological map of the whole of Scotland.[9] The two young men clearly discussed this project together and Griffith imparted to his friend some geological information about the islands of Colonsay, Islay, and Jura gleaned during the course of a recently completed tour of the Hebrides. Necker's map, presumably incorporating Griffith's observations, was eventually presented to the Geological Society of London on 4 November 1808, and this episode perhaps gave Griffith his earliest encounter with the problems of geological cartography. It may be that the incident actually triggered his own thinking. If Necker could compile a geological map of Scotland, then why should Griffith not essay a comparable map of Ireland?

In 1808, with his profession now at his finger-tips, Griffith decided to return home to try his fortune in Ireland, although he afterwards admitted that he had quitted Edinburgh only with the greatest of reluctance.[10] Perhaps this sadness owed something to an affair of the heart because while in Scotland Griffith had fallen 'madly in love' — the words are his own — with a young lady — Maria Jane Waldie (1786-1865) — whose father's favoured residence — Hendersyde Park near Kelso in Roxburgh — was a mere 60 kilometres from Edinburgh. It was to be 21 September 1812 before, in Kelso, the happy couple were united in marriage, but during the interval Griffith showed no sign of being a languid lovelorn swain. On the contrary, he threw himself deeply into public affairs. On St Patrick's Day 1808 he was elected to membership of the Dublin Society upon the proposal of General Vallancey and one Edward Houghton[11]; on 2 February 1809 he was appointed to the Society's Committee for Chemistry and Mineralogy[12]; on 11 May 1809 the Society commissioned him to investigate the Leinster coalfield (see p. 22)[13], and on 28 September 1809 he was named as an engineer to the Bog Commissioners, being credited between that date and 31 December 1813 with that remarkable total of 1300 days of service on behalf of the Commissioners (see p. 24). In addition, between 4 June 1812, when the Dublin Society decided to establish the post of Mining Engineer, and 12 November following, when the election to the new post took place[14], Griffith must have spent a good deal of time not only upon his nuptials, but also upon the far more mundane task of lobbying the Society's members in his own interest. These were indeed busy years for Griffith — busy years during which he began to display that remarkable professional vigour which was to remain one of his outstanding characteristics throughout a long and fruitful life — but they were also the years that saw the genesis of what was to become Griffith's proudest achievement: his quarter-inch geological map of Ireland.

Fig. 2.1. Richard Griffith in 1840 when at the height of his powers. From a pastel portrait in the possession of Colonel A.P. Daniell, by whose kind permission it is here reproduced.

The map in vision

In later life Griffith was to speak frequently about the evolution of his geological map of Ireland, but historical consistency was not one of his fortes and he varyingly quoted 1812, 1811, and 1809 as the year in which he commenced the map's compilation.[15] Indeed, in a letter written in 1821, he would perhaps seem to imply a still earlier date for the origin of the map:

> For many years antecedent to my being elected to the Mining Professorship of the Society [in 1812], I was engaged in making a Geological Survey of Ireland, in the view of ultimately publishing a general Geological Map of the county . . . .[16]

Perhaps any author has difficulty in recollecting that precise moment in time at which he was first moved to undertake some particular project, and in the case of the map of Ireland the difficulty must have been compounded because a second individual was involved in the conception of the work: George Bellas Greenough (1778-1855).

Greenough was a gentleman of means, a former student of Blumenbach's at Göttingen, a member of parliament for the 'rotten' Surrey borough of Gatton from 1807 until 1812, and both a founder member and the first president of the Geological Society of London.[17] He had family connections with the Colthursts who owned estates in the southwest of Ireland, and down to the early years of the present century his geological hammer with a whalebone handle was preserved as a family heirloom at Dripsey Castle in County Cork. Exactly how they met remains uncertain, but Greenough and Griffith early developed a close friendship. Perhaps they became acquainted during Griffith's years of residence in London around 1803, and it could be that Griffith's studies in Cornwall were undertaken at Greenough's suggestion because he had himself made just such a study-tour of the West Country following his return from Germany in 1801. Indeed it was perhaps Greenough's unbounded enthusiasm for geology which gave Griffith's scientific work its particular orientation, and the two men certainly came to share a deep interest in geological cartography. It was in 1808 — the year of Griffith's return to Ireland — that Greenough started work upon the great geological map of England and Wales that was to become his own magnum opus. He placed an early draft of the map before the Geological Society of London in 1812, and the map was finally published in 1820 at a scale of one inch to six miles (1:380,160).[18] It was while he was at work upon this project that Greenough put a proposal to Griffith. Would Griffith undertake the compilation of a geological map of Ireland to stand as a companion to the map of England and Wales that was now engaging so much of Greenough's time? Almost half a century later Griffith must have been remembering this occasion when he admitted that he had begun the map of his native land 'at the pressing instance of Mr. Greenough, one of my oldest and most valued friends'.[19] Exactly how and when Greenough first displayed this 'pressing instance' is not recorded, but the seed from which Griffith's map grew had probably been planted in Griffith's mind earlier than the spring of 1811 because then, on 16 May, he wrote to Greenough saying 'I am collecting materials relative to the general geology of Ireland'.[20] The suggestion had certainly crossed the Irish Sea before the end of that year because in a letter to Greenough dated 25 November 1811 Griffith makes his earliest specific reference to what became his geological map of Ireland.[21] This reference is of particular interest because in the letter Griffith writes of the map as 'our map' and it seems that the two friends were at that stage actually collaborating in its production. Perhaps Greenough then had some hope of producing a geological map of the whole of the British Isles with Griffith assuming chief responsibility for the Irish portion.

The origins of the geological map of Ireland can thus with certainty be traced back to 1811 and that, so far as we know, was the year in which the map entered upon its long gestation period. It is not intended to suggest, however, that Griffith was guilty of misrepresentation when, in later years, he claimed to have begun the map's compilation as early as 1809.

That was the year in which he agreed to execute a survey of the Leinster coalfield, and while a map of the coalfield was clearly then his immediate objective, the geological lines then plotted must have found their place upon the earliest version of the map of Ireland constructed only five years later. In this sense the ancestry of the map can be traced back almost to the time of Griffith's return to Ireland, but between 1809 and 1813 his strenuous labours on behalf of the Bog Commissioners and the Dublin Society can have left him with little time to reflect upon the broader issues of Irish regional geology. Not until late in 1813 did the pressure of his public duties diminish somewhat, and it was only then that he really began to work seriously at his projected map.

In a letter written from his father-in-law's house near Kelso on 1 November 1813 Griffith observed:

> Having finished the field work of the coal district, I employed myself for some time in forming a Geological Map of Ireland, from my own notes, and from those of some of my Mineralogical friends who were kind enough to assist me on the occasion. I shall exhibit this map at my Lectures on Mining, &c. and I have no doubt it will be found generally interesting to the Miner and Agriculturalist.[22]

One of those 'mineralogical friends' must have been Greenough because he had been in Ireland during the summer of that year from 8 July until 2 September, spending the greater part of his time in counties Galway and Mayo.[23] Presumably the two men discussed the projected map of Ireland during this visit, and Greenough can hardly have been surprised when, early in the following January, he received from Griffith a letter containing the following passage:

> I wish to give some idea of the General Geology of Ireland in my mineralogy lectures at the Dublin Society in March. I think Taylor's Map of Ireland is on too small a scale for Public exhibition. I am determined on this account to have one of Arrowsmith's maps coloured for the purpose.[24]

Would Greenough, therefore, Griffith continued, kindly visit the London shop of the cartographer Aaron Arrowsmith to purchase there a good, uncoloured impression of Arrowsmith's map of Ireland in sheets. This, Griffith explained, would serve as his base-map for the projected geological sheet. But Griffith also had a further request: before despatching the map to Ireland would Greenough please pencil in all the geological observations he had made during his Irish tour of the previous summer. In a very real sense it was still 'our map', but Greenough evidently had some doubts about the wisdom of allowing his own Irish investigations to be made public in the manner that Griffith intended, because on 3 February 1814 we find Griffith writing to him as follows:

> With respect to the Geological Map of Ireland I do not mean to make any other use of it than simply exhibiting it one day at my lectures as an imperfect thing but mainly to give some idea of Irish geology. Of course I shall mention the sources of my information. I cannot concur my giving a transcient glimpse of it to some 80 people can have any effect in diminishing its value to the Geological Society.[25]

Griffith's first lecture course at the Dublin Society opened in March 1814[26] and, while there is no proof positive that he did then exhibit his outline geological map of Ireland, there does seem every likelihood that the map did make its first public appearance in the Dublin Society's Hawkins Street lecture-theatre in the spring of 1814.[27] All trace of that particular map has long since disappeared, but presumably in compiling it Griffith had recourse to such earlier maps as those of Fraser, Tighe, Sampson, Townsend, and Stephens and Fitton. Over the greater part of Ireland, however, he had no such precursors, and there he must have leaned heavily upon both his own observations and those of Greenough and other of his friends. The base-map was presumably Arrowsmith's map of Ireland published on 4 January 1811 to a scale of approximately one inch to four miles (1:253,440), although many years later Griffith referred to himself as having coloured Arrowsmith's map not in 1814 but in 1816.[28] This could have been merely a lapse of his memory, or maybe the 1814 map was based upon some sheet other than Arrowsmith's. Lieutenant Alexander Taylor's *A new map of Ireland* published in January 1793 on a scale of approximately one inch to ten miles (1:633,600) was the obvious earlier, if less accurate candidate for such usage. Griffith is certainly known to have added geological colours to the maps of both Arrowsmith and Taylor[29], and his first lines were perhaps plotted upon a Taylor because the Arrowsmith requested of Greenough in January 1814 had failed to arrive in sufficient time for it to be modified for use during the lecture-course beginning in the following March. Whatever the base-map may have been, Griffith was soon displaying concern that his map should be seen far beyond the confines of the Dublin Society's lecture-theatre; in a flush of premature enthusiasm he began to consider the question of publication. On 2 April 1815 he wrote to Greenough:

> I am still working away at the Geological Map of Ireland and have made considerable progress but it is not nearly ready for publication. When it is, I think it will be better to publish it by subscription as really it will be too much to expect from the Geological Society.[30]

Why did Griffith decide to burden himself with the compilation of a geological map of Ireland? Clearly in 1814 his immediate objective was to equip himself with a useful teaching aid in illustration of his Dublin Society lectures, and perhaps he was also not unmindful of that premium for a map of Ireland which the Society of Arts had been offering for so long without success (see p. 21). But there was another, patriotic reason behind his decision. He had firmly in mind the long-term advantages which must accrue to Ireland through a detailed survey of its geological resources. In his Leinster Coal District report of 1814 he observed that

> ... the internal treasures of Ireland remain unexplored, although there is scarcely a county in the Island, which does not exhibit indications of valuable mineral productions.[31]

A geological map could both remedy this unfortunate situation and prevent a squandering of money upon fruitless searches for coal in localities where its presence was stratigraphically impossible. Such utilitarian motives were undoubtedly strong, but we must also suspect that Griffith found a deep intellectual satisfaction in tracing geological boundaries through the Irish countryside as he converted the bald topographical map into a patchwork of geological colours — a patchwork in which each tint tells a story of majestic events that are now countless millions of years distant. His was a satisfaction familiar to any field-geologist who has succeeded in deciphering a fragment of the geological palimpsest, and with both Smith and Greenough known to be at work upon geological maps of England and Wales, was it not natural that Griffith should have sought personal fulfilment through the construction of a companion map of his own native land? There was still a long way to go, but that map displayed at the Dublin Society in 1814 was a beginning — the first essay towards a geological map of the whole of Ireland.

New regional surveys

The return of peace to Europe in the spring of 1814 allowed British travellers to flock to the continent for the first time in many years, but Griffith never showed much interest in geologising abroad. With his inaugural lecture course behind him, and with the Leinster coalfield report safely into print (see p. 25), he spent the summer of 1814 making a major reconnaissance tour of the west of Ireland on behalf of the Dublin Society.[32] His journey took him from the barren plateaux of County Leitrim, across Roscommon and Galway, to the stark karstic limestones of Clare, and thence across Limerick to the majestic Old Red Sandstone mountains of Kerry and Cork. All told he covered a distance of almost 1600 kilometres, and in his report to the Society he observed that the excursion had yielded a great deal of valuable information which he would be embodying into his future lecture courses.[33] On this tour Griffith was not alone; he took with him a young man named John Kelly (1791-1869) from Borrisokane, County Tipperary. Kelly had already been responsible for colouring the 250 published copies of Griffith's Leinster coalfield map — a duty for which he was paid £3.18.1¼[34] — and for the next few years he was in the regular employ of the Dublin Society as Griffith's draughtsman, model-maker, and general assistant. His association with Griffith, however, was to be of much longer duration and it spanned almost forty years, with Kelly playing an important part in many of Griffith's future field investigations.

As a result of the discoveries made during his 1814 tour, the Society instructed Griffith to commence a detailed survey of the so-called Connaught Coal District lying in northwestern Ireland in counties Cavan, Fermanagh, Leitrim, Roscommon, and Sligo. Of this coalfield Griffith had earlier written:

> At present nothing is known, except that the outer edges of several beds of coal have been observed, but they have not been traced to any distance; so that their extent is by no means ascertained.[35]

It was now his task to remedy this state of affairs. He started work in the coalfield during the autumn of 1814 and a further three months of surveying during the ensuing summer brought the study to completion. On 13 June 1816 Griffith reported to the Society that his report on the Connaught coalfield was finished[36] and publication was soon authorised[37], but three years were to elapse before the report was finally ready for distribution. This delay supervened because Griffith insisted upon waiting until a new topographical map of County Roscommon was available to assist him in the preparation of the base-map for the geological sheet which was to accompany the Connaught coalfield report.[38] Griffith had more than a passing interest in this Roscommon map because he and William Edgeworth (1794-1829) (the two men had formerly been colleagues on the bogs survey) were themselves constructing the sheet for the county's Grand Jury, Edgeworth being responsible for the northern portion of the county and Griffith for the southern. The basis of the work was a careful triangulation of the county, and they completed their field operations in 1817 although the map — an excellent one — was not published until 1825.[39] Thus by the close of 1817 Griffith had to hand the missing Roscommon piece of the Connaught coalfield 'jigsaw', and in the following year five hundred copies of the Connaught Coal District report were printed off.[40] At this point Griffith began to display a failing which was to cloud the remainder of his association with the Dublin Society; he seems to have worked strenuously enough in the field, but — like many another geologist — he was dilatory about getting the fruits of his labours into print. His map of the Connaught Coal District is dated 1817 and the title page of the accompanying text is inscribed 1818, but the volume remained unpublished for many months because the map and sections were still only at proof stage even as late as 29 April 1819.[41] Despite its optimistic and misleading dates, it was the summer of 1819 before the Connaught report finally made its appearance.

The new report is in the same general format as that for the Leinster Coal District[42] (Griffith regarded his second report as a supplement to that of 1814) and like the earlier report it presents a somewhat sanguine account of the region's mineral potential. The Dublin Society must have been gratified to receive Griffith's opinion that Connaught 'contains abundance of bituminous Coal of excellent quality, which, from various causes, remains neglected and unwrought'[43], but in reality the Coal District was never to achieve an output any higher than the very modest production attained in the Leinster coalfield. The Connaught report contains two graphical illustrations. Firstly, there is a long east to west cross-section of the coalfield, geologically coloured and with a horizontal scale of one inch to 6720 feet (1:80,640). Secondly, there is a handsome fold-out geological map (size 56 x 49 centimetres) dated 1817 and representing some 2,000 square kilometres of territory centered upon Lough Allen (Fig. 2.2). In character the map is identical to the earlier map of the Leinster coalfield although the engraver

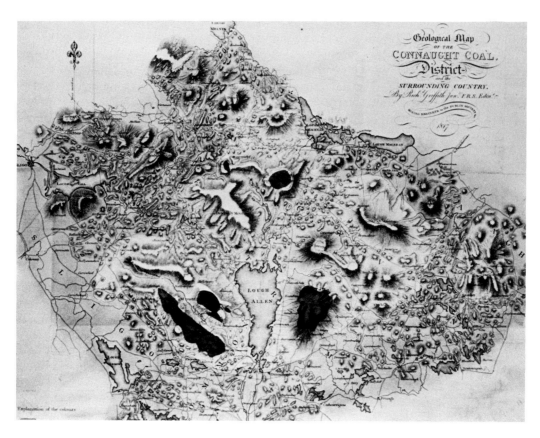

Fig. 2.2. The northern portion of Griffith's map of the Connaught Coal District published by the Dublin Society in 1819. The dark tint represents the 'situation and extent of the different coal fields'. (Reduced in size.)

is now John Taylor of Dublin and the scale has been reduced by a third so that one inch now represents 1.75 miles (1:110,880).

Geologically the Connaught coalfield is very similar to the Leinster coalfield in that they both consist of large outliers of Upper Carboniferous strata rising above limestone lowlands, but topographically the two regions are very different. The rocks of the Leinster coalfield form one expansive plateau, but in the Connaught case the outlier has been carved up by denudation to form a large number of small plateau units disjoined by deep valleys. The steep sides of these valleys provide the geologist with excellent natural cross-sections of the area and this makes the geology rather easier to interpret than is the case in the undissected Leinster outlier.

On his map Griffith shows the following six categories of rock, the categories in most places being separated from each other by engraved lines:

1. 'Granite' (red) – representing the gneiss of Benbo standing to the northwest of the coalfield and forming a part of the Ox Mountain inlier.
2. 'Primary slate' (purple) – washed over the remaining gneiss of the Ox Mountain inlier.
3. 'Old Sandstone' (orange) – employed chiefly to indicate the Old Red Sandstone and the basal Carboniferous sandstones and shale outcropping to the south of the coalfield in the Curlew Mountain inlier.
4. 'Limestone' (blue) – indicating the widespread presence of the Lower Carboniferous limestone around the margins of the coalfield.

5. 'Coal Formation' (uncoloured) — showing the extent of the Upper Carboniferous strata in the main body of the coalfield and in the small Slieve Russell outlier to the east.
6. 'Situation and extent of the different coal fields' (black) — six outliers within the coalfield are differentiated in this manner, and in some of them a dotted line represents the 'outgoing' of a particular coal seam.

As on all Griffith's early maps the tablets in the key are arranged with the oldest rocks at the top and the most recent at the bottom, an arrangement by no means unique to Griffith but nevertheless one that seems strange to a modern geologist familiar with stratigraphical columns that young upwards. Inverted though his key may be, the Connaught map depicts such stratigraphical units as 'Primary slate' and 'Old sandstone', and it therefore represents a further step in the continued movement away from the geognostical type of map and towards a map embodying modern stratigraphical principles. His accompanying memoir certainly makes it clear that he had discerned in the area a stratigraphical sequence rising from the primary granite and mica slate, through the 'first or old floetz sandstone', up to the 'first floetz limestone' and the capping coal formation. The influence of his Wernerian training in Edinburgh is still patent, but, whatever the terminology he might employ, his understanding of Irish stratigraphy was obviously maturing. Indeed, he now felt sufficiently confident to suggest a tripartite division of the Irish limestones grading from the Calp limestone at the base of the pile, through the grey and black crystalline limestones, to upper beds consisting of greyish-blue limestones containing 'a profusion of pertrifactions, but principally of madrepores, chamites, and archites'.[44] This sequence he later had to revise and in any case it was to be twenty years before he first attempted to give cartographic expression to such stratigraphical niceties.

Three years before the publication of his Connaught report Griffith shifted his attention to the third of Ireland's four provinces. It was on 4 July 1816 that the Society instructed him to proceed to Ulster to conduct a survey of the Tyrone and Antrim coalfields along the now well-established pattern.[45] He was shortly authorised to purchase maps of counties Antrim, Armagh, Donegal, and Down to assist him in his investigations[46], and in mining districts those investigations were clearly by no means confined to the surface because in 1817 the Society's accounts reveal Griffith to have spent £1.19.4d. upon the purchase of two safety lamps, one presumably being for himself and the other for Kelly.[47] Griffith worked in Ulster during the summers of 1816, 1817, and 1818 and it was during the autumn of 1817 that he and Kelly undertook a novel project: they started to survey a topographical and geological section across the whole of northern Ireland.[48] Starting on the east coast at Newcastle, County Down, they worked steadily westwards, measuring altitudes by means of one of Edward Troughton's barometers and taking the horizontal distances off existing county maps. Several labourers were employed as assistants at the expense of the Dublin Society[49], but near Castle Saunderson, in County Cavan, Griffith himself met with some kind of accident and Kelly was left to continue the work alone until he reached the west coast at the foot of Benbulbin in County Sligo. The line of section was some 200 kilometres in length — Griffith always referred to it as 'the great section of Ireland' — and it was the western end of the section, where it traverses the Connaught Coal District, which was used to illustrate the Connaught report published in 1819. A copy of the complete section was presented to the Geological Society of London on 15 April 1822.[50] Griffith's manuscript report on the Ulster coalfields was laid before the Dublin Society on 29 April 1819[51], but for the second time publication was delayed, partly because Griffith again lacked an adequate base-map upon which to plot his geological lines.[52] On this occasion, however, the delay was for ten years and when the report was finally published in 1829 it was still without an accompanying map.[53] The Ulster report is thus of little interest in the present context.

By the spring of 1818 only one of Ireland's provinces remained untouched by Griffith — Munster. It seems odd that Munster should have been left to the last because back in 1814 he had clearly entertained high hopes for the province's mineral potential when he observed:

> ... it appears, that very extensive tracts of coal country exist in Ireland; but none, if we except the Leinster district, have been examined; yet the Munster coal district is in extent greater than any in England, and may probably contain inexhaustible beds of coal.[54]

After spending part of the summer of 1818 finishing his fieldwork in Ulster, Griffith was instructed to spend the remainder of the season examining the Coal District lying between Kanturk, County Cork, and Killarney, County Kerry.[55] His manuscript report on this tract of country was laid before the Dublin Society on 29 April 1819[56] and in the following month he was directed to undertake a detailed survey of the entire Munster coalfield.[57] This was to be his last major undertaking on behalf of the Dublin Society and it occupied him for a part of each year down to and including 1824, but it was a project which he was destined never to complete.

Although these various coalfield surveys represent Griffith's main achievements on behalf of the Dublin Society, he did carry out a variety of lesser field-studies at the Society's behest. In 1819, for example, he was despatched to Charleville in County Cork to examine the site of a reported gold strike.[58] In the town he met the discoverer of the supposed lode, one James Carrol, 'who proved to be a poor, deserted school-master, with impaired intellect'. Needless to say there was no gold. In 1821 he was sent upon an equally fruitless mission to search for coal in County Tipperary, but on this occasion his assistance had been called for by five of the county's landowners who thought they could discern 'strong indications of coal' upon their estates.[59] There might be no coal upon the estates in question, but during his visit to County Tipperary Griffith did report upon a deposit that held both high geological interest and modest economic significance – the pipe-clays at Ballymacadam in the Suir valley.[60] The clays are now recognised to be one of the very few confirmed Tertiary deposits in Ireland outside the north-east – the clays are most probably of Upper Eocene age[61] – and while such an identification had to await the modern development of palynology, Griffith did, with remarkable prescience, compare the Ballymacadam clays with those at Bovey Tracey in Devon, a deposit which he had presumably seen during his sojourn in the southwest of England around 1804 and a deposit which is now recognised as being of the same general age as that at Ballymacadam. Under the terms of his appointment as Mining Engineer Griffith was of course retained by the Society for only six months of each year and he seems to have spent at least a part of the remainder of his time in private practice as a geological consultant, his best known work of this type being his involvement in the affairs of Lord Audley's copper mine at Cappagh in County Cork.[62] Thus, whether working for the Dublin Society or at his private practice, Griffith was year by year increasing his knowledge of Irish geology. One would have imagined that all the information so obtained would have been added regularly to the geological map of Ireland which since 1814 had evidently served to illustrate his annual lecture-course, but in 1836 Griffith informed the Geological Society of Dublin that between 1815 and 1835 'no material changes were made in the map'.[63] Whatever the truth of this assertion, there can be no doubt that during those years his geological map of Ireland was slowly progressing, even if for the moment the relevant data remained buried in the pages of Griffith's field-notebooks.

Griffith and the base-map problem

The more expertise Griffith achieved as a geological surveyor, the more exasperated he became at the inaccuracy of the only topographical base-maps available to him. Like all field-geologists, he sought to fix the position of his outcrops by relating them to the topographical detail already present upon the map but, if that detail is inaccurate, then the relative positions of the outcrops are false and the geological boundaries become distorted. In Ireland, as we have seen (p. 19), this problem had first been highlighted by Stephens and Fitton in 1812, and the inaccuracy of their base-maps was a general complaint among Griffith's geological contemporaries. None of them felt more strongly upon this issue than did John Macculloch who at the end of his life (he died following an accident while on his honeymoon in 1835, aged sixty-two) entered a vigorous protest at the defectiveness of Arrowsmith's map of Scotland which he had of necessity used as a base-map while constructing his pioneer geological map of Scotland.[64] That map which Macculloch condemned so roundly was the companion sheet to the map which Griffith was evidently employing as the basis for his own work in Ireland.

The base-map problem remained serious until there became available the carefully surveyed sheets based upon the work of the Ordnance Survey. The earliest of these — a series of four one-inch sheets covering the English county of Kent — appeared in January 1801, and over the next few decades the one-inch map was gradually unrolled across England and Wales. But it was 1824 before the Ordnance Survey commenced operations in Ireland, and when Griffith went down to Munster in the summer of 1818 an Ordnance Survey coverage of Ireland must have seemed a very remote prospect. In the meantime he had to persevere with his field investigations, making do with whatever base-maps were available, and cursing all the while at their inadequacy. On 12 July 1818 he wrote as follows to Greenough:

> I am at present groaning under a weight of County Geological maps, and Notebooks but can compleat [sic] nothing to my mind for want of a general map with some pretensions to accuracy to lay down my work. Unfortunately there is no map of Ireland, at least none deserving the name, Arrowsmith's which is the last is by much the most incorrect & Taylor's & Beaufort's are on too small a scale. I have made an attempt to colour Arrowsmith's map but the positions it gives to many of the towns is so different from the true ones, and the general incorrectness both in the courses of rivers & the situation & extent of mountains renders it impossible to lay down any one geological district in its true form and position. In every part of Ireland that I have examined, I have travelled with the [Grand Jury] county map in my hand & by this means all my observations are laid down with great accuracy. Possessing such materials I cannot be content to debase them by sending forth to the world an inaccurate sketch which is the most that any map of Ireland that has been formed when coloured geologically can pretend to.[65]

In despair, Griffith reached the conclusion that there was no alternative but to embark upon an exercise in cartographic self-help. He had already had some experience of constructing his own base-maps while working for the Bog Commissioners (see p. 20) and while preparing the County Roscommon section of his map of the Connaught Coal District (see p. 41). Now he resolved to adopt the same solution in Munster. It was in that letter to Greenough on 12 July 1818 that Griffith intimated that he was thinking seriously of trying to solve his problem by undertaking the construction of a new base-map designed to meet his own needs. Initially it seems that his objective was limited merely to the preparation of a base-map which would service his geological investigations in the coalfield area of Munster, but the project rapidly blossomed into something larger. What he really wanted was a map which could supersede Arrowsmith's as the basis for his geological sheet of Ireland, and he was soon thinking in terms of a survey of the whole of Munster whence he would continue his lines eastward and northward until the entire island had been covered. In 1821 he wrote to the Royal Dublin Society as follows:

> I have determined, previously to laying down any lines of rock boundaries, to make a perfect Map of Ireland, shewing all mountains and hills, all lakes, rivers and streams, the boundaries of counties and baronies, with all the canals and every road passable for a carriage throughout the country.[66]

His plan was to fix the position of salient features by means of triangulation and he then hoped that it would be possible to fill in the topographical detail by abstracting information from the existing Grand Jury county maps. Griffith's scheme had the blessing of the Royal Dublin

Society, and he and Kelly started work upon the triangulation during the summer of 1819 using an 18-inch reflecting circle and an 8-inch theodolite, the altitudes being measured, as in their 1817 transect of northern Ireland, by means of Troughton barometers.[67] Two base-lines were laid out; one two Irish miles (4.1 kilometres) long near Sneem in County Kerry, perhaps on the boglands to the west of the town, and the other four Irish miles (8.2 kilometres) long somewhere in County Limerick.[68] Haze interfered with some of his observations and another problem was that the local people, suspicious of his activities, sometimes threw down his observational markers.[69] Progress was nevertheless made. When he wrote to Greenough on 1 November 1819 he reported that he had triangulated the whole of Kerry and Cork together with portions of Clare, Limerick and Tipperary, and the greater part of Munster had been covered by the close of the following year. Soon he was looking forward eagerly to the completion of the triangulation of the entire country, and it was perhaps at this stage that he employed a team of draughtsmen to reduce all the county maps of Ireland to the scale of his own projected survey.[70] In 1821 he informed the Royal Dublin Society that his completed geological map would be three feet ten inches (117 centimetres) from north to south and two feet seven inches (79 centimetres) from east to west, its scale being the same as that of Greenough's geological map of England and Wales, namely one inch to six miles (1:380,160).[71] The colour-coding too, Griffith observed, would be identical to that on Greenough's map, and he expressed the hope that his sheet, together with its accompanying memoir, would be fit companions both for Greenough's map and memoir and for the map of Scotland which he understood John Macculloch to be producing. It began to look as though the map's gestation period was nearing its end, but at this point fate took an unexpected turn.

The lean years

In the summer of 1822 famine made one of its regular visitations to the west of Ireland. Faced with an emergency, the government appointed Griffith to be an Engineer of Public Works and despatched him to the southwest to take charge of a programme of relief measures. His area of responsibility comprised counties Cork, Kerry and Limerick, and he was directed to inaugurate a programme of road construction and repair in order to provide employment for the region's starving population.[72] He left for the southwest early in June 1822, taking with him John Kelly who now severed his connection with the Royal Dublin Society and entered Griffith's personal service as an assistant with a salary of £200 per annum. Soon Griffith had seven thousand men working for him on the roads in the southwest and, although the famine quickly passed, he remained in the region to carry through his various civil engineering projects, living for the next seven years not in Dublin but at Ballyellis House near Mallow in County Cork. In the previous century Mallow had achieved some fame as a spa — it had been hailed as 'the Irish Bath' — but by the 1820s its modest glories were fast fading. The Griffiths nevertheless doubtless there enjoyed the usual Anglo-Irish round of dinner parties, hunt balls, and regimental sports-days but, whatever its social attractions, there is no escaping the fact that Mallow was a scientific backwater and Griffith must have found difficulty in maintaining contact with his favourite science during a decade when it was undergoing a vigorous expansion. Even his periodic visits to Dublin can hardly have provided much inspiration because the capital's once bright geological lamp now burned but dimly. Perhaps there was some national disillusionment with a science that had failed to conjure forth a mineral wealth sufficient to trigger off an Irish industrial revolution, but for whatever reason the 1820s was certainly a bleak period in the history of Irish geology. Even the external stimulus once provided by visiting geologists was now largely lost because, after Waterloo, British geological travellers turned their steps towards Auvergne rather than to Antrim, to Campania rather than to Connemara. But maybe Griffith's official duties left him little time for reflection upon such matters because during the eight years following 1822 he was responsible for the construction of some 350 kilometres of new road and for the repair of several hundreds of kilometres of existing highway.[73]

Many of these roads lay within the Munster Coal District and Griffith now tried to combine his road-building for the government with his geological surveying on behalf of the Royal Dublin Society. On 28 June 1822 he wrote to the Society from Newcastle West, County Limerick, explaining that an unusually wet autumn the previous year had prevented his completion of the Munster coalfield survey, but he now offered to complete the investigation without any further cost to the Society — an offer that the Society was naturally pleased to accept.[74] Further, Griffith maintained that the knowledge he had acquired while surveying the region for the Society was now proving invaluable. He was, he claimed, laying out his new roads so that they passed close to the outcrop of hitherto unknown coal-seams, and in this manner he was creating opportunities for the development of important new mines. The coal would be exported from the coal district plateau along the new highways, while as return cargoes carts would bring up limestone from the surrounding lowlands to improve the heavy and acid soils developed upon the strata containing the coal-seams. Frankly, all this reads like a mild exercise in self-glorification — a habit that Griffith regularly indulged in — but his years in the southwest certainly did give him splendid opportunities for the detailed study of the region's geology as, for example, during May 1823 when he surveyed the line of a proposed new road across the Old Red Sandstone of the Boggeragh Mountains from Banteer towards Cork City.

Griffith was now deeply committed to a multitude of activities. For the Royal Dublin Society he was conducting a geological survey of the Munster coalfield, organising a triangulation and geological survey of the whole of Ireland, and giving annual courses of lectures; for the government he was planning and constructing roads in Munster; and for himself there was his continuing private practice as a mining consultant. But he was a man of enormous energy and still his appetite for work remained unsatisfied. Thus in 1825 he seems to have had no hesitation in heaping upon himself yet another onerous duty — a duty connected with a great new national undertaking. It had just been decided that the Ordnance Survey should begin operations in Ireland with the intention of publishing a national map, not on the one-inch scale (1:63,360) as in England and Wales, but at the very much more generous scale of six inches to the mile (1:10,560). The new Survey commenced its work in Ireland in 1824 and it was on 27 August in the following year that Griffith assumed a new office bearing the title Director of the General Boundary Survey of Ireland.[75] In this role it was his duty to supervise the location on the ground of all the Irish county, barony, parish, and townland boundaries so that the lines were ready to be plotted upon the new maps just as soon as the Ordnance surveyors moved into a district. The task proved more difficult than had at first been imagined — there were some 250 baronies, 2,400 parishes and 69,000 townlands to be delimited — and Griffith's staff of boundary surveyors had to be augmented until by the spring of 1828 it numbered about fifty men.[76] Griffith himself was doubtless happy to think that he was facilitating the Ordnance Survey in its production of the first accurate map of Ireland, but for him the chief virtue of his new post was perhaps the freedom it gave to travel throughout Ireland inspecting the work of his staff and steadily adding to his personal store of geological knowledge.

When he became Director of the Boundary Survey, Griffith had already spent some fifteen years tracing geological lines across the Irish countryside, and by the mid-1820s he felt confident that he had sufficient data to allow the preparation of a map for publication. The only hindrance was the lack of an adequate base-map. When he gave evidence before the Select Committee on the Survey and Valuation of Ireland in May 1824 he was questioned about the progress of his map and he responded:

> I can conceive that at this moment I possess in my note books and in county maps a very considerable quantity of the materials which would be necessary to construct the geological map, if a correct map was handed to me, and I should think it necessary where the materials were imperfect, to proceed myself, or employ an assistant, whom I might instruct to fill up the remainder.[77]

Certainly at the Royal Dublin Society it was understood that his geological map had reached an advanced stage, and in February 1825 the Committee of Chemistry and Mineralogy announced that the following summer would probably see the completion of Griffith's geological survey of the whole of southern Ireland.[78] But the summer came and went without the map being published, and in their impatience some members of the Society began to ask awkward questions. Their Mining Engineer was clearly hard at work building roads and surveying boundaries, but exactly what was he doing to earn the salary paid him by the Society? True, he was still giving his annual lecture course on 'Geology and practical mining' — a lecture course which evidently attracted audiences of around four hundred persons[79] — but he had published no mining reports since 1819. His Munster survey was still unfinished, and since 1821 little had been heard of his grandiose scheme for a triangulation of Ireland, although the project was not finally abandoned until 1824 when the beginning of the Ordnance Survey triangulation rendered superfluous Griffith's rather amateurish efforts.[80] In fairness to Griffith it should be said that since 1820 he had never claimed the one pound per day to which he was entitled by way of field expenses[81], but he was now tarred with a suspicion of negligence and in consequence his relationship with the Society became increasingly soured. Between the autumn of 1821 and the summer of 1826 he attended only 15 out of 160 meetings of the Society, and repeatedly he was asked to explain exactly what he was doing during that period of each year when he was supposed to be an employee of the Society. On 29 June 1826 the Society decided to review his position to ascertain whether, in view of his many other commitments, he was still able to perform his duties as Mining Engineer to the Society's satisfaction.[82]

This review in June 1826 elicited a memorandum from Griffith in which he outlined his activities during recent years, but the document's most interesting feature is the request which he now placed before the Society.[83] He asked in future to be employed in the geological mapping of those areas where the Ordnance Survey was then at work so that he would have a series of geological sheets and memoirs ready to be published almost simultaneously with the appearance of the Ordnance Survey's topographical sheets. The scale of Griffith's proposed geological sheets was left vague, but his request clearly demonstrates his recognition that the Ordnance Survey's activities were about to yield those accurate base-maps for which he had craved so long. The request is significant too in another aspect. In 1809 the Dublin Society had abandoned its plan for a general geological survey of County Dublin in favour of a more severely practical survey of the Leinster coalfield. As we have seen (p. 22), this change was in all probability made at Griffith's instigation, but now, in 1826, Griffith was asking to be allowed to travel in the wake of the Ordnance surveyors conducting a general survey of precisely the type which the Society had renounced in 1809. To some extent this change of outlook was probably a result of the evaporation of his earlier optimism about Ireland's mineral resources — an evaporation occasioned by eighteen years' experience of the harsh realities of Irish geology — but perhaps more important was the fact that he was becoming obsessed with an understandable desire to see his map of Ireland completed in the most accurate form possible. His earlier geological sheets had all been intended primarily as aids for the mineral proprietor, but he now saw his map of Ireland from an academic rather than a utilitarian standpoint. The map had become a desirable end in itself.

The Society's Committee of Mineralogy agreed to support Griffith's request and on 3 May 1827 the committee placed before the Society the proposal that Griffith 'should be employed in continuing to collect materials for a geological map of Ireland' and that he should spend the ensuing summer preparing geological maps of County Tyrone and northern County Fermanagh, these evidently being the areas where it was understood that the Ordnance Survey would then be at work.[84] This suggestion met with some opposition and we read that at the Society's meeting on 31 May 1827, with Griffith himself present:

> Doctor Meyler gave notice that on Thursday the 8th day of November next, he would move, "that the salary of the mining engineer should cease from that day".[85]

This motion seems never to have been formally put, and in the event Griffith was allowed to spend the summer of 1827 mapping in an area bounded by a line drawn from Lifford, County Donegal, to Slieve Gallion, County Londonderry, and thence via Clogher and Enniskillen to Pettigo and so back to Lifford. In his subsequent report to the members of the Society — a report dated 30 October 1827[86] — he sought to allay suspicions that they were now supporting pure research by emphasising that during the season he had discovered several important mineral deposits and he observed that when his geological versions of the Ordnance Survey maps were published:

> ...the different proprietors of the mountain estates will be enabled to extend their cultivation by tracing the line of the limestone-beds on the map, and having ascertained the point on the land, may open new quarries where no limestone is at present supposed to occur.

But this was just posturing — just an illustration of what has been termed 'public science'.[87] Griffith was merely employing the utilitarian argument to mantle his deep personal desire to be the author of geologically coloured versions of the Ordnance Survey sheets. Be that as it may, it should be noted that in 1827 the Royal Dublin Society employed Griffith to perform much the same duty as Henry De La Beche undertook in England in 1832 when he was appointed to affix geological colours to the Ordnance sheets of the West Country.

There is just one puzzling feature about Griffith's work in Ulster in 1827. It was quite true that the Ordnance sheets were at long last going to provide him with an accurate base-map upon which to record his field observations, but what map did he employ in the field during his 1827 season? In England De La Beche had the published one-inch map upon which to work, but it was May 1833 before the first of the Ordnance Survey's Irish six-inch sheets appeared. It may be that the Ordnance Survey was able to supply Griffith with early drafts of the sheets covering at least a part of his area (in England the Survey certainly did facilitate certain geologists in this manner) or it may be that Griffith proposed to use the sketch-maps produced by his own Boundary Survey. But whatever the base-maps he had in mind, there is no escaping the fact that in 1827 he was really little better equipped for field operations in Ulster than he might have been in any previous year. Why then was he in such haste to lay aside his studies in Munster in order to turn his steps northwards? Light will be shed on this question by the contents of Chapter Four and it suffices here to suggest that Griffith went to Ulster in order to make a token presence in the Ordnance Survey's field area. Understandably, he seems to have developed a proprietorial feeling for Irish geology, and early in 1826 he must have been disconcerted to learn that the Ordnance surveyors were now under instruction to make geological observations as part of their normal duties. Evidently in Griffith's eyes these 'trespassers' needed a visible reminder that *he* was the leading authority on Irish matters geological and that his own map was already so far advanced as to make their own geological observations unnecessary.

This feeling that the Ordnance surveyors were intruding into his own geological preserve may have been one factor which contributed to the strained relationship that soon developed between the Ordnance Survey and Griffith's Boundary Survey, but clearly Griffith was never an easy colleague even at the best of times. Certainly his association with the Royal Dublin Society continued to be somewhat acrimonious, and during the 1820s the Society increasingly turned for geological insights to its Professor of Mineralogy, Sir Charles Lewis Giesecke (1761-1833). In contrast to Griffith, Giesecke was the Society's blue-eyed boy and the members struck a special gold medal in his honour as early as 1817. In 1825 he was sent to investigate the geology of counties Galway and Mayo, in 1826 he was sent to County Donegal, in 1828 to counties Antrim, Londonderry, and Down, and in 1829 to counties Londonderry and Tyrone.[88] On 5 July 1827 there was even talk within the Society of asking Giesecke to undertake the compilation of a geological map of Ireland.[89] All this criticism, both overt and implicit, stung Griffith into the publication of two new mining reports as evidence that he had not been idle in the Society's interest. The first of these was a short report on the metallic mines of Leinster; it seems that it had originally been laid before the Society in manuscript in May 1821 and it finally got into print in 1828.[90] In his introduction to the volume he promised that it was the forerunner of a series and that the companion study of the metallic mines of Munster would be appearing almost immediately, but in fact no other studies were ever published. The second report was his long-awaited study of the Ulster Coal Districts published in 1829[91] but, since neither of these reports contains a map, they need detain our attention no further.

Griffith was mistaken if he thought that these two reports would shelter him from further criticism. They were both based upon work done more than seven years earlier and the Society's members wanted evidence of his more recent activities on their behalf. Thus on 8 May 1828 he was instructed to lay all his unpublished Society work before the next meeting, the Society specifically asking to see the maps of all those counties which he had surveyed since the announcement of his intention to construct a geological map of Ireland.[92] He complied with this request on 22 May when he displayed the section which he and Kelly had surveyed across northern Ireland in 1817, his reports on the Ulster Coal Districts, and geological maps of counties Antrim, Armagh, Down, Londonderry, Louth and Tyrone, although he had to apologise for the ragged condition of all these sheets by explaining that they had suffered as a result of regular use in the field.[93] In addition, he claimed that he had also completed geological surveys of counties Cavan, Cork, Kerry, Kildare, Limerick, and Roscommon. He thus asked the Society to accept that he had mapped twelve of Ireland's thirty-two counties, the twelve covering some 40 per cent of the country's area. Strangely he laid no geological map of the whole of Ireland before the Society, but the map did not pass unmentioned. He observed:

> I beg also to inform the Society that there is no part of Ireland, in the Geological examination of which I have not made considerable progress. My wishes, I may say the chief public object of my life, has been to complete an accurate Geological map and description of Ireland, and I hope at no distant period, that the work will be accomplished with advantage to the public, and credit to myself.

Even this display of material failed to silence his critics, and on 11 December 1828 a motion was tabled demanding that he state exactly what he had been doing on behalf of the Society since the completion of his Ulster Coal Districts survey more than ten years earlier.[94] The motion was defeated, but on 15 December Griffith did submit a full account of his activities since 1818, and to his earlier claims he now added the observation that he had almost completed a geological map of County Donegal. Thus, he affirmed, he would be in a position to publish a geological map of the whole of Ulster just as soon as the work of the Ordnance Survey was sufficiently far advanced to provide him with an accurate base-map.[95] The passage of time had seemingly brought him neither judgement nor discretion; he was just as buoyantly over-optimistic as ever.

Griffith's remarkable penchant for holding simultaneously a variety of private and public offices was noted earlier, and during the 1820s he added to his hand yet another appointment which, up to this point, has remained unmentioned. In 1827 he was nominated as Commissioner of the General Survey and Valuation of Rateable Property, but the office had remained dormant pending the completion of the earliest of those essential base documents, the six-inch sheets of the Ordnance Survey. Late in 1829 it seemed that the earliest of the new sheets was soon likely to become available and, faced with the task of commissioning a great new undertaking, Griffith decided that the time had come to resign from his post at the Royal Dublin Society. There was also a financial consideration: the salary of his new office as Commissioner of the Valuation was £500 per annum which was substantially more than he was receiving from the Royal Dublin Society, his salary from that source having been reduced to £276.18.5d per annum since 1825. He nevertheless put his decision into effect in an odd manner — a manner that perhaps tells us something about his character. On 12 November 1829 — the seventeenth anniversary of his original election to the post — he allowed himself to be re-elected by ballot as the Society's Mining Engineer, but on the very next day he penned a letter to the Society tendering his resignation.[96] Was he perhaps anxious to establish the point that, after their years of acrimony, it was he who was quitting the post of Mining Engineer and upon his own terms? Nobody was to be allowed to imagine that it was the Society which was dispensing with the services of its officer. Even at this juncture of his life, however, the geological map of Ireland was by no means forgotten. In his letter of resignation Griffith observed:

> ... it is still my intention to continue my researches towards the completion of the Geological Map of Ireland, and I am happy to say, that the peculiar nature of my official duties under the Government, will enable me to effect that object with unusual facility and accuracy and without any further expense to the Society.

But the Society no longer seems to have been interested in such matters. Griffith's resignation was accepted on 19 November 1829 without either comment or regret. Perhaps there is significance in the fact that the vacancy was never filled; Griffith remains to this day the only Mining Engineer ever to have been employed by the Royal Dublin Society.

Griffith's earliest rivals

The years between 1814 and 1830 may have seen geology at a low ebb in Ireland, but this chapter would be misleading if it seemed to imply that during that period Griffith was the sole geologist engaged upon the task of mapping Ireland's rocks. As we will see in Chapter Four, those years saw the inauguration of a full-blown, government financed geological survey of Ireland, but for the moment it is two other geological cartographers who must engage our attention: Thomas Weaver and Alexander Nimmo.

Thomas Weaver[97] was trained at Freiberg. He had experience in the management of mines in County Wicklow and he was responsible for a translation of Werner's *Von den Äusserlichen Kennzeichen der Fossilien* published in 1805.[98] In 1812 he was Griffith's rival for the post of Mining Engineer at the Dublin Society, and in December 1813 he was one of a short-list of four candidates when the Society elected Giesecke to the chair of mineralogy[99] — a surprising election in view of Giesecke's honest admission of his inability to lecture in English! But even setting aside the linguistic issue, the Society surely made a mistake in the 1813 appointment. Weaver was a geologist of the modern stratigraphical school whereas Giesecke represented the older mineralogical tradition. As early as January 1816 Weaver was reported to be 'at present engaged in a Geognostical Map of a great part of Ireland'[100], and on 15 May 1818 he read an elaborate memoir on the geology of southeastern Ireland to the Geological Society of London. When it was published in 1819 this memoir was accompanied by a geological map representing the whole of southeastern Ireland as far westwards as County Limerick.[101] This map was to be the cause of some acrimony between Griffith and Weaver. In later years Griffith was to claim that in 1816, or soon thereafter, Weaver had attended his Dublin Society lectures, that Weaver had seen his map of Ireland on display there, and that in his own work Weaver had merely reproduced the boundaries already depicted upon Griffith's map with the addition here and there of a few minor amendments.[102] Perhaps there was some friction between the two men dating back to their rivalry over the post of Mining Engineer in 1812, and it is now impossible to say whether any plagiarism really did occur. In fairness to Weaver, however, it has to be observed that the text of his paper reveals him to have possessed a very considerable personal knowledge of Irish geology and the text thus does nothing to sustain Griffith's accusation that Weaver was basically a copyist. If Griffith's claim really could be trusted, then it would mean that Weaver's map must hold considerable interest for the student of Griffith's own work. Since all known traces of Griffith's early maps of Ireland have long since disappeared, it follows that if Weaver's sheet really is based upon the map which Griffith displayed around 1816, then we have in Weaver a reflected image of Griffith's long-lost work. In view of this, how accurate is Weaver's map? Errors certainly exist. The large Carboniferous limestone outcrop in southeastern County Wexford is only partially represented, for example, the Slieve Bloom inlier is rather too small, and the Slieveardagh outlier rather too large, while a more adequate distinction might perhaps have been made between the 'Clay Slate' (Silurian) and the 'Sandstone' (Old Red) in the Galty, Knockmealdown, Monavullagh, and Comeragh mountains. In its broad outlines, however, little fault can be found with the map. It is a work of which the author — whoever he may have been — could feel justly proud.

Turning now to our second figure — Alexander Nimmo — we encounter an intriguing little mystery. In the National Library of Ireland there is a copy of Lieutenant Alexander Taylor's *A new map of Ireland* of 1793 (the scale is approximately one inch to ten miles or 1:633,600) dissected, mounted upon cloth, and somewhat crudely colour-washed to represent eight geological categories ranging from 'Granite or other Primary Rock' (coloured lake) to 'Coal formation or Slate clay' (coloured brown) (Fig. 2.3).[103] In addition there are symbols to

Fig. 2.3. The geological map of Ireland which belonged to Alexander Nimmo. (Reduced in size.) Reproduced from the original in the National Library of Ireland by kind permission of the Trustees.

indicate the sites of various mineral deposits, a pattern of ink lines to show 'Flat Peat Bogs', and a number of manuscript comments written in a small running script. Thus in County Down we read 'Between Newt. Ards & Bangor tail of the hill is Granite?', while in the south of Ireland we find 'The Gold of Croghan appears to be in quartz veins penetrating the Grauwacke slate' and 'In Waterford &c Schistus (grauwacke) covered by flat conglomerate'. These notes were evidently based upon field-observations made as the owner of the map travelled around Ireland, and the use of a variety of different inks for the writing suggests that the observations may have been made over a fair period of time. But who was the observer? Upon the reverse of the map, and in an old hand, is the inscription 'This map belonged to Mr. Nimmo the Engineer who Surveyed the Kingdom by Order of the Government'.

Alexander Nimmo (1783-1832)[104] was a Scots civil engineer whose career was in some respects similar to that of Griffith himself. He was educated at the universities of St Andrews and Edinburgh, and in January 1811 he joined the Irish Bog Commissioners as an engineer, having been nominated by no lesser individual than Thomas Telford. Later, following the establishment of the Irish Fishery Board in 1819, Nimmo was employed to conduct fishery surveys and to work on harbour improvement schemes, while during the famine of 1822 he, like Griffith, was despatched to the west of Ireland to take charge of road building and repair as a relief measure. Nimmo is known to have been interested in geology and in October 1823 he read to the Royal Irish Academy a paper on the submarine geology of the English and St George's channels which is the pioneer essay upon the subject of Ireland's offshore geology.[105] The handwriting upon the map in the National Library of Ireland looks very like the handwriting upon the Library's other manuscript plans which can with confidence be ascribed to Nimmo, and there really seems no reason to doubt the map's ascription to him. Granted that the map was in Nimmo's possession and that the manuscript annotations are his, there remains the problem of who affixed the geological colours to the map and under what circumstances. Here there seem to be three possibilities. Firstly, Nimmo may have made the appropriate field observations, may have abstracted details from pre-existing maps, and may then have coloured the new map himself. In this case the map would have to be regarded as being essentially the result of Nimmo's personal field investigation. Secondly, while colouring the various formations represented upon his map, Nimmo may have had access to one of Griffith's maps used to illustrate Griffith's lectures at the Royal Dublin Society. In this case Nimmo's map would have to be regarded as a copy of one of Griffith's early sheets. Finally — and this is the most exciting of the three possibilities — Griffith may actually have presented Nimmo with one of his own early, and perhaps superseded, essays in the mapping of Ireland's geology. Griffith is known to have affixed geological colours to Taylor's map of Ireland (see p. 40) — the very map upon which the National Library of Ireland sheet is based — and Griffith and Nimmo must have known each other through their work for the Bog Commissioners and through their later involvement in famine relief measures in the west of Ireland. This third possibility is thus not mere idle fancy and it presents us with the fascinating situation that in the National Library of Ireland we may have catalogued under 'Nimmo' a map which in reality is one of Griffith's earliest attempts to depict the geology of Ireland. Adjudication between these three possibilities is impossible, but one point can be asserted with confidence: while the map pretends neither to great precision nor to high artistry, it does offer a reasonably accurate sketch of Ireland's geology. All the essentials are present from the basement rocks and the various granite bodies, to the extensive spreads of limestone, the inliers and the outliers, and the basalts of the northeast. It is a remarkable and fascinating map. Someone at some time has clearly been concerned for the map's care and preservation because its dissection evidently post-dates all the manuscript annotations. That care and attention the map richly deserved because today it must be hailed as the oldest surviving geological map of the whole of Ireland. It is to be regretted that such uncertainty has to surround its authorship.

The Battle of Waterloo was one of history's decisive encounters. Among the thousands wounded upon that field was one Lieutenant John Watson Pringle who was later destined for a less hazardous engagement in the field of Irish geology. Aside from that fact Waterloo has no significance for the historian of Irish geological cartography, but the year of Waterloo — 1815 — does nevertheless form a convenient watershed in the history of Irish geology. Before 1815 Ireland was a scene of considerable geological ferment and in the area of geological cartography much had been accomplished; after 1815 Irish geology entered upon a decline of some fifteen years' duration. Griffith's own life epitomises the national trend. Before 1815 he worked strenuously at geological problems on behalf of the Bog Commissioners and the Dublin Society, but after 1815 his attention came to focus increasingly upon his newly assumed public duties rather than upon geological investigation. By 1830 some twenty years had elapsed since he had started work upon his geological map of Ireland, yet the publication of that map must have seemed an event as remote as ever. But fate was to take an unexpected turn, and by 1839 Griffith had achieved his long-cherished ambition — he had published the first detailed geological map of Ireland. The turn of events which allowed him to complete and publish this map forms the subject of the next chapter.

REFERENCES AND NOTES FOR CHAPTER TWO

1. John Davy, *Memoirs of the life of Sir Humphry Davy, Bart.*, London, 1836, vol. 1, 427-432.
2. Some sources for the life of Griffith: Griffith's MS. autobiography dictated 25 August 1869 (the original is in private hands but copies are in the NLI and PROD.); DNB; DSB. *Dublin University Magazine* 83, 1874, 432-437; *Geol. Mag.*, N.S. dec. 2, 5, 1878, 524-528; Gordon Leslie Herries Davies and Robert Charles Mollan (editors), *Richard Griffith 1784-1878*, Royal Dublin Society, 1980, pp. vi + 221; *Trans. Edinb. Geol. Soc.* 3, 1874-1880, 181-184.
3. *Report from the Select Committee on Royal Dublin Society*, H.C. 1836 (445), XII, 204.
4. De Dunstanville to Boulton 7 February 1805, *Boulton Papers*
5. R. Griffith, Inaugural address...19th of February, 1861, *Trans. Instn. Civ. Engrs. Ire.* 6, 1861, 193-221.
6. MS. autobiography, (ref. 2).
7. Richard Griffith, *Geological and mining report on the Leinster Coal District*, Dublin, 1814, xvf; *The fourth report of the Commissioners appointed to enquire into the nature and extent of the several bogs in Ireland, and the practicability of draining and cultivating them*, H.C. 1813-14 (131), VI second part, 177f.
8. MS. autobiography, (ref. 2).
9. V.A. Eyles, Louis Albert Necker of Geneva, and his geological map of Scotland, *Trans. Edinb. Geol. Soc.* 14(2), 1948, 93-127.
10. MS. autobiography, (ref. 2).
11. *Proc. Dubl. Soc.* 44, 1807-8, 95.
12. *Ibid.*, 45, 1808-9, 54.
13. *Ibid.*, 135-139.
14. *Ibid.*, 48, 1811-12, 138-140, 147; 49, 1812-13, 21-22.
15. MS. autobiography, (ref. 2); R. Griffith, Address, *J. Geol. Soc. Dubl.* 1(3), 1837, 141-142; *Q. Jl. Geol. Soc. Lond.* 10, 1854, xx. For earlier but not always accurate histories of Griffith's map see: M.H. Close, Anniversary Address, *J.R. Geol. Soc. Irel.* 5(2), 1878-79, 132-149; A.G. Davis, Notes on Griffith's geological maps of Ireland, *J. Soc. Biblphy. Nat. Hist.* 2, 1943-52, 209-211; J.W. Judd, The earliest geological maps of Scotland and Ireland, *Geol. Mag.*, N.S. dec. 4, 5, 1898, 145-149.
16. Richard Griffith, *Letter to the Royal Dublin Society, describing the proposed plan for constructing an accurate geological and topographical map of Ireland*, Dublin, 1821, 4.
17. DNB; *Q. Jl. Geol. Soc. Lond.* 12, 1856, proceedings xxvi-xxxiv; Horace Bolingbroke Woodward, *The history of the Geological Society of London*, London, 1907, *passim*; M.J.S. Rudwick, Hutton and Werner compared: George Greenough's geological tour of Scotland in 1805, *Br. J. Hist. Sci.* 1(2), 1962, 117-135; Jacqueline Golden, *A list of the papers and correspondence of George Bellas Greenough*, University College London, 1981, pp. 48.
18. George Bellas Greenough, *Memoir of a geological map of England*, London, 1820, 1; Woodward, *op. cit.*, 1907, 57-59, (ref. 17).
19. *Q. Jl. Geol. Soc. Lond.* 10, 1854, xx.
20. *Greenough Papers*.
21. *Ibid.*
22. *Proc. Dubl. Soc.* 50, 1813-14, 17-18.
23. *Greenough's Journal 1811-1815*, *Greenough Papers*.
24. *Greenough Papers*.
25. *Ibid.*
26. *Proc. Dubl. Soc.* 50, 1813-14, 85-86.

27. The earliest published syllabus for Griffith's lectures containing a reference to the display of the map is, so far as I am aware, that for 1827. See Richard Griffith, *Syllabus of a course of lectures on the geology of Ireland, to be delivered in the theatre of the Royal Dublin Society*, Dublin, 1827, pp. 8.
28. *J. Geol. Soc. Dubl.* 3(3), 1846, 176.
29. *Report from the Select Committee on the survey and valuation of Ireland*, H.C. 1824 (445), VIII, 51.
30. *Greenough Papers*.
31. Griffith, *op. cit.*, 1814, i, (ref. 7).
32. *Proc. Dubl. Soc.* 51, 1814-15, 186, 195.
33. *Ibid.*, 195.
34. *Ibid.*, 53, 1816-17, 190.
35. Griffith, *op. cit.*, 1814, xx, (ref. 7).
36. *Proc. Dubl. Soc.* 52, 1815-16, 203.
37. *Ibid.*, 236.
38. *Ibid.*, 203.
39. *A map of the County of Roscommon, laid down from an accurate and minute survey, verified by trigonometrical measurement. Under the orders of the Grand Jury of the County of Roscommon*. The scale is one inch to one and a half miles (1:95,040).
40. *Proc. Dubl. Soc.* 55, 1818-19, 132-133.
41. *Ibid.*
42. Richard Griffith, *Geological and mining survey of the Connaught Coal District in Ireland*, Dublin, 1818, pp. vii + 108.
43. *Ibid.*, iii.
44. *Ibid.*, 4-11.
45. *Proc. Dubl. Soc.* 52, 1815-16, 204, 222.
46. *Ibid.*, 211.
47. *Ibid.*, 54, 1817-18, 14; Chemistry Committee Minutes 1817-1852, *RDS MSS*.
48. J. Kelly, On the subdivision of the Carboniferous formation of Ireland, *J. Geol. Soc. Dubl.* 7, 1855-57, 245f-246f.
49. Chemistry Committee Minutes 1817-1852, *RDS MSS*.
50. *Trans Geol. Soc. Lond.*, series 2, 1, 1822, 434.
51. *Proc. Dubl. Soc.* 55, 1818-19, 132-133.
52. *Ibid.*, 139, 146.
53. Richard Griffith, *Geological and mining surveys of the Coal Districts of the counties of Tyrone and Antrim, in Ireland*, Dublin, 1829, pp. ix + 77.
54. Griffith, *op. cit.*, 1814, xxiv, (ref. 7).
55. *Proc. Dubl. Soc.* 54, 1817-18, 193.
56. *Ibid.*, 55, 1818-19, 133.
57. *Ibid.*, 139, 146.
58. *Proc. R. Dubl. Soc.* 56, 1819-20, 195-196, 217-218.
59. *Ibid.*, 57, 1820-21, 134, 186-187, 213-216.
60. *Ibid.*, 213-216.
61. W.A. Watts, A Tertiary deposit in County Tipperary, *Scient. Proc. R. Dubl. Soc.*, N.S. 27(13), 1957, 309-311.
62. Richard Griffith, *Report on Cappagh copper mine, in the County of Cork, belonging to the Right Hon. Lord Audley*, Dublin, 1822, pp. 9. A copy is in the Haliday Pamphlets in the Royal Irish Academy. See also T.A. Reilly, Richard Griffith and the Cappagh copper mine fraud, pp. 197-210 in Herries Davies and Mollan, *op. cit.*, 1980, (ref. 2).
63. *J. Geol. Soc. Dubl.* 1(3), 1837, 142.
64. John Macculloch, *Memoirs to His Majesty's Treasury respecting a geological survey of Scotland*, London, 1836, 2-16.
65. *Greenough Papers*.
66. Griffith, *op. cit.*, 1821, 4, (ref. 16).
67. *Ibid.*, and *Greenough Papers*, Grffith to Greenough 1 November 1819.
68. Griffith, *op. cit.*, 1821, pp. 7, (ref. 16).
69. *Report from the Select Committee, op. cit.*, 1824, 42, (ref. 29).
70. *Proc. R. Dubl. Soc.* 63, 1826-27, 10.
71. Griffith, *op. cit.*, 1821, (ref. 16).
72. *Employment of the poor, Ireland*, H.C. 1823 (249), X, 3-28; *Report on the Southern District in Ireland*, H.C. 1824 (352), XXI, pp. 9; *Roads, (Ireland)*, H.C. 1829 (153), XXII, 1-5.
73. P.J. O'Keeffe, Richard Griffith: planner and builder of roads, pp. 57-75 in Herries Davies and Mollan, *op. cit.*, 1980, (ref. 2); S. Ó Lúing, Richard Griffith and the roads of Kerry, *Jl. Kerry Archaeol & Hist. Soc.* 8, 1975, 89-113, 9, 1976, 92-124.
74. *Proc. R. Dubl. Soc.* 58, 1821-22, 238-239.
75. *Second report from the Select Committee on the public income and expenditure of the United Kingdom*, H.C. 1828 (420), V, 347-358; John Harwood Andrews, *The paper landscape: the Ordnance Survey in nineteenth-century Ireland*, Oxford, 1974, *passim*.
76. *Second report, op. cit.*, 1828, 351-358, (ref. 75).
77. *Report from the Select Committee, op. cit.*, 1824, 56, (ref. 29).
78. *Proc. R. Dubl. Soc.* 61, 1824-25, 77.
79. *Report from the Select Committee on the Irish miscellaneous estimates*, H.C. 1829 (342), IV, 189, 194.
80. *Proc. R. Dubl. Soc.* 63, 1826-27, 10.
81. *Report from the Select Committee, op. cit.*, 1829, 194, (ref. 79).
82. *Proc. R. Dubl. Soc.* 62, 1825-26, 215.
83. *Ibid.*, 63, 1826-27, 8-10.
84. *Ibid.*, 140-141.
85. *Ibid.*, 154, 160.
86. *Ibid.*, 64, 1827-28, 4-6.
87. F.M. Turner, Public Science in Britain, 1880-1919, *Isis* 71, no. 259, 1980, 589-608.
88. *Proc. R. Dubl. Soc.* 62, 1825-26, 109-110 and appendix to the proceedings of 2 March 1826; Charles Lewis Giesecke, *Second account of a mineralogical excursion to the counties of Donegal, Mayo, and Galway*, Dublin, 1828, pp.

23; *Proc. R. Dubl. Soc.* 63, 1826-27, 56 and appendix to the proceedings of 14 December 1826; *ibid.*, 66, 1829-30, 4; 67, 1830-31, appendix no. 1, *Account of a mineralogical excursion to the County of Antrim*, Dublin, 1829, pp. 22.
89. *Proc. R. Dubl. Soc.* 63, 1826-27, 178.
90. Richard Griffith, *Report on the metallic mines of the Province of Leinster in Ireland*, Dublin, 1828, pp. 29. See also *Proc. R. Dubl. Soc.* 57, 1820-21, 205.
91. Griffith, *op. cit.*, 1829, (ref. 53).
92. *Proc. R. Dubl. Soc.* 64, 1827-28, 138.
93. *Ibid.*, 142-143.
94. *Ibid.*, 65, 1828-29, 31, 35-36, 40-41, 133.
95. *Ibid.*, 145-147 and appendix to the meeting of 18 December 1828.
96. *Ibid.*, 66, 1829-30, 23, 30-31.
97. DNB; *Q. Jl. Geol. Soc. Lond.* 12, 1856, xxxviii-xxxix.
98. Abraham Gottlob Werner, *A treatise on the external characters of fossils*, Dublin, 1805, pp. xx + 312.
99. *Proc. Dubl. Soc.* 50, 1813-14, 33. The four short-listed candidates were Robert Bakewell (1768-1843), Giesecke, James Millar (1762-1827), and Weaver.
100. James Ogilby to Robert Jameson 22 January 1816. *Pollock-Morris Papers.*
101. T. Weaver, Memoir on the geological relations of the east of Ireland, *Trans. Geol. Soc. Lond.* 5, 1819, 117-304.
102. *J. Geol. Soc. Dubl.* 3(3), 1846, 176.
103. NLI 16 B 3(15). See also G.L. Harries Davies, The oldest surviving geological map of Ireland, *Ir. Nat. J.* 17(12), 1973, 397-399.
104. DNB; Sheila Devlin-Thorp (editor), *Scotland's cultural heritage*, vol. 1, Edinburgh, 1981.
105. A. Nimmo, On the application of the science of geology to the purposes of practical navigation, *Trans. R. Ir. Acad.* 14, 1825, part 1, 39-50. See also *The Dublin Philosophical Journal, and Scientific Review*, 1, 1825, 152-161.

3

GRIFFITH'S UNOFFICIAL SURVEY

1830-1857

It was the Irish capital's impressive response to the Crystal Palace exhibition — the Great Dublin Industrial Exhibition of 1853, organised by the Royal Dublin Society and located upon the Society's premises at Leinster Lawn. On the morning of Wednesday 31 August 1853 Queen Victoria and Prince Albert paid their second visit to the exhibition. They were greeted in rousing style as a band played Dr Robert Stewart's specially composed Exhibition Grand March (by permission the piece was dedicated to the Prince himself) and now the royal party was moving through the Great Hall, past the equestran statue of Her Majesty by Baron Marochetti, past the scientific instruments displayed by Thomas Grubb of Rathmines and the carved bog-oak from the house of Cornelius Goggin in Dublin's Nassau Street (the Queen herself was actually wearing one of his bog-oak brooches set in Wicklow gold), and past the splendid array of cast-iron ware from the Coalbrookdale Company in England. Now it was that the Lord Lieutenant directed Prince Albert's attention to a striking map displayed upon Stand 39. It was Richard Griffith's magnificent quarter-inch geological map of Ireland. The visitors moved across to examine the map more closely and by the stand they found Griffith himself in attendance eager to demonstrate the salient features of Ireland's geology in both map and sections. The Queen, we are told, was deeply interested in what Griffith had to say, while Prince Albert evidently revealed himself to be by no means ignorant of the basic facts of Ireland's subterranean structure. Before the royal party moved away, Griffith presented to the Queen a small-scale version of his great map and this Her Majesty was graciously pleased to accept. It was perhaps the proudest moment of Griffith's life, but how did it come about? In 1829, when he resigned from the office of Mining Engineer to the Royal Dublin Society, the evidence of his past performance gave little hope that his much talked-of map would ever really achieve publication. Yet by 1853 not only had the map been published, but it was widely known and held in such esteem that the Lord Lieutenant singled it out for the particular attention of his royal guests. This chapter tells the story of how Griffith's map underwent its transformation from being merely a manuscript teaching-aid in the lecture-theatre of the Royal Dublin Society to being a familiar and highly regarded public document.

The Valuation Survey
The year 1831 gave Griffith the mid-point of his long life and it was the four years centred upon 1831 which were to prove crucial in the development of his scientific career. Four events during those years were of outstanding importance.

Firstly, there was Griffith's change of residence. By 1828 his engineering responsibilities in the southwest were well in hand and he decided to transfer himself and his family from Mallow and back to Dublin. Their new home in the metropolis was number 2 Fitzwilliam Place, a substantial house in a recently constructed Georgian terrace, and this was to be Griffith's Dublin residence until he died there in 1878. This removal brought Griffith back to the city which was the focus of Irish scientific endeavour. It was the city of such international luminaries of science as Edmund Davy, William Rowan Hamilton, Humphrey Lloyd, and Robert Mallet, and these men were the members of that stimulating circle into which Griffith was now to be drawn.

Secondly, there was the establishment of a new Irish society. After languishing for some fifteen years, Irish interest in the earth-sciences underwent revival in the 1830s and at a meeting held in the Provost's House of Trinity College on the evening of 29 November 1831 it was resolved to found the Geological Society of Dublin 'for the purpose of investigating the mineral structure of the earth, and more particularly of Ireland'.[1] Griffith was one of the eighteen gentlemen present that evening and he was soon deeply involved in the affairs of the new society — a society which for the next fifty years served as the forum for a lively Irish school of geology. Griffith helped to find the society its earliest premises; he purchased specimens for its museum and designed the museum's display-cases; between 1831 and his death in 1878 he was the society's president or vice-president for a total of 37 years (during the remaining ten years he was a member of the council); and to the society he presented some twenty research papers devoted to aspects of Irish geology.

Thirdly, there was the foundation of the British Association for the Advancement of Science at York in 1831. Griffith was evidently not present at that inaugural meeting, but by 1833 he was a member of the Association and he was certainly present in Edinburgh the following year when the Association assembled for the fourth of its annual meetings. Thereafter he attended the Association's peripatetic gatherings with fair regularity and to the Association he read many a paper, the last of them — a paper on the Irish eskers — being presented in 1871 when Griffith was in his eighty-seventh year.[2] His presence at these annual 'wisdom meetings' allowed him to enter the circle which contained such eminent geologists as Sir Charles Lyell, Sir Roderick Murchison, and Adam Sedgwick, and through the Association Griffith earned his membership of a stimulating and influential 'Invisible College'.

Fourthly — and here we encounter a development of paramount importance for Griffith's geological career — there was his new appointment in the public service. In 1827 the Lord Lieutenant had named him to be Commissioner for the General Valuation of Rateable Property. For three years this office remained dormant because the Ordnance maps which were to be the basis of the valuation were insufficiently far advanced, but by 1830 the Ordnance Survey was ready to make the first of its sheets available to the valuators. It was this fact which prompted Griffith's resignation from the post of Mining Engineer at the Royal Dublin Society in November 1829 (see p. 50) and he immediately set about the task of organising and staffing his new department. He continued to be involved in road engineering in the southwest, and he retained his responsibility for the Boundary Survey, which in 1830 still had to examine the entire southern half of the island, but it was now the Valuation Survey which came to have the chief call upon his attention. As Commissioner for the Valuation he was responsible for supervising the valuators who for the next thirty-five years were to travel throughout the length and breadth of Ireland valuing and re-valuing land and property as the basis for new rating systems, and the valuation carried out under his direction between 1852 and 1865 — officially 'the Tenement Valuation' — is still known throughout Ireland as 'the Griffith Valuation'.[3] The valuation surveys were remarkable achievements in their own right, but in the present context attention must be focussed upon the implications that the inauguration of the Valuation Survey held for Griffith's geological studies. Here there are three points to note.

Firstly, Griffith's new office gave him the freedom to travel throughout Ireland at public expense. Ostensibly he journeyed in order to direct the activities of his valuators — a duty he performed conscientiously and with great ability — but he doubtless availed himself of every opportunity of adding to his store of geological information. The Rev. Maxwell Close (1822-1903), who knew Griffith well in later years, observed that Griffith was pleased to hold the Boundary and Valuation survey posts concurrently because the two surveys were quite independent and he thus had the opportunity of visiting the same area in two different capacities at wide intervals of time.[4] In this way he could allow his initial geological opinions to mature before he returned to an area to make his final judgements. Near the end of his life Griffith reminisced about this period for the benefit of Archibald Geikie:

> ... during the entire period I never lost sight of the Geology of Ireland and carried it on in connection with all my public duties, which were very numerous and required my presence in every Parish, and almost in every large farm more than once....[5]

Certainly his travels now became very extensive and he once claimed — though perhaps with some exaggeration — that during these years he commonly journeyed from the north of Ireland to the south up to forty times each year, moving regularly by night so as to avoid the loss of a day in the field.[6]

Secondly, geology had a direct relevance to his new task because in their valuations his staff were involved in the assessment of soil quality. Each valuation party contained a Spadesman whose task it was to dig up soil samples for appraisal, but Griffith had no intention of allowing his men to confine their investigations to the superficial mantle; he insisted that in performing this part of their duties they should have due regard for the nature of the underlying solid rock. A modern pedologist would hardly agree with Griffith in ascribing a vital pedogenic role to the underlying solid strata, particularly in heavily glaciated and drift-encumbered Ireland, but Griffith gave instruction that attention was always to be paid to what he regarded as the underlying parent material. Indeed in the 1850s, at the start of the 'Griffith Valuation', he sought to assist his valuators in their task by issuing each of them with a colour-printed map of the country's solid geology at a scale of about 1:1,152,000 — a map which, printed by Forsters of Crow Street, Dublin, was the first colour-printed geological map of the whole of Ireland.[7] Prior to 1830 Griffith had perhaps sought to convince his contemporaries of the importance of geological mapping through the argument that such studies might reveal the presence of mineral resources which could become the basis for an Irish industrial revolution, but the employment of such a thesis made it difficult to justify detailed mapping outside the coalfields and a few other areas of obvious mineral potential. Now, after 1830, geology acquired for him a new significance as the science which he could claim was basic to the work of the Valuation Office. Privately, in his own mind, Griffith of course saw his geological map as a desirable end in itself, but the relationship of geology to agriculture through the medium of soils now afforded him with the overt utilitarian excuse he needed to justify the production of a detailed geological map of the entire country.

Thirdly, not only did Griffith now have an excellent excuse to justify his personal involvement with geology, but in his new office he was provided with a large staff which could be employed to assist him in a programme of geological surveying. The faithful John Kelly joined the valuation staff in 1830 as Griffith's chief assistant, and within a few years Griffith had about one hundred valuators under his control, all of them instructed in the making of geological observations. In his official reports he was understandably reluctant to disclose how much of his men's time was being spent upon purely geological investigations, but later he was rather more forthcoming in his correspondence with Archibald Geikie:

> I required my numerous land surveyors, and other assistants to mark on the large county maps and on the Ordnance Survey where it had been completed each quarry marking the dip of the strata in each case and collecting a small specimen of the rock.[8]

Apart from Kelly, there were three members of the valuation staff whose geological investigations seem to have been of particular value to Griffith: Patrick Ganly (1809?-1899) from Dublin, Patrick Knight from County Mayo, and Samuel Nicholson from Dumfriesshire in Scotland. Some of Griffith's men were to become proficient geologists in their own right. Knight, for instance, read a paper on the geology of County Mayo to the Geological Society of Dublin as early as May 1833[9]; in 1838 Ganly was the discoverer of the fact that cross-bedding can be used to demonstrate whether a stratum has been inverted[10]; and Kelly himself was to become an officer of the Geological Survey of Ireland, a vice-president of the Royal Geological Society of Ireland (this was the new title granted to the Geological Society of Dublin in March 1864), and the author of a number of geological papers together with a book in which he joined so many other Victorian authors in seeking a reconciliation of *Genesis* with geology.[11] Perhaps valuators such as these needed little encouragement from Griffith in order to persuade them to pursue geological investigations during the course of their other duties, but Griffith was soon allowing geology to get quite out of hand. The measurement of dips and the collection of fossil specimens is hardly necessary for the assessment of soil quality, and we must be highly sceptical of Griffith's oft-repeated claim that his men only made such studies upon a voluntary basis and 'after the official business of the day had terminated'.[12]

The Valuation Survey commenced its operations at Coleraine, County Londonderry, on 1 May 1830, and it was there that Griffith displayed his geological map of Ireland for the instruction of his trainee staff[13], that map presumably being a coloured copy of Arrowsmith's map such as had been used to illustrate Griffith's Royal Dublin Society lectures. From that moment onwards Griffith was in reality running a rudimentary and quite unofficial geological survey of Ireland, and it is worthy of note that the inauguration of this survey pre-dates by two years the appointment of Henry Thomas De La Beche to affix geological colours to the Ordnance maps of southwestern England. Had the authorities been aware of what was afoot in Ireland, then Griffith must surely have faced the accusation that he was guilty of diverting public funds into a private enterprise. But in fact the authorities remained quite oblivious to the existence of a geological cuckoo within the Valuation Office nest. If they had discovered what was afoot, then we may be confident that the devious, cunning but enormously able Griffith would have had some ready rejoinder. He had located the headquarters of the Valuation Office within his own residence at 2 Fitzwilliam Place and there he succeeded in getting his public and private duties inseparably intertwined. The authorities were even in doubt as to how far the hall-porter at 2 Fitzwilliam Place was one of the Griffith family's domestics and how far he was an employee of the Valuation Office![14]

With the resources of the Valuation Survey at his command, Griffith now began to think seriously again about his geological map of Ireland. On 20 March 1833 the council of the Geological Society of Dublin asked him to sanction their copying his geological map for the use of the society, and he agreed, although he asked for the matter to be deferred for one year so that he might introduce certain corrections in the map's representation of Connemara.[15] This particular incident would seem to belie Griffith's claim that he made no significant alterations to the map between 1815 and 1835 (see p. 44), but it was certainly 1834 before he began to undertake a major revision of the map. In September of that year the British Association met in Edinburgh and there it was resolved that the following year's meeting should be held in Dublin. Griffith was present in Edinburgh and he found himself nominated to serve as the president of Section C (Geology and Geography) for the forthcoming Dublin meeting. He must have been delighted with the honour, and in a moment of euphoria he made the Association a promise: when the members reconvened in Dublin in 1835 he would put on display his geological map of Ireland to provide the visitors with an introduction to the structure of his native land.[16] This was to submit the map to its most critical test yet, and with the coming ordeal in mind he began to revise the map just as soon as he returned to Dublin from Scotland. On 3 December he wrote to G.B. Greenough:

> I am looking over all my note books and County maps, as a commencement of my new edition of the Geological Map of Ireland. I have some doubts about the colouring and wish to have your assistance on this point.[17]

Very fittingly it was Greenough who conveyed Griffith's promise to the Geological Society of London. The veteran geologist was now enjoying his third term in the society's presidential chair and in his anniversary address on 20 February 1835 Greenough observed:

> The coloured copy of Arrowsmith's map of that portion of the United Kingdom which Mr. Griffith has undertaken to lay before the British Association in August next, will bring within our reach an abundant supply of geological information, which though it has been in his possession for many years past, a natural repugnance to combining geological correctness with geographical inaccuracy has hitherto induced him to withold.[18]

By the following summer Griffith was ready. The Dublin meeting was held between 10 and 15 August and it was generally acclaimed as an outstanding success. At the conclusion of the meeting the Assistant Secretary, John Phillips, wrote to his sister:

> Everything has worked very well, each Section thinks itself to have been eminently successful.... The weekly festivities were incredibly beautiful particularly the Zoological Fete, and (as I hear for I could not go to it) the Dublin Garden Fete. The most elegant dinner I ever saw was given to 300 members on Saturday in Trinity College when the fine quire of the College sang delightfully....[19]

Such was the enthusiasm that more than three hundred local applicants had to be refused membership because of the pressure of numbers. Section C met in the lecture-theatre of the Royal Dublin Society in Leinster House (the building now accommodates the Irish parliament) and there Griffith presided over meetings attended by such notables as Louis Agassiz, Greenough, Phillips, Murchison, Sedgwick, James Smith of Jordanhill, William Whewell, and the famed William Smith himself. Rarely can one room have contained so distinguished a band of geological cartographers, and Griffith must have been delighted at the calibre of his audience when, on the first morning of the meeting, he vacated the chair in favour of Sedgwick and then rose to address the section on his favourite subject — his geological map of Ireland.[20]

In his opening remarks Griffith briefly outlined the map's history, paying due acknowledgement to such assistance as he had received from the works of Berger, Buckland, Conybeare, and Weaver. In the continued absence of any Ordnance Survey map of Ireland, the map on display was still only a coloured Arrowsmith and Griffith proceeded to remind his audience of the problems attendant upon using inaccurate base-maps for geological purposes. Finally he launched into a discourse on the geology of Ireland illustrating his various points by reference to both the map and the section which he and Kelly had surveyed across northern Ireland in 1817 (see p. 43). His paper evidently aroused much interest, because the discussion had to be adjourned on the Monday morning and it was resumed on the Wednesday. After the meeting Sedgwick wrote to Charles Lyell:

> Griffith exhibited a geological map of the whole of Ireland, coloured from his own observations. It is a thousand pities he did not publish it fourteen years since. As far as I have examined the demarcations, they appear to be very well laid down — much more correctly I think than in our friend Greenough's first edition. The description and discussion of the map took two mornings.[21]

Greenough and Murchison both took the opportunity of discussing with Griffith the principles which should govern the colouring of geological maps[22], and Phillips secured Griffith's permission to copy the map of Ireland for incorporation into a geological map of the British Isles which he was then compiling. When it was published soon after (probably in 1837)[23] this map of Phillip's became the first published version of Griffith's work, but it is only on the very small scale of approximately one inch to thirty miles (1:1,900,800).[24]

The Railway Commission

Prior to 1835 Griffith's map of Ireland can have been seen by very few of the leading British geologists of the day, but now, as a result of his Dublin exposition, the map was well and truly launched upon the geological scene. He must have been encouraged by the favourable reception accorded to the map, but there still remained that old thorny problem: how could the map be published in the absence of an accurate base-sheet? At its Dublin meeting the British Association had itself forwarded to the government a resolution pointing out 'that the advancement of various branches of science is greatly retarded by the want of an accurate map of the whole of the British Islands'[25], and at the beginning of 1836 a solution to Griffith's base-map problem must have seemed as remote as ever. But in reality the answer now lay just around the corner. On 28 October 1836 there was established a small commission 'to inquire into the manner in which Railway Communications can be most advantageously promoted in Ireland'. Griffith was a member of the four-man body, and again he demonstrated his subtle ability to direct affairs towards his own geological ends. His three fellow commissioners were Peter Barlow F.R.S. the mathematician, Colonel John Fox Burgoyne the distinguished Peninsula veteran who was now chairman of the Irish Board of Works, and Thomas Drummond, the able Under Secretary of State for Ireland. By no stretch of the imagination could men such as these be regarded as pliable light-weights, yet Griffith does seem to have been able to urge their thinking in precisely that direction he most desired. Perhaps there was already some sympathy for his geological bias because two of his fellow commissioners — Barlow and Burgoyne — were members of the Geological Society of Dublin, but no matter how it was achieved, we can surely discern Griffith's influence being reflected in two ways within the commission's reports.

Firstly, Griffith laid his newly completed manuscript map before the other commissioners and satisfied them that a knowledge of geology was basic to the planning of Ireland's future railway system. Were the Carboniferous limestone regions not both the seat of the country's richest agricultural land and the site of most of the major towns and cities, and were they not therefore the areas which would generate the majority of the nation's future rail-traffic? Equally, were the regions of pre-Carboniferous rocks not areas of poor soils, low population density, and subsistence agriculture? Railways were thus hardly likely to be economic if they were allowed to stray down the stratigraphical column into regions consisting of rocks of Devonian or earlier age. Even more directly there was the influence which geology was able to exert upon the railway engineer through the structural control of topography. As the commissioners observed:

> ...we found that the lowest and most level lines throughout the country, are almost exclusively confined to the carboniferous limestone, and the moment we passed the boundaries of that rock we encountered difficulties which it was desirable to avoid.[26]

Secondly, the commissioners became convinced — and it must have been at Griffith's insistence — that there existed no map of Ireland sufficiently accurate to be employed in railway planning at even the outline national level. The commissioners therefore approached the Ordnance Survey and it was agreed that the Survey would construct a new map of Ireland expressly for the commission's use.[27] The map was to be based upon the primary national triangulation which the Survey had finished in 1832, and upon the completed six-inch survey of the nine northern counties of Antrim, Armagh, Donegal, Down, Fermanagh, Londonderry, Louth, Monaghan, and Tyrone. Elsewhere the Ordnance Survey would have to abstract its details from the old county maps as Griffith had once hoped to adjust the same maps to fit his own triangulation. The construction of the new map was entrusted to Lieutenant Thomas Larcom who was in charge of the Ordnance Survey Office in Phoenix Park, Dublin, and the drawing of the map was completed in 1837 to a scale of one inch to four miles (1:253,440).[28]

Griffith had played his cards well. He had prevailed upon his fellow commissioners to extract from the Ordnance Survey that very map which for so long had been his heart's desire — an accurate and large-scale topographical map of the whole of Ireland. But further than this, the commissioners were now so convinced of the importance of geology that they decided to feature it prominently in their final report. They determined to add to the report a substantial appendix entitled *Outline of the geology of Ireland*[29] (needless to say Griffith was its author), and they resolved to publish Griffith's geological work in its entirety upon the quarter-inch map that Larcom had prepared. The engraving of that map, however, took time, as did the transference to it of Griffith's geological lines, and the map was still incomplete when the commissioners published their final report in 1838. A smaller scale 'provisional' version of the geological map was therefore prepared to accompany the report and this map is today perhaps the best-known of all Griffith's geological sheets (Fig. 3.1).

Sheets of Many Colours 63

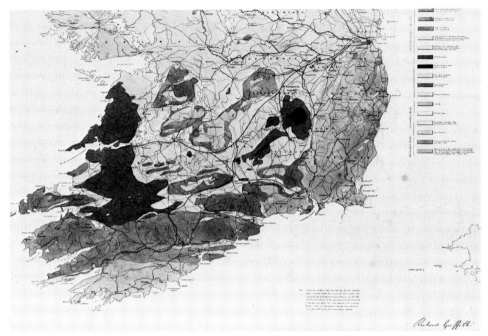

Fig. 3.1. The southern half of Richard Griffith's map for the Railway Commissioners' Atlas *of 1838. (Reduced in size.) The note to the left of the facsimile of his signature in the bottom right-hand corner explains that reduction to the scale of the original has necessitated the omission of much of the detail which has been mapped in the field.*

The 'provisional' sheet is contained in an atlas of six maps published in 1838,[30] the other maps depicting phenomena such as Ireland's population density and internal trade, and some of the maps have recently attracted attention because of the employment of cartographic techniques that were remarkably sophisticated for their day.[31] Two thousand copies of each of the six atlas maps were produced[32] and the geological map bears the following title:

GEOLOGICAL
MAP OF
IRELAND
to accompany the Report of the
RAILWAY COMMISSIONERS
1837

It carries a printed facsimile of Griffith's signature and the inscription 'Dublin, April 28th 1838'. The scale is approximately one inch to ten miles (1:633,600) and the engraved area is 63.5 x 79 centimetres. The topographical detail was supplied by Larcom, but the actual engraving was the work of James Gardner of Regent Street, London. Geology is represented by a series of twenty different colours, and in conception the map is thoroughly modern in that the stratigraphical approach has finally replaced the geognostical. Thus we now find represented, in descending stratigraphical order, 'Tertiary Formation' (the Lough Neagh Clays), 'Lyas, Greensand and Chalk', 'New Red Sandstone and Red Marl', 'Coal Formation', 'Millstone grit', 'Carboniferous or Mountain Limestone', 'Yellow Sandstone and Conglomerate', and 'Old Red Sandstone and Sandstone Conglomerate'. Of the post-Old Red Sandstone systems now known to exist in Ireland, only the magnesian limestones and breccias of the Permian are absent from the map, but for this omission Griffith can hardly be blamed because the Permian system was not conceived until Murchison's Russian journey of 1841. In any case, the Irish

Permian rocks are extremely limited in their outcrop and it was to be 1852 before they were first dated positively.[33]

Among the rocks older than the Old Red Sandstone, Griffith's classification is of course much more rudimentary because Lower Palaezoic stratigraphy was then in its infancy and Murchison's seminal work *The Silurian System* was not published until 1839.[34] Griffith does nevertheless make a clear distinction between the rocks which he terms 'Clay Slate, Graywacke, and Graywacke Slate' – rocks now known to be of Silurian, Ordovician and Cambrian age – and the 'Mica Slates' of northwestern Ireland which were to become the Dalradian Schists of a later generation of Irish geologists. True, the map does have its geological shortcomings and it has to face its most serious criticism in the far south in the very region which Griffith must have known so well during his years of residence at Mallow in the 1820s. There he has placed the extensive tracts of what are really Old Red Sandstone not into the Old Red Sandstone category, but into their own unique category – the 'Old Conglomerate' – placed stratigraphically directly beneath the Old Red Sandstone. Two closely related facts led him into this error. Firstly, the Old Red Sandstone in the south is generally more deformed than the equivalent rock elsewhere in Ireland and this caused him to regard the southern sandstones as older than those farther to the north. Secondly, the Old Red Sandstone of the south locally contains beds of quartzite and slate and, as he explained in 1840, this reminded him of lessons learned at Robert Jameson's feet thirty years before:

> From the Wernerian character of my geological education, I found it almost impossible to conceive the quartz-rock and clayslate could be newer than the old red sandstone, or even belong to that series.[35]

This mistake he compounded with another: he mapped what are today known as the Carboniferous Cork Beds of southern County Cork as 'Clay Slate' which he regarded as older than the 'Old Conglomerate' (Old Red Sandstone) lying immediately to the north. With such a faulty stratigraphy in mind, he in 1838 can have had little conception of the true nature of the region's structure. Other lesser points of criticism can also be found. The Carboniferous outcrop near Wexford town is inadequately represented, for example, the Limerick volcanics are poorly delimited, and many quartzite masses have been overlooked in counties Galway, Mayo, and Donegal. But all such weaknesses do little to detract from the fact that the basic outlines represented upon this map are remarkably good. The map certainly represents a considerable advance upon the geological sheet of southwestern Ireland by Thomas Weaver which the Geological Society of London had published in the previous year[36], and Griffith's 1838 map really suffers little when compared with its modern counterpart, the 1:750,000 sheet of the Geological Survey of Ireland. Indeed, from the aesthetic standpoint, the older, hand-coloured map has a quality which is entirely lacking from its somewhat garish descendant. In short, Griffith's atlas map is both an attractive piece of cartography and an excellent representation of Ireland's geology. It was the first published geological map to show the whole of Ireland upon a reasonably large scale, but its significance was even greater than this because almost one fifth of the country had never before been depicted upon *any* published geological map save for such very small-scale representations as that compiled by Phillips around 1837 (see p. 61).

A cartographic masterpiece

The atlas map of 1838 was merely a trial-run in anticipation of the much more ambitious map which was to follow. Larcom had completed the construction of the new quarter-inch base-map in 1837, and during 1837 and 1838 the map was with the engravers. The prospect of at last being handed an accurate topographical map of Ireland now threw Griffith into a flurry of activity. This new geological map clearly had to be prepared with care and precision. Unlike his earlier maps of Ireland, the new map was not merely a teaching aid to be put upon fleeting display in some lecture-theatre. Rather was it a map which would be in the hands of other geologists — skilled geologists — as they walked the Irish countryside observing, checking, criticising. Griffith was a proud man; he wanted to be confident that his map would survive any test that his peers might choose to apply. At this late stage he therefore resolved to confirm and improve many of his geological lines by further field investigation. He himself can have done little of this work; he was far too deeply involved in public duties to be able to spend weeks searching for geological outcrops. As he wrote to Greenough on 13 November 1838:

> I have done little in Geology since we parted at Newcastle [in August 1838]. I am so much absorbed with the detail of business connected with the Ordnance Survey and the Valuation of Ireland that I have not one leisure moment left to entertain even a geological thought.[37]

Almost all of this work of revision was left to the valuators, with the chief burden falling upon Patrick Ganly. At the time Ganly was in theory a draughtsman in the Valuation Office, but in reality between 1837 and 1846 he held a roving commission to explore Ireland's geology under Griffith's direction. This particular episode in the history of Griffith's map receives illumination from the *Ganly Papers*, most of which are now lodged in the library of the Royal Irish Academy.[38] The papers consist of over six hundred letters dated between June 1837 and April 1846, and written by Ganly in the field to either Griffith or Kelly back in Dublin. Personal detail is almost entirely lacking from the letters; they take the form of a series of reports written twice a week and recording Ganly's day by day activities for the benefit of his Dublin superiors. Whatever Griffith may have claimed about his men only conducting geological investigations during their off-duty hours, these letters prove conclusively that during the years from 1837 to 1846 Ganly was working hard and long doing little but geology. Indeed, Griffith clearly kept him under considerable pressure and he was certainly allowed no close season during the final winter before the publication of Griffith's great map; he was kept hard at work in West Cork and Kerry right through the rigorous months of December, January, and February of 1838 to 1839. It was there, at Dunmanway in County Cork, that he must have experienced that legendary event of Irish folk memory, the Big Wind of the night of Sunday 6 January 1839.

By the close of 1839 six-inch Ordnance Survey sheets had been published covering seventeen of the thirty-two Irish counties, and these were the maps with which Ganly was if possible provided as a basis for his field investigations. As early as 25 August 1837 he wrote to Griffith from Manorhamilton, County Leitrim, recording the despatch to Dublin of thirteen six-inch maps of County Cavan and fourteen similar sheets of County Fermanagh, all of them evidently bearing geological field annotations. Four of his equivalent six-inch field-sheets from County Galway — all dating from the early 1840s — have only recently come to light in University College Galway.[39] From his field-sheets he transferred the geological lines to the county index maps (mostly drawn to the half-inch scale, 1:126,720) which accompanied the six-inch survey, and to these index maps he subsequently added geological colours. Both his letters and his few surviving field-maps show him to have been an extremely thorough and perceptive geologist, and the following extract, taken from a letter he wrote to Kelly from Drumkeeran, County Leitrim, on 27 July 1837, is a typical example of the content of Ganly's reports:

> Monday, 24 July. Traced lower Sandstone of Lugnaquilla, by eastern base of Mullahirk mountain, to northern shore of Loughmacnean; also, also marked junction of this sandstone with the great overlying shale.
> Tuesday, 25 July. Removed from Sun Inn to this place. The day very wet. Attempted to make a section through the upper sandstone of Lugnaquilla with a view to determine the position of the coal beds, but failed in consequence of rain.

Wednesday, 26 July. Traced outgoing of the upper sandstone of Lugnaquilla from Tubber, in a western direction to head of the river Ayle; and thence, in a southern direction to Altun near the southernmost point of Dowbally mountain.

During the course of all these field investigations Ganly was expected to collect geological specimens for Griffith and this commission he executed in his usual conscientious manner. Thus on 15 October 1838 we find him at Foynes, County Limerick, recording the despatch of five boxes of fossils to Dublin by the canal company's boat. Griffith had been slow to appreciate the vital stratigraphical importance of palaeontology — a slowness which can in part be attributed to his semi-withdrawal from geology during the 1820s at a time when stratigraphical palaeontology was developing apace — but now in the 1830s he rapidly amassed a large fossil collection. Recognising his own weakness as a palaeontologist — on New Year's Day 1839 he wrote to Greenough saying 'I know scarcely anything of the matter myself & am too old to learn'[40] — he about 1840 took into his personal employment a precocious, red-headed young naturalist named Frederick McCoy (1823-1899) to examine and classify all the fossil material. Of McCoy, the son of a distinguished Dublin physician, we will hear more later, but the early fruit of his labours was seen in the publication of synopses of the Irish Carboniferous fossils (1842 and 1844) and of the Irish Silurian fossils (1846)[41], all the studies being based upon the contents of what Griffith persisted in regarding as his personal fossil collection despite the fact that much of it had been assembled through the activities of the valuators.

Griffith played the role of a proud father to his geological map, and in his enthusiasm he determined to give the scientific world every opportunity of inspecting the work even during those intimate final stages in its birth. Having displayed the map in embryo to the British Association in Dublin in 1835, he two years later returned to the Association — it was now assembled in Liverpool — to exhibit what is described as 'his large geological map of Ireland', although it is difficult to see how this can have been anything more than either a revised version of his 1835 Arrowsmith or a copy of the recently published Railway Commissioners' atlas map. Perhaps what the members of Section C chiefly remembered of Griffith's contributions that year was not so much the map as a veritable geological anaconda that Griffith had brought over from Ireland — a geological section at a horizontal scale of six inches to the mile (1:10,560), extending fifty miles from the graywackes near Butler's Bridge, County Cavan, northwestwards across the Carboniferous rocks of the Cuilcagh Plateau Country, to the sea at the foot of Benbulbin.[42] The topographical detail of the section was based upon the maps of the Ordnance Survey, and the geology seems to have been derived largely from Ganly's field investigations. That geology was important because it gave Griffith what he came to adopt as his type sequence for the Irish Lower Carboniferous strata: the Yellow Sandstone at the base, then the Carboniferous Slate, the Lower Limestone, the Calp (alternating beds of dark shales and impure limestones) and finally the Upper Limestone. This interpretation of the Irish Lower Carboniferous sequence was destined for a long life; in its essentials it continues to this day to feature upon the most widely circulated map of the Geological Survey of Ireland, namely upon the 1:750,000 geological map of Ireland (see p. 182).

When the Association re-assembled at Newcastle-upon-Tyne in August 1838 Griffith was again present to display various geological sections, this time of southern Ireland, but his *chef-d'oeuvre* was indeed a map which he now proudly brought before the public for the first time.[43] It was a proof impression of Larcom's quarter-inch map of Ireland to which Griffith had added geological colours so that members of the Association could see what the new geological map would look like when it was eventually published. This map exhibited at the Newcastle meeting evidently bore the manuscript inscription '1838' and it was seemingly the only copy of Griffith's quarter-inch map ever to display that date. Later, as we will see (p. 67), this particular map came into the possession of the Geological Society of Dublin, but unfortunately the map's present whereabouts are unknown. Its rediscovery would certainly be a matter of much interest because that 1838 map pre-dates the earliest published version of the map by some nine months. In southwestern Ireland, at least, the map of 1838 must have differed considerably from the map which was published in the following year, because throughout the winter of 1838 to 1839 Ganly was engaged in his field-revision in counties Clare, Limerick, Kerry, and Cork, and his amendments continued to stream back to Dublin. Larcom's new map was printed in six sheets and in this field-work Ganly was now checking a proof copy of the southwestern sheet against the field evidence. Such was Griffith's confidence in Ganly's work that early in 1839 Ganly was instructed to send his revisions not to Griffith at the Valuation Office, but direct to the Ordnance Survey where the new lines were to be engraved immediately by Larcom's staff. The hardworking Ganly was told to place his final amendments to the geological lines of the southwest into the hands of the engravers by 28 January 1839 at the latest and, like the conscientious employee that he was, he met the deadline by despatching his last revisions to the Ordnance Survey from Macroom, County Cork, on 26 January.[44]

Two months later, on 28 March 1839, Griffith inspected and signed a still unfinished proof copy of his new map — a proof copy which survives in the archives of the Irish Ordnance Survey — and over the next few days the final amendments must have been added to the copper plates from which the map was to be printed.[45] Griffith tells us that it was also during March 1839 that he added geological colours to the two maps which were to serve as the colourists' master-copies, one for Hodges and Smith of College Green, Dublin, and the other for James Gardner of Regent Street, London.[46] The map's publication date may be regarded as 22 May 1839 because upon that day Griffith displayed to the Geological Society of London one of the first copies of the map to come from the colourists.[47] Colouring the maps was a laborious task — we are told that it took from seven to eight days to colour a copy of William Smith's geological map of England and Wales[48] — and several months therefore elapsed before copies of Griffith's map became generally available. It was August 1839 before Charles William Hamilton, the Secretary of the Geological Society of Dublin, was able to acquire a copy from Gardner's[49] and even Griffith himself had difficulty in obtaining copies. When he addressed the Geological Society of Dublin on 12 June 1839 no copy of the new map was to hand and he therefore had to give the society the early proof copy dated 1838 which he had displayed before the British Association at Newcastle-upon-Tyne in the previous August.[50]

The new map (Fig. 3.2) bears the following title:

A GENERAL MAP
OF
IRELAND
to Accompany the Report
of the
RAILWAY COMMISSIONERS
shewing the
Principal Physical Features and Geological Structure
OF THE COUNTRY

The map's engraved area is 147.5 x 182 centimetres and the base-map is Larcom's fine quarter-inch (1:253,440) hachured map, often referred to as 'the Railway Commissioners Map', and printed, as we have seen, in six equi-sized segments. Sometimes the six segments are found dissected, separately folded and contained in a slip-case, but in other examples the segments have been joined together and mounted upon rollers to form an impressive wall-map. The price of the map was twenty shillings uncoloured, and these uncoloured copies, complete with all their printed geological lines and information, were widely used as a base-map for the plotting of non-geological phenomena because for many years Larcom's was the only accurate topographical map available depicting the whole of Ireland. Two thousand copies of the map were produced to accompany the report of the Railway Commissioners and by the close of 1846 a

Fig. 3.2. Part of Sheet 3 of Griffith's quarter-inch geological map of Ireland of 1839. (Reduced in size.) From a dissected copy of the map in the collection of the Geological Survey of Ireland, reproduced by kind permission of the Director.

total of 5406 copies of the map had been distributed.[51] Only a small fraction of that number were ever coloured to become real geological maps, however, and it seems that geological versions were produced by Hodges and Smith and by Gardner's chiefly in response to orders received from individual customers. On the geological versions of the map, rock-type is represented by colour washes set within geological boundaries which in most places are represented by engraved lines. In the key there are twenty-six colour boxes, and in addition a set of symbols is employed to locate and differentiate the various types of mine. As on all Griffith's earlier maps, the key is upside down in the sense that the colours representing the Primary rocks are at the top while the Tertiary beds are down near the bottom. At the foot of the map numerous acknowledgements are made concerning the sources from which its topographical detail has been derived (one acknowledgement is to Alexander Nimmo for certain submarine information) but where the geology is concerned there is merely the bald statement: 'The geological lines have been laid down by Richd. Griffith Esqr.' If Ganly, Kelly, and the others had hoped that their work would receive some small recognition, then they were doomed to disappointment, and in later years Kelly, for a variety of reasons, did harbour a sense of grievance against the man whom he had served for so long.[52]

The use of a quarter-inch base-map presents any geologist with both a challenge and an opportunity. On the one hand, such a scaled base-map demands careful attention to field detail. Nothing can be slurred over. Even relatively small inliers and outliers must be discovered and plotted. On the other hand, a quarter-inch map gives the conscientious field-geologist the opportunity of presenting his field-observations with a minimum of generalisation, and most of those sinuous boundaries traced out so laboriously upon the ground can find a place upon the eventual completed map. This challenge and opportunity Griffith readily accepted. He and his staff had carried through a painstaking field-survey, and now they had the satisfaction of seeing the great geological jigsaw assembled for the first time upon a really satisfying scale.

The map makes no attempt to depict Ireland's ubiquitous Quaternary deposits; earlier in his life Griffith had spent years mapping the Irish peat-bogs and later he was to turn to study the Irish eskers, but for the moment his attention was firmly fixed upon Ireland's more solid foundations. Among those foundations he has very satisfactorily delimited the higher courses — rocks belonging to the Tertiary and Mesozoic eras — but such strata underlie only a small proportion of Ireland's surface and it was amidst the far more extensive Palaeozoic rocks that he encountered his principal challenge. At the top of the Palaeozoic pile Permian rocks of course still pass without title, although in the key, opposite the New Red Sandstone colourbox, is the prescient observation: 'In one instance lower beds alternate with magnesian limestone'. The Carboniferous System receives nine colour-boxes in the key, these ranging from the Coal Formation at the top of the sequence to the Yellow Sandstone at the bottom. The distinction that Griffith introduces here between regions of workable and unworkable coal, between Millstone Grit with and without coal-seams, serves to remind us of the origins of his map in a series of coal district surveys a quarter of a century earlier. Those nine Carboniferous colour-boxes do nevertheless convey a false sense of the stratigraphical refinement that Griffith had been able to introduce into his actual map. In September 1837 he had suggested a tripartite division of the Irish limestone into a Lower Limestone, a Calp Limestone, and an Upper Limestone (see p. 66), but the Irish Carboniferous limestones as a whole underlie almost forty thousand square kilometres and between 1837 and March 1839 there had been insufficient time for all the limestone terrain to be re-examined in the field so that the new subdivision could be introduced into the map. As a result, the map was published with most of the limestone depicted as Lower Limestone; only a few areas of Upper Limestone are represented and no Calp Limestone whatsoever is indicated. Indeed, upon some early issues of the map even the Calp colour-box is left empty of its tint.

Still within the Carboniferous sequence, but in the far south of Ireland, Griffith had revised his views on what we now know as the Cork Beds. In 1838 he had mapped the Cork

Beds as 'Clay Slate' and he had placed them stratigraphically beneath the sandstones and conglomerates — what he had termed the 'Old Conglomerate' — lying immediately to the north (see p. 64). In 1839 he wisely placed the Cork Beds into the new 'Carboniferous Slate' category and this he now recognised to be really a younger formation than the sandstones and conglomerates to the north. With this stratigraphical succession correctly represented, he was in a far better position to appreciate the folded structure of Munster's Ridge and Valley province. The sandstones and conglomerates — the 'Old Conglomerate' of 1838 — nevertheless still presented him with a problem. He was now satisfied that in eastern Munster the sandstones and conglomerates, despite their relatively high degree of deformation (see p. 64), really belonged to the Old Red Sandstone and as such were they represented upon the 1839 map. But was the 'Old Conglomerate' of western Munster also Old Red Sandstone? Here Griffith encountered the problem of how far the newly designated Silurian System might be present in Ireland. He had doubtless been in Leinster House in August 1835 to hear Murchison and Sedgwick read their paper on the Silurian and Cambrian systems[53], and he had certainly corresponded with Murchison, 'the King of Siluria', upon his newly discovered realm. The first indication that Silurian rocks extended into Ireland came in 1837 when Captain Joseph Ellison Portlock, an officer of the Ordnance Survey, announced the discovery of a small patch of Silurian strata near Pomeroy in County Tyrone.[54] That patch had not featured as Silurian upon Griffith's 1838 map — the patch had then been coloured as 'Clay Slate and Graywacke Slate' — but upon the 1839 map the patch was differentiated by a green tint as being the only known area of Silurian rocks in Ireland. Perhaps it was a mark of Griffith's want of familiarity with the new system that in the key, opposite the appropriate green colour-box, the title of the system is misspelled as 'Silurean' upon early printings of the map. But, as the ideal geological cross-section at the foot of the 1839 map serves to emphasise, Griffith was aware that elsewhere in Ireland there might be Silurian strata intervening between the Old Red Sandstone and what he termed 'the Transition Clay Slate'. With this problem in mind he had obtained fossils from Dunquin, Ferriters Cove, and other sites in the Dingle Peninsula of western Munster, and from Knockmahon in County Waterford, and these fossils he had despatched to London for examination by William Lonsdale, by James de Carle Sowerby, and by Murchison himself. Their verdict was unanimous; the fossils were of Silurian age. The result came too late for the appropriate change to be made in the colouring of the 1839 map; the rocks were left bearing the 'Transition Clay Slate' tint, but Griffith added to the map a pasted-on label dated in Dublin on 20 May 1839 and explaining that certain areas of the map now depicted as Transition Clay Slate were probably founded upon rocks belonging to the Silurian System. He hazarded the suggestion that further investigation would reveal the graywackes to be of Silurian age both in western Munster and in northern County Dublin. In the latter area his surmise was to prove correct, but in the case of western Munster it was eventually to emerge that he was wrong, although the true age of some of that region's rocks is still engaging the attention of geologists even today almost a century and a half after the first appearance of Griffith's map. It nevertheless has to be admitted that the problems of Munster geology led Griffith to publish what must be the worst of all the geological boundaries to appear upon the 1839 map. If, as he supposed, the uplands of eastern Munster were developed in Old Red Sandstone while those of western Munster were developed in Transition Clay Slate of possible Silurian age, then clearly there had to be drawn some boundary between the two regions. The boundary chosen is essentially a topographical one separating the Munster plateaux country in the east from the more rugged, mountainous terrain to the west, the boundary following the southeastern and eastern foot of the Sheehy, Derrynasaggart, and Boggeragh mountains. In reality the country on both sides of this line is underlain by Old Red Sandstone and nothing like Griffith's line finds expression upon the modern geological map. His line is, however, strangely reminiscent of the line which Horatio Townsend had drawn upon his map of County Cork in 1810 to separate his 'Red or Brownstone district' in the east from the 'Secondary Mountn. & high

Moorland' in the west (see p. 15).

Some imperfections Griffith's map of 1839 may have possessed, but it was nevertheless a very fine map. Back in 1821 he had expressed the hope that his eventual map might be found a fit companion to take its place alongside Greenough's England and Wales and Macculloch's Scotland (see p. 46). His hope was now amply fulfilled, and whatever the merits of Griffith's map vis-à-vis that of Greenough, there can be no doubting the fact that Griffith's pioneer geological map of Ireland is far superior to Macculloch's equivalent map of Scotland published in 1836, only three years before Griffith finally launched his masterpiece.

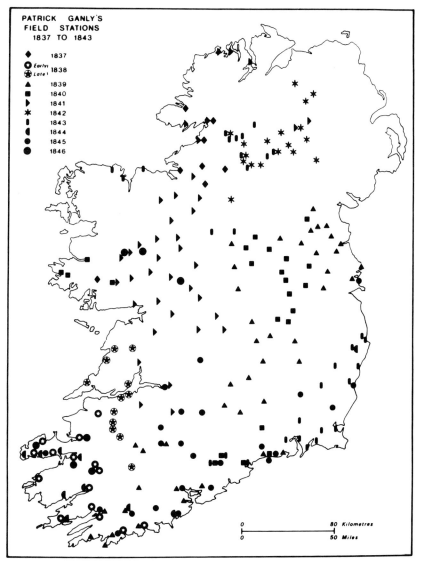

Fig. 3.3. Patrick Ganly's field-stations from June 1837 to April 1846. Ganly visited many of the stations represented more than once within a given year. Based upon information in the Ganly Papers *in the Royal Irish Academy and upon further documents in private hands. It should be noted that Ganly appears to have worked in West Cork and Kerry, in County Donegal, and in County Monaghan before the writing of the earliest item in the* Ganly Papers *in June 1837.*

The map's later history

The publication of Griffith's quarter-inch map in 1839 by no means signalled the end of Griffith's geological activities. He now embarked upon a policy of continuous revision for the map. He went on making his own geological investigations as he travelled around Ireland in the course of his public duties; he persisted in the employment of his valuators in the recording of geological observations; and he incorporated into his map discoveries made by other Irish geologists — discoveries which had commonly been announced at meetings of the Geological Society of Dublin. But in this revision there was one individual above all others upon whom Griffith relied — Patrick Ganly. The publication of the map in May 1839 did nothing to release Ganly from his intensive field investigations upon Griffith's behalf. From 1839 until 1846 he was kept hard at work collecting fossils and rock samples, checking geological boundaries, re-mapping formations, and refining stratigraphical classifications. The *Ganly Papers* show that during these years he was employed in most parts of Ireland save the northeast (Figs 3.3 and 3.4), and it needs to be emphasised again that in Ganly Griffith had found a most astute and perceptive field-geologist who worked with remarkable industry under what must often have been rigorous conditions of both climate and terrain. The four recently discovered six-inch field-sheets used by Ganly in County Galway early in the 1840s are fascinating documents, three of which reveal that he then achieved a more satisfactory subdivision of the Lower Palaeozoic strata southward of Killary Harbour than the subdivision devised by the officers of the Geological Survey of Ireland some twenty years later.[55] Sometimes Ganly was left to work out his own field programme; sometimes he was given very specific instructions, as in August 1842 when John Kelly wrote to Griffith:

> I have desired Mr Ganly proceed to trace the boundary of the Calp of the Caledon district towards Dungannon — Connecting the Carnteel Sandstone with the others there abouts.[56]

Ganly himself was now often remarkably forthright in his letters to Griffith. On 19 December 1840 he wrote to Griffith from Clifden, County Galway:

> In my examination of your present geology of this country, I have found the granite outline to be *good*, and the junction of the metamorphic district with the ordinary mica slate, *middling*; but your representation of the quartz rock is decidedly *bad*.[57]

Three months later, on 26 March 1841, he wrote to Griffith from Dromahair, County Leitrim, this time referring to the Connaught Coal District mapped by Griffith himself around 1815:

> Your present boundary of the coal shale south of Dromahaire [*sic*] is quite wrong but I have in some measure now corrected it.

In the following month Ganly was in northern Donegal and from Carrickart he reported upon his activities on 15 April:

> Examined the country between Carrickart and Glen Lough and found it, in general, to be quite wrong and much in want of correction.[58]

Indeed so blunt is Ganly in much of his criticism of Griffith's map — so often does he assume an air of geological superiority — that one is left wondering about the precise nature of the relationship existing between the two men. Was Ganly just tactlessly exuberant and Griffith just benignly tolerant, or does the lack of forelock-touching subservience perhaps indicate between the two men a relationship that was in some way more intimate than the overt relationship to be expected between the Commissioner for the Valuation and a relatively junior employee in the Valuation Office?

The results of all this field-revision were added regularly to maps in Griffith's possession, and from time to time the colourists in Dublin and London were provided with corrected master copies so that their customers might be assured of obtaining up-to-the-minute versions

Sheets of Many Colours　　73

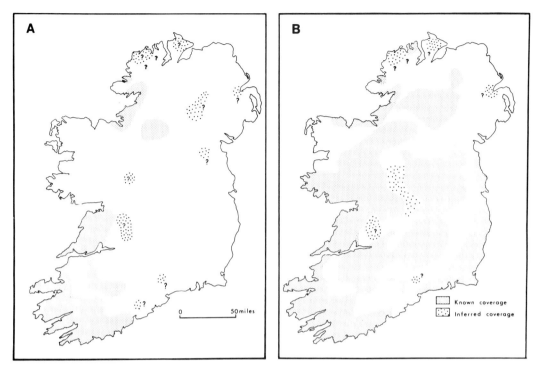

Fig. 3.4. Map A indicates the extent of Patrick Ganly's known contribution to Griffith's quarter-inch geological map of Ireland published in May 1839. Map B indicates the extent of Ganly's known contribution to the new version of the quarter-inch map published by Griffith in 1855. Reproduced, after slight amendment, from the map by Dr J.B. Archer in Gordon Leslie Herries Davies and Robert Charles Mollan (editors), Richard Griffith 1784-1878.

of the map. Indeed, there even seems to have been a scheme whereby the owners of 1839 issues of the map could return their sheets to the colourists to have new detail added.[59] A multitude of minor varieties of Griffith's map must therefore exist, and some single copies produced to order during periods of slack demand could well prove to be unique in some respect or other. As early as 6 June 1840 Griffith drew up a long list of amendments made to the map during the previous twelve months, these involving changes in no less than twenty of Ireland's thirty-two counties.[60] Four days later he presented to the Geological Society of Dublin a revised version of the map upon which the most important modification was that presaged by the inscription upon the label which had been affixed to early copies of the map: Griffith now transferred the rocks of Lambay Island and Portrane, County Dublin, and the rocks of western Munster, from the Transition Clay Slate to the Kingdom of Siluria. But so far as Munster was concerned, this was an extension of his realm that the 'King of Siluria' himself refused to welcome. After the somewhat unsuccessful British Association meeting at Cork in August 1843, Murchison, accompanied by John Phillips and Griffith, went to examine the rocks of the southwest, and he and Phillips concluded that the greater portion of the rocks which Griffith was now representing as Silurian were in reality a westward extension of the Old Red Sandstone beds of eastern Munster. They were, as Phillips put it, 'very, very old red, grown grey with age'[61] (Fig. 3.6). But Griffith remained unmoved; after 1839 he never represented the ancient rocks of western Munster as being anything other than Silurian, and he thus continued to give prominence to a slightly modified version of that entirely specious boundary drawn at the southeastern and eastern foot of the Derrynasaggart and Boggeragh mountains (see p. 70).

As the years passed by, the colourists of Griffith's map increasingly found themselves

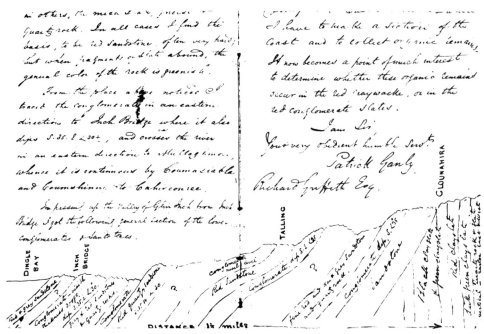

Fig. 3.5. Part of a letter written by Patrick Ganly to Richard Griffith and dated at Anascaul, County Kerry, on 17 April 1838. This was evidently Ganly's second period spent investigating the rocks of the Dingle Peninsula and he again returned to the region in the summer of 1838 and in the spring of 1844. The section above extends from the shores of Dingle Bay northwards across the Old Red Sandstone and onto the Lower Palaeozoic rocks of the Anascaul inlier. Reproduced from the Ganly Papers *by kind permission of the Royal Irish Academy.*

Fig. 3.6. 'Phillipsia attacking Griffithides' as depicted in a drawing made by Captain Henry James in about 1843. Phillips, on the left, had claimed that the rocks of southwestern Ireland, which Griffith had represented as Silurian, were really Old Red Sandstone. The referee is Murchison and the spectator outside the ring immediately to the right of Murchison would appear to be Adam Sedgwick. Dr J.B. Archer suggests that the 'trilobite' peering over Griffith's left shoulder may be the only known pictorial representation of Patrick Ganly! The original drawing is in the archives of the Institute of Geological Sciences, London, and it is here reproduced by kind permission of Mr F.W. Dunning, Curator of the Institute's museum.

under instruction to disregard certain of the printed boundaries upon the map; before reaching for their tints they now had to face the tedious task of inserting a multitude of new manuscript boundaries. Most of these new boundaries had to be added to regions of Carboniferous limestone because early in the 1840s Ganly had devoted much time to the field examination of the limestones so that the Calp and Upper Limestone could be differentiated from the Lower Limestone (Figs 3.7 and 3.8). As early as the Glasgow meeting of the British Association in September 1840 Griffith was able to display a map incorporating many of the new intra-Carboniferous boundaries[62], and by the time he displayed a fresh version of his map to the Geological Society of Dublin on 12 January 1842, his subdivision of the Irish limestones was almost complete.[63]

Detailed discussion of the many changes introduced into the map after 1839 is here impossible, but there is one point that should be noted: all the versions of the map produced down to July 1852 carry Griffith's printed signature and the printed date 'Dublin March 28th 1839', that being the date upon which he had signed the original proof copy.[64] Not unreasonably, some of his contemporaries protested at this persistence of the 1839 date.[65] The maps dated 1839 but published at various times down to 1852 incorporate discoveries made by geologists other than Griffith subsequent to March 1839 and it seemed that Griffith was trying to establish a false priority for himself by allowing the maps to appear in a back-dated form. In his own defence Griffith claimed as early as 1841 that the persistence of the old date was the result of an engraver's oversight[66], but it was a lame excuse, the more so since Griffith did nothing to correct the 'error' until the early 1850s.

The offending date at last disappeared from the map in August 1852. It was then that a new edition of the map was issued, printed, at Treasury expense, from fresh copper plates upon which the many manuscript boundaries introduced since 1839 finally became engraved lines. This new edition carries a wealth of detail beyond that present upon the map in 1839, and the increased complexity of the 1852 edition is indicated by the fact that the key now contains forty-six colour boxes in place of the original twenty-six. But Griffith has now gone even further in refining his geological classification; he has added to his colour-coding a complex system of alphabetical notations whereby, for instance, A represents Mica Slate, B represents Quartz Rock, C represents Azoic Limestone, D represents the Cambrian Clay Slate Group, and E represents the Silurian. Sub-groups are indicated by a lower case letter so that Ab is talcose mica slate, Bc is slaty quartz, Ca is gray slaty Azoic limestone, Db is greenish gray clay slate, and Ee is chloritic quartzite. Also, in the map's maritime margins is a series of six geological cross-sections (four of them are labels pasted onto the map) together with a large and elaborate 'Synoptical view of the principal fossils characteristic of Irish strata'. It was a specially prepared version of this new edition — a version bearing the printed date 1853 — which Griffith displayed to Queen Victoria and Prince Albert at the Great Dublin Industrial Exhibition of 1853. The Queen herself was somewhat short in stature, and if her gaze became fixed upon the lower part of the map she may have been pleased to note that Cobh in County Cork was now become Queenstown in honour of her visit in August 1849.

Two years later the holding of another great exhibition was instrumental in bringing Griffith's map into its final form. That exhibition was the Exposition Universelle staged in Paris in 1855. Griffith was invited to display there his now famed map — an invitation that he was of course pleased to accept, although he later felt somewhat aggrieved upon discovering that his map was not to be exhibited in the main concourse.[67] This new edition of the map, prepared in response to the Paris exhibition, is essentially the same as the 1852 edition, but there are now eight geological cross-sections in the map's maritime margins and all of them are integral parts of the map, being printed directly from the copper plates. The map bears the inscription 'The geology revised and improved in 1855' together with a facsimile of Griffith's signature and the date April 1855, but it was probably some months after that before copies of the map became generally available. It was 1 August 1856 before Griffith signed a presentation

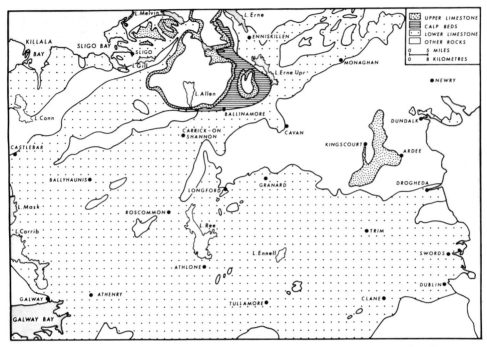

Fig. 3.7. The subdivision of the Carboniferous Limestone of the northern Irish Midlands as it appeared upon Griffith's quarter-inch map of 1839.

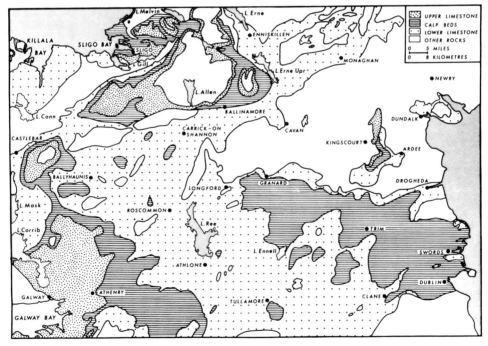

Fig. 3.8. The subdivision of the Carboniferous Limestone of the northern Irish Midlands as it appeared upon the special version of the quarter-inch map prepared for the use of the Geological Survey of Ireland in the spring of 1846 (see p. 133). The map is today in the archives of the Geological Survey of Ireland, and almost all the subdivision that it depicts was the work of Patrick Ganly.

copy of the map destined for Murchison[68]; it was 1 January 1857 before Griffith presented a copy of the map to the Geological Society of Dublin[69]; and it was 8 November 1858 before he presented a copy to the Royal Irish Academy (Plate 2).[70] With the copy presented to the Geological Society of Dublin there seems to have been something wrong because the society's council minutes for 18 November 1857 record that:

> Professor Haughton read a letter from Dr Griffith presenting another copy of his Map of Ireland to the Society in exchange for the one now hanging in the Society's Room.[71]

Griffith, it will be noted from that quotation, was now 'Dr Griffith'. He had received an honorary LL.D. from Trinity College Dublin at a commencements ceremony held on 20 February 1849. Very appropriately this was the same degree as Trinity College had conferred upon William Smith in August 1835 while he was present in Dublin for that year's meeting of the British Association. Trinity College can thus number among its alumni both the distinguished author of the first geological map of England and Wales and his no less distinguished Irish counterpart. In connection with Griffith's commencements ceremony there is, nevertheless, an odd coincidence. At the very ceremony where he received his LL.D. degree, there was conferred upon Patrick Ganly the B.A. degree which had been awarded to him in Michaelmas Term 1846 following a course of university study as an external student. It is singularly appropriate that these two individuals, who between them had done so much to bring into existence the geological map of Ireland, should have appeared to take their respective degrees upon the very same day. But was this nothing more than a happy coincidence? Surely it is more likely to have been a planned event, and as such it must again pose for us the question of the nature of the relationship existing between these two men. Of one thing we may be certain: in deciding to confer an LL.D. upon Griffith the University was honouring a man who by 1849 was become one of Ireland's leading public figures, but we can be confident that the University must have regarded the geological map of Ireland as by no means the least of Griffith's claims to academic distinction. In its edition of 1839 Griffith's map represents a fine achievement; in its edition of 1855 it is on every count — cartographic, aesthetic, and geological — a truly superb map. Not only is it the most attractive geological map of Ireland ever produced, but for some parts of Ireland the map remains to this day the most accurate representation of the geology ever achieved. It is understandable that in his old age Griffith should have wished to leave to posterity a photograph of himself displaying one of the various geological maps of Ireland with which his name has been so intimately associated (Fig. 3.9).

Who made 'Griffith's Map'?

The triumphant note just sounded would have made a convenient conclusion to the present chapter but such it cannot be allowed to become. We now have to consider the question of who actually made the quarter-inch geological map of Ireland which has for so long been hailed as 'Griffith's Map'. In so designating the map posterity has implied not merely that Griffith compiled the map — not merely that Griffith stimulated the publication of the map — but rather that Griffith was responsible for identifying upon the ground the actual geological boundaries represented upon the map. This was certainly the sense that Griffith himself intended to convey when he referred to 'my map'. He did of course admit that in drawing his geological lines he had paid due regard to the work of others, and with great diplomacy he seems to have varied the style of his acknowledgement to accord with the character of the audience he might be addressing. Thus when he spoke to the British Association in Dublin in 1835 his audience contained some distinguished Fellows of the Geological Society of London and Griffith therefore emphasised the assistance that he had received from such stalwarts of that society as Berger, Buckland, Conybeare, and Weaver.[72] When he addressed the Geological Society of Dublin in 1836 he stressed the contributions made to the map by three members of the staff of the Valuation Office 'acting under my direction', these three being Patrick Knight in County

Fig. 3.9. Sir Richard Griffith, Bart, aged in his eighties. The geological map which he brought to the photographer's studio for this sitting is the map prepared in 1853 to assist the staff of the Valuation Office. It was the first colour-printed geological map of the whole of Ireland (see p. 59). The portrait first appeared in Dublin University Magazine, *83, 1874, and it is derived from a photograph taken in Dublin by Thomas Cranfield of 115 Grafton Street.*

Mayo, Samuel Nicholson in northern County Antrim, and John Kelly whose investigations had ranged over many parts of Ireland.[73] Again, when he appeared before the British Association in Belfast in 1852 it was such local geologists as James Bryce (1806-1877) and James MacAdam (1801-1861) that he singled out for particular mention.[74] But while Griffith might regularly give the appropriate acknowledgements when talking of his map, the message that he left with his audiences was always clear enough: 'The geological lines upon my map are essentially my lines painstakingly traced out upon the ground over many years during such intervals of time as I have been able to snatch from my onerous public duties'. This is the view of Griffith's map which was widely accepted by his contemporaries. On 17 February 1854 when Griffith received the Wollaston Medal, the premier award of the Geological Society of London, the president (Edward Forbes) explained that Griffith was receiving the distinction:

> ... for the valuable services rendered by you to geological science, and particularly for your Geological Map of Ireland, the result of your own laborious and judicious researches.[75]

Archibald Geikie received an account of the map's history from Griffith himself in the 1870s and later, in his highly influential book *The founders of geology*, Geikie referred to the map as 'one of the most remarkable geological maps ever produced by a single geologist'.[76]

Perhaps this chapter has already given sufficient indication that the traditional view of Griffith's map — the view enshrined in those words of Geikie's — is in need of substantial revision. In reality Griffith's relationship to the map changed through time. Down to the 1830s Griffith's geologically coloured versions of the maps of Arrowsmith and Taylor were doubtless Griffith's maps in the strictest sense — they were the result of his own essentially unaided field observations. In 1830, however, Griffith began to run his unofficial geological survey of Ireland from within the Valuation Office and from that point onwards there is a progressive decline in the proportion of work upon the map which can be attributed to Griffith's personal field investigations. Perhaps the geological lines upon the atlas map of 1838 were still in essence Griffith's lines, but by the time the first edition of the quarter-inch map appeared in 1839 there were wide tracts of country where the mapping was now Ganly's rather than Griffith's. Throughout the 1840s, as the *Ganly Papers* make so abundantly clear, this process continued, and when the map assumed its final form between 1852 and 1855 it was a document which owed far more to Ganly's field activities than to those of Griffith. If the map has to be an occasion for eponymy, then perhaps in its 1852 to 1855 versions the map should take its name from Ganly rather than from Griffith. In retrospect it does now seem surprising that Griffith's claims to the principal authorship of the map should for so long have passed unquestioned because there were always circumstantial grounds for suspecting that Griffith's public declarations upon the subject represented rather less than the whole truth. To bring the map from its outline condition of 1839 to its remarkably detailed state of 1852 to 1855 must clearly have involved the expenditure of many thousands of hours upon meticulous field investigation, but during the years in question Griffith himself could never have found the time necessary for such field-studies; he was far too busy a public figure. In the 1840s, in addition to being Director of the Boundary Survey and Commissioner for the Valuation, he was also one of the Shannon Improvement Commissioners and the Deputy Chairman of the Board of Works, in which capacity he found himself deeply involved in measures to alleviate the distress of the Great Famine. For his devotion to these and other public duties he in March 1858 became a baronet but, while his public services may have secured his social advancement, they can hardly have left him with much opportunity to tramp the Irish countryside, collecting-bag upon the shoulder and hammer in the hand. We have Griffith's own admission that as early as the autumn of 1838 he was having difficulty in finding the time even to think about geology (see p. 65); for the time-consuming task of conducting geological field-work he clearly had to rely upon somebody other than himself, and it is fortunate that the survival of the *Ganly Papers* removes all the doubt that might have existed as to the identity of the figure whom Griffith used as his principal field investigator.

The somewhat enigmatic character of the relationship existing between Griffith and Ganly has already been noted (pp. 72 and 77) but, as Jean Archer has convincingly demonstrated,[77] there can be no doubt about Ganly's important role in bringing the quarter-inch geological map of Ireland to that state of remarkable excellence which it had achieved by 1855 (Fig. 3.4). It needs to be re-emphasised that most of the lines appearing upon the versions of the map issued between 1852 and 1855 had been traced out upon the ground not by Griffith but by Ganly. In talking about his map, as we have seen, Griffith regularly acknowledged the assistance he had received from others, but among those acknowledgements we search in vain for Ganly's name. During the years between 1837 and 1846 — the years when Ganly was so deeply involved in Irish geology upon Griffith's behalf — Griffith evidently made only one public reference to Ganly: in addressing the Geological Society of Dublin on 12 July 1839 he mentioned Ganly in the very menial role of having collected fossils for him at Knockmahon in County Waterford.[78] Not until thirty years later, in giving evidence before the Select Committee on the Valuation Survey of Ireland on 13 May 1869, did Griffith make his one and only reference to the fact that in compiling his map he had relied heavily upon Ganly's services.[79] Why did Griffith publicly admit his indebtedness to such other of his Valuation Officer staff as Kelly,

Knight, and Nicholson, and yet conceal his far greater indebtedness to Ganly? Two answers spring to mind. Firstly, Ganly was supposed to be a draughtsman in the Valuation Office yet for more than ten years Griffith had him working as a field-geologist constructing an entirely unauthorised geological map of Ireland. As was suggested earlier (p. 60), to reveal the extent of Ganly's involvement in geology was to raise questions about a misappropriation of public funds. Secondly, and this was surely a more weighty factor, Ganly's contribution to the making of the geological map of Ireland was far too extensive for Griffith to acknowledge with equanimity. Griffith was proud of his map; its completion was the obsessive ambition of his life; it was to be his memorial in science. To admit that he had only been able to complete the map by enlisting the services of another — to concede that the map was largely the work of one of his staff — these were revelations that Griffith could not bring himself to make.

It was not only upon the quarter-inch map that Griffith subsumed Ganly's work as his own. It seems that many of the geological communications that Griffith made to both the Geological Society of Dublin and the British Association were based not, as he implied, upon his personal field observations but rather upon the studies conducted by Ganly. In short Griffith's acquaintance with Irish geology in the raw was far less intimate than he liked to suggest. There is an interesting fact that may here have relevance. Griffith presented many a paper to the Geological Society of Dublin but the papers were commonly read for him in his absence. When he reported upon the latest changes introduced into the quarter-inch map — reports presented to the society on 10 June 1840, 8 December 1841, and 12 January 1842 — Griffith was absent upon all three occasions, and the first report was read for him by the president and the second two were read for him by John Kelly. Later, between 1854 and 1860, Griffith presented nine papers to the society but on six of the occasions he was definitely absent and for the three remaining occasions the records leave us in doubt as to whether or not he was actually present at the meetings in question.[80] At least one of his contemporaries noticed this odd behaviour; on 22 February 1842 Charles William Hamilton wrote to Murchison:

> ...we have a Geological Society from which Mr Griffith has personally withdrawn although he sends his clerk to attend & make announcements.[81]

A charitable comment upon Griffith's behaviour would be that he was a busy man whose public duties made it difficult for him to attend society meetings. A less charitable explanation would be that Griffith absented himself in order to avoid exposure to the kind of well-informed questioning that might have revealed his own want of familiarity with the field evidence under review. Since that field evidence had so often been observed not by Griffith, but by Ganly, it would seem that in debate Griffith's credibility might indeed have proved somewhat fragile.

We must nevertheless strive to be fair to Griffith. It is certainly not intended to imply that he was an imposter and charlatan in science. He really was very widely experienced in Irish field-geology and as a field-geologist he was noted for his powers of endurance. Even in his seventies he was still to be found attired in a swallow-tailed coat and striding over the Dingle mountains during torrential downpours of rain which severely taxed the endurance of men but half his age.[82] Equally, there was nothing deficient in his powers of geological observation and he evidently shared with William Smith one of the most valuable faculties that a geologist can possess: the ability to carry in his mind for many years a detailed mental picture of the character of the rocks as they are displayed at crucial geological sites. Griffith was a widely experienced and very competent field-geologist; his weakness was that his overwhelming ambition to be the author of the first detailed geological map of Ireland led him to pretend to the possession of a knowledge of Irish geology far more detailed than his onerous public duties could in reality ever have allowed him to achieve. In Ganly he found a most able accomplice who, for whatever reason, was prepared to acquiesce in the deceit by allowing his own astute and painstaking field-studies to be presented to the world as Griffith's. When the field-mapping was completed would Griffith have felt more at ease with Ganly out of the way and is that why in February 1851 he gave Ganly every encouragement to emigrate to the United States?[83] How far Griffith's contemporaries were aware of the true character of the quarter-inch geological map of Ireland in its later versions must remain an open question. There is certainly evidence that a coolness existed between Griffith and some of his fellow members of the Geological Society of Dublin but it is difficult to know whether or not this arose from a suspicion that Griffith was being less than honest with them. There is, however, one positive fact suggesting that Griffith's scientific credentials may have been seen as slightly suspect: unlike so many other of Ireland's geologists, Griffith was never elected to Fellowship of the Royal Society of London. He associated with men such as Sir Henry De La Beche, the Earl of Enniskillen, Humphrey Lloyd, Sir Charles Lyell, Sir Roderick Murchison, and Adam Sedgwick — all of them entitled to the dignity of an F.R.S. — yet Griffith's own name seems never to have been brought before the society in anticipation of his election to Fellowship. Perhaps not all of Griffith's contemporaries were convinced that the quarter-inch geological map of Ireland deserved to be known as 'Griffith's map'.

Sir Richard Griffith died peacefully at his Dublin home on the evening of Sunday 22 September 1878, just two days after entering his ninety-fifth year. At the time of his death the house at 2 Fitzwilliam Place must surely still have contained a great wealth of material relating to Griffith's studies in Irish geology over a period of seventy years. His fossil collection had already been dispersed, most of it going to the museum of the Royal Dublin Society, but in 1878 there presumably lay upon Griffith's shelves the note-books in which he had recorded his field-observations ever since those far-off days when as a young man he had first plied his hammer to the rocks of the Leinster Coal District. Maybe some of his drawers contained drafts for the quarter-inch map in its various states. Perhaps he still possessed the geologically coloured copies of Arrowsmith's map which he had exhibited during his Royal Dublin Society lectures and which had featured so prominently at the British Association meeting of 1835. And can we seriously suppose him to have destroyed the geological sections — there were twenty-one of them — which he had displayed so proudly to the Queen and Prince Albert in 1853? He certainly still retained the six hundred letters of the Ganly correspondence; they were bound into four volumes and they stood upon his library shelves as invaluable works of detailed local reference. Somewhere, too, he must have had the small-scale county index maps compiled by Ganly in the light of his six-inch field-mapping. And what of Ganly's six-inch field-maps themselves? It seems likely that Griffith possessed a large number of the six-inch Ordnance sheets of Ireland (a full set would have contained almost two thousand sheets) most of them carrying Ganly's field annotations. In short, 2 Fitzwilliam Place must have contained a remarkably detailed and extensive archive of Irish geology, a great deal of it assembled through the activities of Griffith's valuators. But it is an archive of which there is now but little trace. In the first week of November 1878 many of Griffith's effects came under the hammer in the D'Olier Street auction rooms of John Fleming Jones.[84] Among the lots were his geological texts — texts over which he must have pored while compiling the map of Ireland. There were surveying instruments — some of them perhaps the very instruments which he and the faithful Kelly had taken to the summits of the Munster peaks during their attempted triangulation of Ireland sixty years before. Lot 1133 was the four volumes of the Ganly correspondence. Lot 185 is described as 'Ordnance Survey Index maps of Ireland, coloured, on Linen in four Morocco Cases'. Were these colours geological colours, and were these the index maps to which Ganly had transferred all his six-inch field observations? We shall probably never know the answer to that question, but if the maps were geological in character, then they were the only such maps from Griffith's Irish geological archive to reach the saleroom. The fate of the remainder of the archive — the note-books, the sections, the maps — is unknown. Perhaps they were simply destroyed as Griffith's executors removed from the house the accumulated residue of his fifty years of residence.

This evident destruction has seriously distorted posterity's assessment of Griffith's geological achievement. We have come to think of Griffith merely as the man who compiled a remarkably fine quarter-inch geological map of Ireland; we have come to see him as the pioneer who first traced the broad outlines of Irish geology and thus laid the foundations upon which his successors have built through their exploration of Irish geology at levels of detail far beyond those ever attained by Griffith himself. Such intepretations do far less than justice to Griffith. It needs to be emphasised that the quarter-inch geological map of Ireland was by no means the alpha and the omega of Griffith's geological activities. Rather is that map merely the published tip of a great mountain of geological information gathered by Griffith during seventy years of research into the geology of his native land. Far from being seen as representing the full extent of Griffith's geological knowledge, the quarter-inch map is more truly to be viewed as Griffith's synopsis of Irish geology derived from detailed field-studies conducted upon scales far larger than the quarter-inch. To this point we will return in the next chapter when we encounter the strenuous efforts that Griffith made in 1844 to bring his unofficial geological survey of Ireland out into the open and to secure for it recognition as the official national survey (see p. 115). In this ambition Griffith failed; the extensive geological archive assembled by his unofficial survey was seemingly largely destroyed, and posterity speedily forgot its very existence.

But can we be certain that Griffith's geological archive really was largely destroyed? Anglers tell stories of the large fish that escaped; archival historians recount tales of elusive documents. It is said that as recently as the 1960s the mews house at the rear of 2 Fitzwilliam Place still contained 'old maps and documents', and perhaps somewhere else there may even yet survive relics of Griffith's unofficial geological survey. Past events do afford some ground for modest optimism. Three volumes of the Ganly correspondence were found in the Dublin Valuation Office in 1944; four of Ganly's six-inch field-sheets were discovered in the Mitchell Museum of University College Galway in 1971; and the fourth volume of the Ganly correspondence was located in private hands in 1977. Who knows what may yet await discovery?

REFERENCES AND NOTES FOR CHAPTER THREE

1. G.L. [Herries] Davies, The Geological Society of Dublin and the Royal Geological Society of Ireland 1831-1890, *Hermathena* no. 100, 1965, 66-76.
2. R. Griffith, On "The Boulder Drift and Esker Hills of Ireland", and "On the position of erratic blocks in the country", *Rep. Br. Ass. Advmt. Sci.*, Edinburgh, 1871, part 2, 98-100.
3. J. Lee, Richard Griffith's land valuation as a basis for farm taxation, pp. 77-101 in Gordon Leslie Herries Davies and Robert Charles Mollan (editors), *Richard Griffith 1784-1878*, Royal Dublin Society, 1980; W.E. Vaughan, Richard Griffith and the Tenement Valuation, *ibid.*, 103-122.
4. *J. R. Geol. Soc. Irel.* 5, 1879-80, 137, and *Scient. Proc. R. Dubl. Soc.*, N.S. 2, 1878-80, 196.
5. *Geikie Papers*, Gen. 524, letter dated 26 November 1874.
6. *Report from the Select Committee on General Valuation, &c. (Ireland)*, H.C. 1868-69, (362), IX, minutes of evidence 44.
7. *Geological map of Ireland to accompany the instructions to valuators appointed under the 15th. & 16th. Vic. cap 63*, Dublin, August 1853; Richard Griffith, *Instructions to the valuators and surveyors appointed under the 15th and 16th Vict., cap. 63, for the uniform valuation of lands & tenements in Ireland*, Dublin, 1853, 14-15.
8. *Geikie Papers*, Gen. 524, letter dated 26 November 1874.
9. Patrick Knight, Notice of the general geology of Erris, County of Mayo, *J. Geol. Soc. Dubl.* 1(1), 1833, 45-51; *Erris in the "Irish Highlands", and the "Atlantic Railway"*, Dublin, 1836, pp. viii + 178.
10. P. Ganly, Observations on the structure of strata, *J. Geol. Soc. Dubl.* 7, 1855-57, 164-167; R.C. Simington and A. Farrington, A forgotten pioneer: Patrick Ganly, geologist, surveyor and civil engineer, (1809-1899), *J. Dep. Agric. Repub. Ire.* 46, 1949, 36-50; J.B. Archer, Patrick Ganly: geologist, *Ir. Nat. J.* 20(4), 1980, 142-148.

11. John Kelly, *Notes upon the errors of geology illustrated by reference to facts observed in Ireland*, London, 1864, pp. xvi + 300.
12. VOLB, 16 December 1843 - 27 May 1845, Griffith to the Chancellor of the Exchequer 25 March 1844.
13. *Report from the Select Committee, op. cit.* 1868-69, minutes of evidence 44, (ref. 6).
14. *Ibid.*, minutes of evidence 186.
15. RGSIP, *Minutes of meetings of council, 1831-1863*.
16. *J. Geol. Soc. Dubl.* 1(3), 1837, 142.
17. *Greenough Papers*.
18. *Proc. Geol. Soc.* 2, 1833-38, 155.
19. J.M. Edmonds and P.A. Beardmore, John Phillips and the early meetings of the British Association, *Advmt Sci.* 12, 1955-56, 102. For the meeting of the Royal Dublin Society financed the compilation of a booklet about the geology of the Dublin region and this work contains a *Geological map of the vicinity of Dublin*. See [John Scouler], *Memorandum of objects of geological interest in the vicinity of Dublin*, Dublin, 1835, pp. 26.
20. R. Griffith, On the geological map of Ireland, *Rep. Br. Ass. Advmt. Sci.*, Dublin, 1835, part 2, 56-58.
21. John Willis Clark and Thomas McKenny Hughes, *The life and letters of the Reverend Adam Sedgwick*, Cambridge, 1890, vol. 1, 447-448.
22. R. Griffith, On the principle of colouring adopted for the geological map of Ireland, and on the geological structure of the south of Ireland, *J. Geol. Soc. Dubl.* 2(1), 1839, 78.
23. J.A. Douglas and J.M. Edmonds, John Phillips's geological maps of the British Isles, *Ann. Sci.* 6, 1948-50, 361-375.
24. John Phillips, *An index geological map of the British Isles; constructed from published documents, communications of eminent geologists, and personal investigation.*
25. *Rep. Br. Ass. Advmt. Sci.*, Dublin, 1835, part 1, xxvi.
26. *Second report of the commissioners appointed to consider and recommend a general system of railways for Ireland*, H.C. 1837-38, XXXV, 34.
27. *First report of the commissioners appointed to inquire into the manner in which railway communications can be most advantageously promoted in Ireland*, H.C. 1837, XXXIII, 6.
28. John Harwood Andrews, *A paper landscape: the Ordnance Survey in nineteenth-century Ireland*, Oxford, 1975, 183-185; *History in the Ordnance map: an introduction for Irish readers*, Dublin, 1974, 48-49.
29. *Second report of the commissioners, op. cit.*, 1837-38, appendix 1, 1-23, (ref. 26). The appendix was reprinted and published in Dublin in 1838 as a 23-page booklet containing Griffith's atlas map as a fold-out.
30. *Atlas to accompany 2^D report of the Railway Commissioners Ireland 1838*.
31. Arthur Howard Robinson, *Early thematic mapping in the history of cartography*, Chicago and London, 1982, *passim*; The 1837 maps of Henry Drury Harness, *Geogrl J.* 121, 1955, 440-450.
32. *Railways (Ireland)*, H.C. 1839, (88), XLVI, 1.
33. W. King, On the Permian fossils of Cultra, *Rep. Br. Ass. Advmt. Sci.*, Belfast, 1852, part 2, 53. See also J. Bryce, On the magnesian limestone and associated beds which occur at Hollywood in the County of Down, *J. Geol. Soc. Dubl.* 1(3), 1837, 175-180.
34. Roderick Impey Murchison, *The Silurian System, founded on geological researches in the counties of Salop, Hereford, Radnor, Montgomery, Caermarthen, Brecon, Pembroke, Monmouth, Gloucester, Worcester, and Stafford*, London, 1839, pp. xxxii + 768.
35. R. Griffith, Reply to that part of Mr. Weaver's paper relative to the mineral structure of the south of Ireland, *Lond. & Edinb. Phil. Mag.* 17, 1840, 162.
36. T. Weaver, On the geological relations of the south of Ireland, *Trans. Geol. Soc. Lond.*, second series, 5(1), 1838, 1-68. See also T. Weaver, On the mineral structure of the south of Ireland, with correlative matter on Devon and Cornwall, Belgium, the Eifel, &c., *Phil. Mag.* 16, 1840, 276-297, 388-404, 471-476.
37. *Greenough Papers*.
38. The three volumes of Ganly letters in the Royal Irish Academy cover the years 1837-39, 1839-41, and 1841-43. A fourth volume for 1844-46 is in private hands. See also Simington and Farrington, *op. cit.*, 1949, (ref. 10).
39. J.B. Archer and G.L. Herries Davies, Geological field-sheets from County Galway by Patrick Ganly (1809?-1899), *J. Earth Sci. R. Dubl. Soc.* 4(2), 1982, 167-179.
40. *Greenough Papers*.
41. Richard Griffith, *Notice respecting the fossils of the Mountain Limestone of Ireland, as compared with those of Great Britain, and also with the Devonian System*, Dublin, 1842, pp. 25. Richard Griffith and Frederick McCoy, *A synopsis of the characters of the Carboniferous Limestone fossils of Ireland*, Dublin, 1844, pp. viii + 207 + 29 plates. A later edition was published in London in 1862. Richard Griffith and Frederick McCoy, *A synopsis of the Silurian fossils of Ireland*, Dublin, 1846, pp. 72 + 5 plates. A later edition was published in London in 1862.
42. R. Griffith, On the leading features of the geology of Ireland, *Rep. Br. Ass. Advmt. Sci.*, Liverpool, 1837, part 2, 88-90. Part of the section is reproduced in *J. Geol. Soc. Dubl.* 7, 1855-57, plate x.
43. R. Griffith, On the geological structure of the south of Ireland, *Rep. Br. Ass. Advmt. Sci.*, Newcastle-upon-Tyne, 1838, part 2, 81-84.
44. *Ganly Papers*, Ganly's letter from Macroom 27

January 1839.
45. G.L. [Herries] Davies, Notes on the various issues of Sir Richard Griffith's quarter-inch geological map of Ireland, 1839-1855, *Imago Mundi* 29, 1977, 35-44.
46. *J. Geol. Soc. Dubl.* 2(2), 1843, 167-168.
47. *Proc. Geol. Soc.* 3, 1838-42, 136-138.
48. John Phillips, *Memoirs of William Smith, LL.D.*, London, 1844, 77.
49. *J. Geol. Soc. Dubl.* 2(2), 1843, 139f.
50. *Ibid.*, 167-169.
51. Railways (Ireland), *op. cit.*, 1839, 1, (ref. 32); *Trigonometrical surveys*, H.C. 1847, (171), XXXVI, 4. The quarter-inch map continued to feature in Ordnance Survey catalogues until the 1930s still at its original price of one pound for the six uncoloured sheets. The copper plates for the map are still in the Ordnance Survey office in Dublin and copies of the map have been struck from these as recently as the 1970s.
52. See, for example, *The Atlantis* 2, 1859, 242f.
53. A. Sedgwick and R.I. Murchison, On the Silurian and Cambrian Systems, *Rep. Br. Ass. Advmt. Sci.*, Dublin, 1835, 2, 59-61. See also R. Griffith, Address delivered at the fifth annual meeting of the Geological Society, *J. Geol. Soc. Dubl.* 1(3), 1837, 142-166.
54. Thomas Frederick Colby, *Ordnance Survey of the County of Londonderry. Volume the first. Memoir of the city and north western liberties of Londonderry, Parish of Templemore*, Dublin, 1837, 5-6. See also J.E. Portlock, On a small tract of Silurian rocks in the County of Tyrone, *Rep. Br. Ass. Advmt. Sci.*, Newcastle-upon-Tyne, 1838, part 2, 84.
55. Archer and Herries Davies, *op. cit.*, 1982, (ref. 39).
56. *Ganly Papers*, a pencilled note at the foot of Ganly's letter of 22 August 1842.
57. *Ibid.*
58. *Ganly Papers*, Ganly to Griffith 17 April 1841.
59. *J. Geol. Soc. Dubl.* 2(2), 1843, 170-172.
60. VOLB, 8 December 1838–24 September 1841, Griffith to Apjohn 6 June 1840.
61. *J. Geol. Soc. Dubl.* 3(2), 1845, 156. See also Archibald Geikie, *Life of Sir Roderick I. Murchison*, London, 1875, vol. 2, 20; *Monteagle Papers*, Murchison to Monteagle, 29 August 1856.
62. R. Griffith, On the Yellow Sandstone, and other points of the geology of Ireland, *Rep. Br. Ass. Advmt. Sci.*, Glasgow, 1840, part 2, 110.
63. This map survives in the Department of Geology, Trinity College Dublin.
64. Before they left the Ordnance Survey maps were commonly inspected, signed, and dated by one of the engravers. These MS dates, appearing in the bottom right margin of a map, have much chronological value but it should be remembered that the geological colours may have been added to a particular map years after it left the Ordnance Survey.
65. *J. Geol. Soc. Dubl.* 2(2), 1843, 159-160, 167-172.
66. *Ibid.*, 167-169.
67. VOLB, vol. 79A, Griffith to Richard Thompson 13 June 1855.
68. This copy of the map is now in the Institute of Geological Sciences in London.
69. *J. Geol. Soc. Dubl.* 7, 1855-57, 210.
70. *Proc. R. Ir. Acad.* 7, 1857-61, 107.
71. RGSIP, *Minutes of meetings of council 1831-1863*.
72. R. Griffith, *op. cit.*, 1835, (ref. 20).
73. R. Griffith, Address, *J. Geol. Soc. Dubl.* 1(3), 1837, 142.
74. Notices of the geology of Ireland, *Rep. Br. Ass. Advmt. Sci.*, Belfast, 1852, part 2, 47.
75. *Q. Jl. Geol. Soc. Lond.* 10, 1854, xix.
76. Archibald Geikie, *The founders of geology*, London, 1905, 455.
77. J.B. Archer, Richard Griffith and the first published geological maps of Ireland, pp. 143-171 in Herries Davies and Mollan, *op. cit.*, 1980, (ref. 3).
78. R. Griffith, On the principle of colouring adopted for the geological map of Ireland, and on the geological structure of the south of Ireland, *J. Geol. Soc. Dubl.* 2(1), 1839, 85.
79. *Report from the Select Committee, op. cit.*, 1868-69, minutes of evidence 49, (ref. 6).
80. RGSIP, *Proceedings of general meetings 1831-1875*.
81. SP, ID.142.
82. *Geol. Mag.*, N.S. dec. 2, 5, 1878, 524-528.
83. VOLB, 3 October 1850 - 29 March 1853, Griffith to Ganly 27 February 1851.
84. *Catalogue of books... including the library of the late Sir Richard Griffith, Bart.*, Dublin, 1878, pp. 54. A copy is in NLI.

4

ORDNANCE SURVEY GEOLOGY

1824-1845

The Richmond National Institution stood at 37 Upper Sackville Street, Dublin, and there industrious blind folk were taught skills so that they might be the better able to support themselves. But the motives of the gentlemen who gathered in the Institution on the afternoon of Wednesday 10 February 1836 were hardly philanthropic. They were the members of the Geological Society of Dublin come to attend their society's fifth annual general meeting in the rooms which they rented from the Institution at a cost of £60 per annum. It was a long meeting — members returned to their seats in the evening after a dinner adjournment — and among the business transacted was a transfer of the presidency. Richard Griffith had completed his term of office in the society's chair, and as his successor the members chose a military gentleman whose deeply scarred forehead seemed to belie the fact that in truth he had never heard a cannon roar in anger. He was Lieutenant-Colonel Thomas Frederick Colby of the Royal Engineers. Griffith and Colby were well known to each other. Colby was in charge of the Ordnance Survey, and Griffith, as Director of the General Boundary Survey of Ireland, was charged with the responsibility of locating all the Irish administrative boundaries in advance of Colby's surveyors. The relationship between the two men had not always been harmonious, but now, as Griffith handed over the chair, he referred to 'my valued friend Colonel Colby' and on both sides there were doubtless uttered all those pleasantries appropriate to such an occasion. Behind this deferential posturing there nevertheless lay a rivalry. Griffith and Colby were each responsible for different and competing schemes designed to yield a geological map of Ireland. Theirs was in essence a race for scientific priority. By February 1836 Griffith already had a good lead over Colby — as if to emphasise this fact Griffith devoted part of his retirement address to outlining the history of the geological map which he had displayed to the British Association the previous August — but the eventual result was still by no means a foregone conclusion. Fortune, however, was to smile upon Griffith rather than upon Colby. Over a period of twenty years it was Colby's fate to encounter a succession of frustrations which thwarted him in the attainment of his geological objective. Those frustrations loom large in the present chapter. It is a chapter concerned not so much with the niceties of a branch of thematic cartography, as with the gyrations of public figures eager to exploit geology as a vehicle for their personal ambitions.

Colby's strategy

Thomas Colby was born in 1784, the son of an officer in the Royal Marines.[1] Educated at the Royal Military Academy, Woolwich, he in 1801 took a commission in the Royal Engineers, and the following year he joined the Ordnance Survey. For more than forty years thereafter he served in all parts of the British Isles, his chief adversary being the elements rather than some foreign foe. Throughout his life he displayed both a selfless devotion to all aspects of the Survey's work, and a passion for precision in surveying. Despite the loss of a hand in 1803, following the bursting of an overloaded pistol — the same explosion was responsible for his scarred forehead — he achieved high skill as an instrumental observer. Year after year he was to be found, come rain come storm, upon the most exposed of British and Irish mountaintops, conducting the triangulation which was the basis for the entire mapping programme. In 1820 the Duke of Wellington, then the Master-General of the Ordnance, appointed Colby to succeed Major-General William Mudge as the Superintendent of the Ordnance Survey, and in this capacity he soon found himself deeply involved in Irish affairs. It was in 1824 that the decision was taken to extend the Survey's operations across the Irish Sea, and it fell to Colby to plan and implement the new survey which, it will be remembered, was designed to yield a six-inch (1:10,560) coverage of the entire island. It was a challenging project — nowhere else in the world had a national map been attempted upon so large a scale — but Colby was outstandingly successful in executing the task, and the magnificent first edition sheets of the six-inch survey of Ireland are surely the finest memorial that any surveyor could possibly desire. Happily he remained in office sufficiently long to see the completion of his great work; he retired from the Survey upon attaining the rank of Major-General in 1846, the very year in which the last of some two thousand six-inch sheets of Ireland was published.

Colby took his personal responsibilities in Ireland very seriously; in 1828 he married an Irish lady — Elizabeth Boyd from Londonderry — and for ten years thereafter he resided in Dublin rather than in London. From the outset he concluded that the Ordnance Survey should in Ireland conduct its operations in a manner far more lavish than that which had been the custom in England. He resolved that instead of confining itself merely to the production of six-inch maps, the Survey should conduct a thorough scientific, historical, economic, statistical, and social investigation of the entire country. One of his officers later observed of the scheme:

> Colonel Colby felt that such a noble opportunity of connecting with a topographical survey, all those collateral inquiries, by which the springs of national wealth are discovered and directed into their proper channels, ought not to be lost.[2]

The maps themselves were envisaged as a kind of geographical index to the compendium of

Fig. 4.1. Major-General Thomas Colby. He was elected to Fellowship of the Royal Society on the same day — 13 April 1820 — as William Whewell the future Master of Trinity College Cambridge. From Charles Close's The early years of the Ordnance Survey.

information which it was hoped the mapping parties would be able to collect as a by-product of their normal surveying duties. Admittedly the Survey had been at work in England for over thirty years without ever engaging in such comprehensive collecting activities, but Colby could argue that English precedent had no relevance in the Irish context. The English counties, for example, were already fully described in the works of many a topographer, whereas the literature devoted to the Irish counties was still extremely meagre. Again, in England the Survey's activities had been geared to the production of sheets upon the relatively small one-inch (1:63,360) scale, but in Ireland the Survey would be proceeding slowly with all that meticulous attention to detail which is imperative for the production of maps at a scale of six-inches to the mile. Thus, it was hoped, the Irish survey parties would be afforded ample opportunity for general observation and collection. Finally, there was the contrasted economic condition of the two nations. England was enjoying an unprecedented economic boom as she converted herself into the workshop of the world; Ireland was struggling in a morass of rapidly growing population, social unrest, recurrent famines, and stagnating industry. Surely the Ordnance Survey's activities should be designed to provide a basis both for an understanding of Ireland's ills and for the planning of appropriate remedies. Colby could therefore claim that this was no collection for collection's sake; the motives to which he admitted were utilitarian just as much as academic. But in emphasising the practical significance of his scheme was he being entirely frank? Was there not a strong element of personal satisfaction involved? Colby had but recently assumed control of the Ordnance Survey and he may well have felt a need to flex his new muscles both to satisfy himself of their existence and as a public display of his new authority. How better to satisfy such urges than by leading the Ordnance Survey into a new venture undreamed of by his predecessor — a venture which had the additional advantage of extending Colby's suzerainty far beyond the confines of mere cartography and into the almost limitless terrain represented by historico-statistical inquiry.

The history of Colby's scheme has been traced in admirable detail by Professor J.H. Andrews[3], and it is only necessary here to focus upon the geological aspects of Colby's programme. Although he himself never made any really original contributions to geology, Colby had more than a passing interest in the science. He joined the Geological Society of London as early as 1814 and he served on the council of the society three times between 1815 and 1825. Later, during the years of his Irish residence, he took an active part in the affairs of the Geological Society of Dublin, and, as we have seen, he served as the society's president during the year 1836-1837. In view both of his own inclinations and of the science's obvious economic importance, it comes as no surprise to find that geology loomed large in Colby's scheme for a comprehensive survey of Ireland. In his report on the progress of the Irish survey dated 2 February 1826 and submitted to Sir Henry Hardinge, then the Clerk of the Ordnance, Colby wrote:

> The general Disposition of the Minerals of a Country is so important in every Branch of Domestic and Political Economy connected with its improvement that I have thought it right to direct very particular attention to this subject. The outline Plans of Ireland will furnish the best Basis ever given in any Country for a Geological and Mineral Survey and the Ordnance will I hope be able ultimately to accompany their Map of Ireland with the most minute and accurate Geological Survey ever published.[4]

This was an ambitious objective; how was it to be attained? When he appeared before the Select Committee on the Survey and Valuation of Ireland, in March 1824, Colby was asked whether he considered that a geological survey of Ireland should proceed concurrently with the topographical survey. He replied:

> I should apprehend that it would be more convenient for the geological survey to follow the survey of the country, than to be connected with it in the first instance.[5]

Faced with the realities of the Irish situation he changed his mind, and by 1825 he had decided that the two surveys should proceed in parallel with the geological sheets appearing as a by-product of the topographical mapping. Nowhere in the surviving records are his plans for geology outlined with any precision, but a reading between the lines suggests that his proposal was simple in the extreme. Survey parties in the field were evidently to collect samples of rocks as they went about their topographical work, carefully recording upon maps the precise location of the points sampled. Then, since very few members of the field-parties possessed any training in geology, the samples and their accompanying maps were to be despatched to the Ordnance Survey's headquarters at Mountjoy House in the Phoenix Park, Dublin, where, seemingly, it was anticipated that an expert would identify the samples. What was to happen next remains uncertain. Perhaps it was intended that, having named the samples, the expert would then interpolate the appropriate geological boundaries onto the six-inch maps so as to create geological versions of the topographical sheets. Clearly, the drawing of such boundaries would have owed more to the expert's powers of divination than to his knowledge of field realities! It may therefore be that at this stage Colby aspired to nothing more than petrographic maps displaying only point-information — maps upon which the character of the sample taken from each collecting site would be recorded by means of a symbol. Although there is no real indication of the type of map that Colby had in mind around 1825, some insight into his thinking is perhaps to be obtained from another, but related, aspect of the six-inch sheets. Relief, like geology, is a continuous and three-dimensional phenomenon but, upon the six-inch sheets, Colby was about to represent the configuration of Ireland by means of spot-heights, bench-marks, and trigonometrical stations. In other words, relief was being represented not as a continuous phenomenon, but in the form of point-data. It seems likely that Colby at this stage proposed to treat geology in the same manner. Certainly the report of the 1824 Select Committee on the Survey and Valuation of Ireland contained a set of symbols deemed suitable for the representation of geological point-information upon what was then being termed 'The atlas of Ireland'.[6] Many of the symbols (gold, silver, copper, iron, lead, etc.) are the old alchemical symbols, and had Colby relied upon a scattering of such signs to constitute his geological maps, then those maps must have belonged in the same generic class as Sampson's County Londonderry maps of 1802 and 1814.

The topographical survey of Ireland started in earnest in 1825 when the surveyors commenced operations in northern Ireland, and it was there that the collecting of geological samples started in accordance with Colby's system. The Survey's letter registers contain various references to consignments of rocks being despatched to Dublin by the field-parties, and Colby soon felt it necessary to make one of his officers directly responsible for the supervision and co-ordination of all the Irish Survey's geological activities. The officer he selected for the post was Captain John Watson Pringle (c1793-1861) who on 14 November 1826 was appointed to be Superintendent of the Geological Survey.[7] Pringle was a veteran of the Peninsula (he had fought at Nivelle, Nive, and Orthes) and of Waterloo (where he had been severely wounded) but after the war his interests had turned to science. He had studied mineralogy under Friedrich Mohs at the Freiberg School of Mines, and in the 1830s he was to become a member of the council of the Geological Society of London. It is said that Colby secured Pringle's appointment to the Ordnance Survey expressly because of his mineralogical training[8], and perhaps from the outset Colby had regarded him as a potential director for the geological work in Ireland. Pringle was to grapple manfully with the task assigned to him, but it is sad that Griffith should have been nominated to the Boundary Survey instead of being brought in to take charge of the Ordnance Survey's geological operations. Under Griffith's direction, Ordnance Survey geology might have had a greater hope of success, but Colby almost certainly never considered such an appointment. As a professional soldier he would hardly have countenanced the idea of a civilian geologist directing the field-activities of the Survey's commissioned officers. The Boundary Survey to which Griffith was appointed had an exclusively

> Geometrical elevation on a line passing through the parishes of Aghanloo Drumachose and Balteagh exhibiting the strata at the north western limits of the basaltic district.

The second diagram — the work of Lancey alone — is dated 6 January 1828 and entitled:

> Geometrical elevation on the north end of the baseline Magilligan parish, Londonderry.

Neither Fenwick nor Lancey has produced a real geological section; their two diagrams are merely crude panoramas to which a scattering of geological information has been added. Perhaps during their off-duty hours in Mountjoy House the two lieutenants had been thumbing through the Survey's newly acquired set of the *Transactions of the Geological Society of London* wherein they must surely have lighted upon the 'Sectional view of the N.E. coast of Ireland' between Glenarm and Magilligan Head published in 1816 by Buckland and Conybeare.[19] Maybe that particular panorama served as Lancey and Fenwick's inspiration but, if such was the case, then the quality of their own work fell far short of that of their mentors. Even their most sympathetic critics must admit that the two subalterns seem to have possessed only the haziest notion of what constituted a geological section, yet, despite all their imperfections, the two panoramas were lithographed by the Survey, presumably so that copies could be circulated to all the field-parties as examples of the type of work required for each parish. Fenwick and Lancey were among the more enthusiastic geologists upon the Survey's staff, and one can only shudder at the thought of the so-called sections that must have been returned to Mountjoy by those officers who begrudged every minute that they were required to spend upon geological investigation.

Two other documents preserved in the Dublin office of the Ordnance Survey, and dating from Pringle's era, must be mentioned at this point. The first of these is a printed six-inch topographical map of the parish of Aghanloo, County Londonderry, dating from about 1828 and to which manuscript geological annotations have been added in red ink drawing attention to the presence in various localities of deposits of basalt, amygdaloid, 'blue shell limestone', 'mulatto stone', gravel, and alluvium (Fig. 4.3). Certain outcrops and river sections have been picked out in colour and the Plateau Basalts in the northeast of the parish have received a mottled carmine wash. No geological lines have been plotted (the carmine is left with an indeterminate edge) but there are marked some twelve sites at which rock samples had been collected for despatch to Dublin. The second document is another map closely related to the sheet just described. It is a rather crudely lithographed version of the Aghanloo parish map probably dating from 1828 or early 1829 and bearing in its margin the inscription 'by order of Capt. Pringle RE, Wm Lancey'. All the manuscript geological notes appearing upon the earlier sheet have been lithographed, save for those relating to drift deposits, and geological boundaries have been sketched in, although whoever plotted the boundaries clearly recognised that he was interpolating lines based upon far too little field evidence. In many places it was just guesswork and he has therefore added notes such as 'supposed line of junction' and 'line of junction very indefinite'. As in the case of the panoramas by Fenwick and Lancey, this Aghanloo map was evidently lithographed to serve as an exemplar for the other field-parties, and in the margin of the map Pringle has caused to be added the lithographed instruction:

> In the Geological Maps of Parishes all Topographical Remarks must be written in Black Ink, but those referring to the Geology are to be inserted in Red.

Whatever their failings — and they were many — these panoramas and maps produced under Pringle's direction do show that by 1828 the Ordnance Survey was making a serious attempt to conduct a comprehensive geological survey of Ireland, and it will be remembered that it was around this time that Griffith pricked up his ears and began to take note of the Survey's activities. His request to be employed in the geological mapping of northern Ireland was placed before the Royal Dublin Society in June 1826 and he certainly made no secret of the motives behind his proposal (see p. 48). He wanted to be ready to rush into print with a

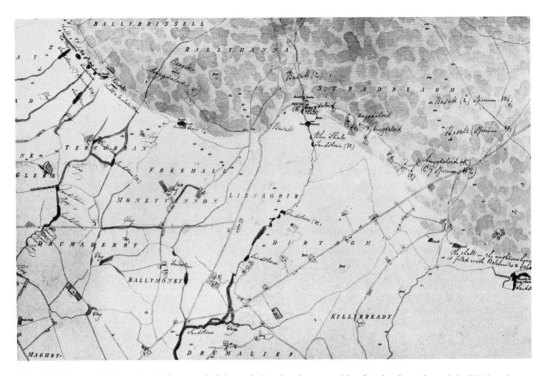

Fig. 4.3 A part of the geological map of the north Londonderry parish of Aghanloo, the original being drawn to a scale of six inches to one mile and most probably dating from around 1828. The base-map is a printed parish map and all the geological annotations have been added in red ink, many of them referring to numbered geological specimens collected by the field-parties. A blotched carmine wash has been added to show the Plateau Basalts in the north and northeast, and smaller, linear washes indicate places where the surveyors had noted the presence of continuous road and river sections. Reproduced from the map in the archives of the Ordnance Survey of Ireland by kind permission of the Director.

series of geological maps and regional memoirs just as soon as the appearance of the first batch of six-inch Ordnance sheets gave him the accurate base-maps that he had for so long desired. His request granted by the Royal Dublin Society, he passed the summer of 1827 mapping in and around the districts where the Ordnance Survey was then at work and, as was suggested earlier, Griffith's field-programme that year was probably intended to remind the Ordnance surveyors that they were interlopers in what he liked to regard as his own geological preserve. Perhaps, too, he hoped to embarrass the Survey through a demonstration of its own ponderous geological impotence when compared with the incisive activities of just one trained and experienced geologist. Griffith must have been acutely aware that if Colby's scheme succeeded, then his own work of seventeen years was wasted, and there was frustrated his long-cherished ambition to be the author of the first geological map of Ireland. But all Griffith's fears were to prove groundless; Colby's geological project was heading for troubled waters, and Griffith was destined to far outpace the official survey.

Colby's first defeat

Geology must have been the most popular science in Britain during the first half of the nineteenth century, but one suspects that its vogue did not extend throughout the ranks of the Ordnance Survey. The construction of detailed topographical sheets was in itself a taxing responsibility, and many officers and men probably resented having in addition to engage in geological duties. Some perhaps felt that the plying of a hammer to an Irish rock-face was hardly a fit duty for one of His Majesty's commissioned officers, while in the ranks there must have been many who were indignant at having to trudge long distances weighed down under the burden of rock-filled knapsacks. The Ordnance Survey letter registers of the period convey the impression that the field-parties were less than enthusiastic about their geological work; Pringle certainly had to issue regular reminders about the importance of sending an adequate number of rock-samples up to Dublin.[20] Exactly what was to happen to the specimens when they reached Dublin remains cloudy; certainly no geological expert had been appointed to the staff to sort and name the material. Perhaps Pringle himself proposed to perform these duties, although there is one reference to some specimens collected by Lieutenant Bordes being sent over to London 'that their names may be ascertained'.[21] It may be that the question of naming the specimens never became a serious issue because, despite Pringle's best efforts, the field-parties remained recalcitrant, and even as late as 18 April 1828 he had to admit to Colby that very few specimens had as yet arrived at Mountjoy.[22]

In 1828 doubts began to be entertained in official circles in England about the manner in which Colby was conducting the entire survey of Ireland, and as a result the Master-General sent a senior officer to Dublin to conduct an on the spot enquiry. That officer was Sir James Carmichael Smyth, a distinguished military engineer; his task was to examine Colby's methods with a view to discovering whether they involved any undue expense or delay. Sir James arrived in Ireland on 24 July and he immediately began a thorough investigation of the Survey in all its many facets. Pringle's geological department came in for especially close scrutiny because it clearly contributed nothing towards the completion of the six-inch topographical maps, and it was therefore an obvious area in which economies might be effected.[23] In this geological sector Colby was immediately thrown onto the defensive. 'What is the cost of the geological survey', asked Sir James, 'and what proportion of the surveyors' time is devoted to geological investigations?'. Colby pleaded that it was impossible to answer such questions. In particular, independent costing was impracticable because the geological work was so integral a part of the entire survey programme. A man might be chaining distances and collecting rock-samples more or less simultaneously. 'Have you ever received any specific authorisation to permit your inauguration of a geological survey of Ireland?', pressed Sir James. Colby admitted that he had not, but he pointed out that in his annual reports he had freely divulged that a geological survey was in progress so there could be no accusation of his having left his superiors in ignorance of what was afoot. 'How much delay in the progress of the six-inch map may be imputed to the diversion of staff into geological studies?', inquired Sir James. Initially a good deal of delay had resulted, Colby confessed, partly because the men had been geologically untrained, and partly because the geology of County Londonderry had proved unexpectedly complex. But now, he emphasised, the men were more experienced and the geology of the districts currently being mapped was simpler, so any future delay would only be on an insignificant scale. This became his chief line of defence. He played geology *pianissimo*. He represented it as consuming only a fraction of the Survey's total expenditure of time and money, and the following is, for example, part of his response to one of Sir James's probes:

> It was never my intention to do more on the subject of Geological and Mineralogical researches than to make those primary examinations, whilst the Officers were on the ground, which might facilitate future operations and render the Survey more complete in conformity with the spirit of the report of the Committee to which I was referred. I considered it as a matter of very subordinate detail . . . .

This may have been a prudent disclaimer for such a moment of crisis but it was hardly to be reconciled with those ambitious proposals which Colby had outlined in his annual report of 2 February 1826.

As the officer directly responsible for the geological survey, Pringle soon became involved in Smyth's searching inquiry. He was asked to report upon the extent of the country which had been examined geologically, and his reply must evoke both our surprise and suspicion because he claimed to have received rock collections, geological cross-sections, or geological plans from almost three quarters of the parishes which had been mapped from the topographical standpoint. This statement is difficult to accept because only four months earlier Pringle had admitted to Colby that very few samples had as yet reached Mountjoy, and we must suspect that Pringle was guilty of somewhat exaggerating his department's achievements. He must surely have been nonplussed had Sir James asked to see this splendid array of geological material! When invited to estimate the proportion of their time which officers were devoting to geology, Pringle responded that the figure lay somewhere between one tenth and one sixteenth, and he re-emphasised the point made by Colby; geology was only interfering marginally with the topographical survey. In defending himself against the charge that his staff was in any case not qualified to undertake geological investigations, he protested that of the Survey's thirty officers in the field, no less than five had some geological qualifications, these being Lieutenant Portlock, whom he described as fully trained, and lieutenants Bordes, Fenwick, Lancey, and Robe, all of whom he recorded as partly trained. When he was asked how long he thought a geological survey of the whole of Ireland would take if conducted in accordance with the principles laid down in his *Observations*, he replied that the investigation of the geology of each of the anticipated 1500 six-inch sheets (the idea of parish maps had now been abandoned) of Ireland would occupy three days, that the entire survey could be completed by 1834, and that the total cost would be only £5500 made up as follows:

	£
Lieutenants commanding Divisions extra pay 14/-, 3 days to each sheet, 1500 sheets	3150
Draftsmen 5/- per diem, one day to each sheet	375
Labelling packing and carriage of specimens, about 10 specimens to each sheet (being sent in duplicate), 5/- per sheet	375
Hammers, test bottles for acid, haversacks, etc. etc.	100
Drawing paper, tracing paper, etc. etc.	300

The remaining £1200 was to be allocated to Pringle himself as travelling expenses and additional pay spread out over the years 1828 to 1834. It was a wildly over-optimistic estimate and it only serves to emphasise that the Ordnance Survey had no appreciation of the magnitude of the geological task which Colby has so gaily undertaken.

By no means all of the Survey's staff shared Colby's belief that geology could be pursued without interference to the topographical mapping. The Ordnance Survey letter registers record, for example, one communication which arrived at headquarters on 29 August 1828:

> Captain Wright explaining facts as to time consumed in Geological Survey of Parishes in opposition to Capt. Pringle's Estimate.[24]

Captain Wright was evidently no friend of geology, but then neither, it transpired, was Sir James Carmichael Smyth. He was empowered by the Master-General to issue instructions to Colby for the future conduct of the Survey, and on 1 September 1828 Sir James acted.[25] His first order was decisive. All the Survey's geological investigations were to cease forthwith and the men were to be returned immediately to their normal surveying duties. The order was passed to the staff in the field on 4 September[26], and perhaps the final fruit of this abortive episode in the history of Irish geological mapping was Lieutenant Rimington's geological plan of Island Magee, County Antrim, which arrived at Mountjoy on 9 September.[27]

Thus the first attempt to conduct an official geological survey of Ireland had failed without yielding so much as a single published sheet. Perhaps Griffith alone found satisfaction in the Ordnance Survey's discomfiture. Failure it certainly was, but this should not be allowed to obscure the significance of what Colby had attempted. His had been a very early venture in the organisation of an official national geological survey. True there had been previous attempts to develop such surveys elsewhere. As early as 1811 John Macculloch had received a roving commission to explore the geology of Scotland on behalf of the Board of Ordnance and, in the United States, North Carolina and South Carolina possessed state surveys which commenced respectively in 1823 and 1824.[28] But Colby's Irish project envisaged a geological survey far more lavish than those being attempted elsewhere — far more lavish both in the sense of the manpower being deployed and in Colby's expressed ambition (see p. 87) that the eventual maps would be 'the most minute and accurate Geological Survey ever published'. The geological survey of Ireland has admittedly had a somewhat interrupted career since 1825, but, if we overlook a few small breaks, it would seem that of all the world's national geological surveys only that of France is possessed of a pedigree longer than that of the Irish survey. The French survey was planned by the Corps Royal des Mines in 1822 but actual field operations were not commenced until 1825[29], the very same year in which Colby's men began their own investigations amidst the schists and basalts of northern Ireland.

The Colonel's second sally

Colby's geological plans may have gone awry, but he was not a man to be diverted from what he saw as a just and important objective. He doubtless believed that Carmichael Smyth had been misguided in ordering the abandonment of the geological survey, and Colby was certainly determined to see the order rescinded at the earliest opportunity. The Nelsonic 'blind-eye approach' was of course quite out of the question, but surely the harshness of Sir James's verdict would mellow away with the passage of time, allowing the case for geology to be heard again under rather more favourable circumstances. This happened sooner than might have been expected. Sir James had issued his prohibition on 1 September 1828, but a resumption of the geological work was being discussed at Mountjoy as early as the November of the following year[30], and soon thereafter Colby somehow secured the Master-General's approval for a revival of the geological investigations, the news being conveyed to the field-parties in an order issued from Mountjoy on 13 January 1830.[31] For his part Colby must have been well satisfied at the thought of getting the Survey's hammers and collecting bags into use once more, but one wonders how his men reacted to the news that their two years of respite from geology were now at an end.

Pringle was no longer available to superintend the geological survey and the man appointed in his stead was Lieutenant Joseph Ellison Portlock who was promoted to captain's rank on 22 June 1830. Like Pringle, Portlock came of naval stock.[32] He was born at Gosport on 30 September 1794, the son of a naval officer who had sailed with Cook on that final voyage which ended so tragically at Kealakekua Bay in Hawaii. Today Portlock Point is a favourite surfing locality on the neighbouring island of Oahu, and Portlock junior must often have been regaled with his father's tales of the exotic Pacific world. The young man nevertheless chose for himself a military rather than a naval career, and in 1813 he accepted a commission in the Royal Engineers. In May of the following year he was sent to Canada where he served with some distinction during the final operation of the Anglo-American war of 1812-1815, and following the conclusion of hostilities he participated in a number of exploratory expeditions in the vicinity of the Great Lakes where his own name is perpetuated in Portlock Harbour on Lake Huron. Not until September 1822 did he finally return to Britain, and soon thereafter he transferred from the regular duties of his corps to the Ordnance Survey. Here he early made his mark, and in July 1824 Colby nominated him as one of the officers who were to be responsible for the triangulation of Ireland. From 1827 onwards Portlock was in sole charge of this vital work, and in order to complete the observations he and his men spent long periods encamped upon some of Ireland's most exposed mountain-tops where their hardships must have been severe. What, for instance, were Portlock's memories of Christmas 1827? His team spent the festive season atop remote Slieve League in County Donegal, their wind-blown tents perched at no great distance from the edge of a stupendous cliff over which the surveyors could gaze at the incessant foaming of the Atlantic breakers almost 600 metres below.

Living so, in the intimacy of the field, it might have been imagined that the distinctions of rank and social class would have been forgotten for the moment but, for whatever reason, his men found Portlock to be cold, humourless and withdrawn. He addressed them but sparingly and then only in the course of duty, and they for their part never presumed to tender the uninvited comment. Yet the captain held their unanimous respect, not least because of a personal toughness which perhaps owed a good deal to those years spent in the Canadian backwoods. We are told that for long periods he was capable of maintaining a steady pace of four miles per hour, leaving his companions trailing and exhausted. Quite apart from the elemental rigours of life amidst the Irish mountains, Portlock must have been plagued with boredom as he sat awaiting those all too infrequent spells of clear visibility which were essential for his observations. Perhaps he whiled away the hours of enforced idleness by examining the quartzites, granites and sandstones which form Ireland's highest mountain peaks. He was certainly a devoted geologist by the time he assumed his responsibility for the Irish triangulation. Having joined the Geological Society of London in 1825, he was taking an active interest in the Survey's geological operations as early as January 1826[33], and two years later Pringle returned Portlock as the only fully trained geologist upon the Survey's Irish staff. In November

Fig. 4.4. Major-General Joseph Ellison Portlock aged in his sixties and perhaps soon after his term as president of the Geological Society of London (1856-1858). Reproduced from a photograph in the archives of the Institute of Geological Sciences, London, by kind permission of Mr F.W. Dunning, Curator of the Institute's museum.

1831 Portlock became a founder member of the Geological Society of Dublin, and he addressed the society no less than twenty-one times between April 1832 and April 1841.[34] In addition, he was to become an active member of the British Association, serving on a number of occasions as the Secretary of Section C (Geology), and his geological reputation was sufficient to earn his election to the presidency of the Geological Society of London in 1856. In Portlock, Colby had found no ersatz geologist, but the genuine article; the captain was a full-blown enthusiast, well versed in the most modern stratigraphical approach to the subject, and very much a part of that rapidly developing community which was British geology of the mid-nineteenth century.

Exactly how Portlock proposed to conduct the geological survey at this period is nowhere made clear, but presumably he had little alternative but to employ the methods originally devised by Pringle. Indeed, the sole known surviving copy of Pringle's *Directions* of 1827 contains manuscript annotations by Portlock, suggesting, perhaps, that he may once have considered the reissue of the work in revised form. For two reasons Portlock was less than happy about his new responsibility for the geological survey. Firstly, he still remained responsible for the continuing activities connected with the primary triangulation, and he therefore had to divide his time between geology and squatting upon mountain-tops. Secondly, he had no power to order the field-parties to make geological observations; in his own words 'geology had, in fact, been permitted, but not commanded'.[35] He could thus impose no sanctions upon any field-parties which refused to shoulder their share of the geological burden. Portlock nevertheless had the firm support of Colby, and it was Colby who on 9 April 1830 signed a document, for circulation throughout the Survey, listing the colours and letters that were to be used to differentiate the various rock-types upon all the Survey's geological maps and sections.[36] One of the document's instructions now reads rather strangely. Opposite 'sap green' is the direction:

> This Tint to extend over those parts where depth of Soil has precluded investigation of the Substratum, and where wells or other excavations do not afford the requisite facilities.

Colby's mind thus seems still to have been focussed upon petrographic maps with their point-information, rather than upon maps in which evidence might be extrapolated between exposures and beneath an overburden. Perhaps such extrapolation introduced too strong an element of the subjective to satisfy a surveyor who expected every feature of the landscape to be plotted to the highest level of instrumental accuracy.

Between 1830 and 1832 geological investigation proceeded in a leisurely and perhaps rather aimless manner, the areas examined lying in and around County Londonderry where the Survey was still engaged upon its topographical mapping. In 1832, however, the pace of the work suddenly quickened. During that year Portlock completed the primary triangulation of Ireland and, although he continued to have responsibility for the secondary triangulation, he did now have more time to devote to the supervision of the geological work. Also it was now made clearer exactly how the Survey's geological studies were to be published, and this must have reassured the field-parties that the fate of any geological plans they prepared would not merely be to gather dust in the basements of Mountjoy. There of course always had been the notion that some kind of statistical and scientific account of Ireland would accompany the six-inch map, but until the early 1830s the whole project had been left delightfully vague. Now the scheme took firm shape in the mind of Lieutenant Thomas Aiskew Larcom who in 1828 had become the officer in charge at Mountjoy. In 1832 Larcom proposed to Colby that a series of detailed parish memoirs should be prepared for the whole of Ireland, each memoir offering a comprehensive conspectus of its region and including an account of the parish's geology. Colby accepted the proposal, and in July 1833 the publication of the memoirs was formally resolved upon. As soon as the scheme had received Colby's approval, Larcom prepared a thirty-seven page document entitled *Heads of enquiry* which was circulated to all the officers in the field so as to assist them in the collection of information for the memoirs.[37] The pamphlet is really a skeleton memoir listing the topics that officers were to bear in mind, and ranging in its scope from natural features and gentlemen's seats, to fairs and the quality of grazing grounds. There is even advice upon the making of rubbings from monumental inscriptions! In the realm of geology the pamphlet is unfortunately not very informative. Under that head we read on page 12:

> *Geology* — On this subject separate tables, &c, will be furnished.

Presumably Portlock was responsible for these additional documents, but sadly no copies seem to have survived.

The memoir scheme having at last given firm purpose to the Survey's geological investigations, something like a wave of enthusiasm for the science swept through the Irish Survey during the 1830s with such members of Portlock's department as lieutenants Fenwick and Stotherd appearing before scientific societies from time to time to read geological communications. We must note, however, that this new-found enthusiasm was by no means unique to the Irish branch of the Survey; Colby's men in England and Wales were also now undertaking geological investigations as a strong sideline to their topographical work. About 1831, for example, John Robinson Wright and Henry Mclauchlan, two of the Survey's civilian employees, had made a geological map of the Welsh Borderlands, and this was presented to the Geological Society of London by Colby himself.[38] Later, Roderick Murchison made use of Wright's mapping around Ludlow and found it to be 'a model of accuracy'.[39] Mclauchlan displayed a geological map of the Forest of Dean to the Geological Society in February 1833[40], and J.R. Wright's brother, Romley Wright (another of the Survey's employees), was the author of a paper on Brown Clee Hill in Shropshire read to the same society in the following December.[41] It should also be remembered that De La Beche's offer to produce geological versions of the Survey's one-inch sheets of southwestern England had been accepted, upon Colby's recommendation, in April 1832. Thus Colby's encouragement of geology was by no means exclusive to Ireland and some of his letters written in the 1830s shed interesting light upon the reasons for his concern with the science. Firstly, he had a professional motive for wishing to stimulate geological studies as is shown in this extract from a letter written to J.R. Wright on 28 February 1832:

> An intimate knowledge of the stratification and of the Geological features of a Country, is highly useful in the delineation of the forms of hills; and I have always been of opinion that this knowledge might be collected in passing over a country for the formation of a map, without any sacrifice of time.[42]

Secondly, there were utilitarian and financial motives involved because, in his letter of 9 April 1832 to the Master-General, recommending acceptance of De La Beche's offer, Colby wrote of:

> ... the one, the utility of making the Ordnance Maps subservient to the most perfect illustrations of the Geological features of Great Britain as a matter of general political economy for the benefit of the Empire. The other, the increased profit arising from the additional Sale of the Ordnance Maps, resulting from their increased utility, and the knowledge of their value becoming more general in the Country.[43]

It is interesting to note that in the new Master-General who assumed office in May 1835 both Ireland and geology had a good friend. In agreeing to place De La Beche's geological survey of England and Wales upon a formal footing the Master-General, in June 1835, added the following comment to the Board of Ordnance's minute:

> At the same time I must observe that all the arguments contained in the Report of the Geologists as applicable to the advantages to be derived to England from a correct Geological Map, apply in a still stronger degree both to Ireland & Scotland & most especially

to the former — a country abounding in minerals & the Geology of which is infinitely less known than that of England. I would therefore propose that as fast as persons fully qualified are found to carry on a Geological survey some of them should be attached to the Irish Survey for the purpose also of a Geological Map of that Country, and the same also with respect to Scotland when proceeding with that Survey.[44]

The Master-General concerned was Sir Richard Hussey Vivian who had just returned from being commander of the forces in Ireland, bringing with him, as his second wife, Letitia Webster from County Longford. Perhaps this explains his Irish proclivities.

The memoir in its heyday

The Ordnance Survey had decided to produce a series of memoirs descriptive of the Irish parishes, but the country contains 2400 such units; where was a start to be made? In 1833 the Survey had published its six-inch sheets for County Londonderry — the first Irish county to be completed — and it was therefore only natural that Colby and Larcom should have turned to that county in search of the parish which was to become the subject of their earliest published memoir. The choice fell upon the parish of Templemore which includes most of the city of Londonderry, and it was perhaps just fortuitous that it was a parish with which Mrs Colby's family — the Boyds — had strong connections. In 1834 Portlock was formally asked to prepare contributions for the forthcoming work, and the entire project was pressed ahead so rapidly that early in the summer of 1835 a provisional edition of the memoir was printed off at Treasury expense.[45] This edition was expressly intended for circulation among the British Association members who would be attending the Dublin meeting in the August of that year, and Colby and Larcom were anxious to test the mettle of their memoir against the visiting scientific steel. They need have had no fears; the Association was much impressed.[46] On 15 August 1835 John Phillips, the Association's assistant general secretary, wrote to the Lord Lieutenant expressing the Association's thanks for the copies of the memoir which had been distributed among the sections, offering his congratulations upon the appearance of the work, and expressing the hope that other parishes would soon be accorded similar treatment.[47]

Although the memoir had passed its first test with flying colours, the Ordnance Survey was still in no haste to get the volume into the bookshops, and it was not until November 1837 that the work was finally published. Fifteen hundred copies were printed and the volume of some 350 quarto pages was sold for twelve shillings.[48] Only in detail does it differ from the volume shown to the British Association two years previously, and the memoir is a remarkable, if somewhat indigestible, compendium of information, its scope extending from physiography and natural history, via gentlemen's seats, antiquities and history, to commerce and 'productive economy'. The title page bears Colby's name, but the memoir was really the work of many Survey pens, Portlock being responsible for both the natural history section (geology, botany, and zoology) and that dealing with the parish's industry and commerce. Portlock's geological contribution is brief — only seven pages — but it is accompanied by a coloured geological map of the parish on a scale of one inch to one mile (1:63,360). The geological mapping of Templemore was a relatively simple task because the parish is underlain almost everywhere by the Dalradian Schist which Portlock has coloured pink as representing what he terms varieties of mica slate, the various varieties being represented by letters. Thus 'G.m.q.' is mica slate passing into quartz slate. Some dip directions are shown and a few drift deposits are indicated by colour washes plus a descriptive comment such as 'A Chain of Sand and Gravel Hills in the Boggy Bottom'. One interesting feature is the use of a deeper tint of colour to represent those locations from which rock-samples had actually been taken, but members of the British Association who received copies of the 1835 version of the map must have been puzzled by those deeper splashes of colour because Portlock had forgotten to add a note to his key explaining their significance. This oversight was remedied in 1837 and the resulting addition to the marginal information is the only significant difference between the map of 1835 and its slightly younger successor. The two maps are both small — their coloured areas are little more than 100 square centimetres — but they nevertheless merit a special place in the history of Irish geological cartography because they were the first official geological maps to be published for any part of Ireland. It is very appropriate that their scale should have been the same as that adopted for the official Geological Survey maps which were to become so familiar in later years.

By 1837 Ordnance Survey geology was riding the crest of a wave. In England and Wales, De La Beche's geological department had been put upon a secure formal footing[49], while in Ireland Portlock and his men found the intellectual climate entirely favourable to their geological studies. In large measure this was because geology was enjoying a quite remarkable popularity among the British middle classes — indeed a geological mania was sweeping the land — but the Irish clutching at the science was also to some extent a clutching born of despair. Between 1791 and 1831 the country's population had almost doubled to reach a total of nearly eight million, but there had been little accompanying expansion of the nation's economy and population pressure had become acute. More than ever before the obvious palliative seemed to be an Irish industrial revolution based upon native mineral resources. The 1830s therefore saw a resurgence of that buoyant notion that Ireland somewhere *had* to contain a mineral treasure-house sufficient to rival the rich deposits being exploited so successfully elsewhere in the United Kingdom. Geology seemed to be the magical divining rod which would bring those hidden riches to light and yield a quick profit to the impoverished nation. These are Colby's own words from an address to the Geological Society of Dublin on 15 February 1837:

> The situation and the productions of Ireland are favourable to manufactures and to commerce; and a detailed geological survey which shall exhibit its natural resources, cannot fail to give profitable employment to its population.[50]

With the current of Irish opinion running strongly in his favour, Portlock was in 1837 instructed to open what was termed 'a geological office', but in reality this was a misnomer. The scope of the new establishment actually extended far beyond the realm of geology to encompass a soils laboratory, botanical and zoological departments, and a statistical bureau.[51] This comprehensive organisation was located not in Dublin, but in the infantry barracks at North Queen Street in Belfast, and by May 1838 Portlock had based there a staff of no less than thirty-five geologists, botanists, zoologists, and collectors of statistical data, all of them being civilian employees of the Ordnance Survey.[52] Among Portlock's geologists there were some who were destined to achieve eminence in their chosen field. His chief assistant, for example, Thomas Oldham (1816-1878), was successively to hold the Local Directorship of the Geological Survey of Ireland and the Superintendency of the Geological Survey of India, while George Victor Du Noyer (1817-1869), Portlock's artist, was to become one of the Irish Geological Survey's senior field-staff. On a more humble plane, James Flanagan, one of Portlock's fossil collectors, later filled the same office with the Geological Survey of Ireland, and it is interesting to note that another of Portlock's men — Patrick Doran — left the team in order to become first a fossil collector with Griffith's unofficial geological survey, and later a dealer in geological specimens. The task of the geological office was of course to collect information for the proposed series of parish memoirs, and the establishment represented a remarkably lavish public investment in the preparation of regional studies. Certainly Ordnance Survey geology had come a long way since the dark days of the Carmichael Smyth inquiry of only ten years earlier.

By the close of 1839 the Ordnance Survey had carried its six-inch topographical mapping across more than two thirds of Ireland's surface, and in the December of that year the surveyors were at work in counties Clare, Limerick, Tipperary, Kilkenny, Waterford and Wexford.

But no geological hammers and collecting bags were to be seen among those field-parties down in Leinster and Munster; all the geologists were still back in Ulster chipping away at the very same schists and basalts as had engaged the attention of Pringle and his men more than a decade earlier. Gone was the old pretence that geology could be a by-product of the topographical survey, and Portlock was now unquestionably running a geological survey comparable in every way with that being conducted under De La Beche in England and Wales. Both surveys were under Colby's control, both were staffed by civilian employees of the Ordnance Survey, and the parallel even extended to their both possessing associated museums, Portlock's in the infantry barracks, Belfast, and De La Beche's in Craig's Court, London.

For reasons which will shortly become clear, Portlock's survey left to posterity very little in the way of a cartographic memorial, but in the office of the Geological Survey of Ireland there are twelve manuscript geological sections plotted by Oldham and Du Noyer while they were working for Portlock and dated between March 1838 and September 1841. All are drawn to a horizontal scale of six inches to one mile (1:10,560) and they represent portions of counties Donegal, Fermanagh, Londonderry, and Tyrone. Of the actual methods used by Portlock's men in the field we know little[53], but the following order issued from Mountjoy on 24 April 1837 does give some indication of one of the temptations which might befall a Survey geologist in an age when geological specimens of quality could command a high price on the collectors' market:

> Capt. Portlock authorized to remove any person engaged in collecting specimens connected with Geology who may be discovered to have retained, given or sold specimens without his knowledge.[54]

The precise mode in which the results of the geological investigations were to be published for the national benefit was for the moment left very uncertain. Presumably there would be other memoirs on the Templemore pattern to which the geological department would be contributing, but Portlock can hardly have viewed with equanimity the prospect of preparing a series of two and a half thousand parish maps and memoirs. Such fragmented studies would clearly render difficult any appreciation of the broader regional structures of Ireland, while the accompanying parish memoirs would obviously afford little scope for the discussion of those wider theoretical issues in which Portlock was so keenly interested. He must surely have reflected, therefore, upon the possibility of publishing the geological work outside the cramping framework of the memoir project, the more so since in England and Wales Colby was leaving De La Beche free to publish a steady stream of large geological maps on the one-inch scale (1:63,360) — maps that were rapidly becoming exemplars for the rest of the geological world.[55] But at this point there must have returned to haunt Portlock that very same problem as had confronted every Irish geologist from Fraser in 1801 down to Griffith in the 1830s: upon what base-map should he publish the results of his field investigations? Hitherto the difficulty had been simply one of securing a base-map of sufficient precision, but in Portlock's case the problem assumed a slightly different guise. He had at his disposal the beautiful and accurate six-inch productions of the Ordnance Survey, but unfortunately Colby had neglected to compile any medium-scale map of Ireland to serve as a base for the geological sheets which Portlock would doubtless have liked to publish. All the field observations made by Portlock's department were recorded upon six-inch maps but, on mature reflection, the publication of six-inch geological sheets was out of the question, partly because they were too expensive and partly because upon such a scale sheet after sheet might appear in a single uniform hue as a result of some regions lacking any geological diversity. What Portlock needed was a national map upon a scale smaller than the six-inch, but no such document existed until he had descended the scale-ladder down to the level of the county index maps compiled to illustrate the sheet lines of the six-inch survey. These county indexes were mostly on a scale of half an inch to the mile (1:126,720), and this was surely rather a smaller scale than Portlock would have wished for, not least because it was only twice the scale of the map which Larcom had just constructed to serve as a base for Griffith's geological map of 1839. The public had a right to expect that Portlock's well-staffed department would produce a geological map of Ireland which was rather more than just a glorified Griffith map. But Portlock had no choice; county indexes it had to be. In some ways the indexes were suitable enough for the purpose, and they certainly had the advantage of representing the structure of a large area within the compass of a single sheet, but they did have one serious practical drawback. The counties in which Portlock had set his men to work — Donegal, Londonderry, Tyrone, and Antrim — are among the largest of the Irish counties (varying in area from 2115 to 4847 square kilometres) and the use of the indexes as base-maps meant that the geologists had to examine a very large piece of country before they could have so much as a single sheet ready for publication. Perhaps, too, the use of the six-inch map as a field-sheet encouraged a leisurely approach to mapping operations. On the six-inch map a day's field-work can appear as a triumph of achievement; on a smaller scale map the same piece of territory withers into insignificance.

On the opposite side of the Irish Sea such problems were quite unknown to De La Beche. His department was both recording its field-observations and publishing upon the one-inch map (1:63,360), which at that time constituted the Ordnance Survey's basic coverage of England and Wales, and even a full sheet in this series depicted no more than 2080 square kilometres of territory. But many of the sheets incorporated large expanses of sea and they thus contained far less than 2080 square kilometres of mappable country. For example, of the eight one-inch sheets which De La Beche completed for the Ordnance Survey in 1835, only two were full sheets and no less than five were at least 50 per cent water. In this manner the use of the one-inch map gave De La Beche the opportunity of pleasing his superiors by maintaining a steady output of completed work, and between 1835 and January 1840 he published no less than thirteen one-inch geological sheets.[56] In Ireland Portlock was denied the same opportunity of placing his department in a favourable light. But perhaps the most fundamental difference between the geological departments located on either side of the Irish Sea was that Portlock possessed little of that initiative, that administrative drive, and that influence in high places which enabled De La Beche to create in the flourishing Geological Survey of England and Wales an archetype for geological surveys the world over. Maybe, as a member of an Ordnance Survey which was essentially military in character, the soldier Portlock experienced constraints that were unknown to the civilian De La Beche but, whatever the mitigating factors, Portlock must himself shoulder much of the blame for the fact that between 1835 and 1840 his department succeeded in publishing no geological sheets other than the small map contained in the Templemore memoir. It was a disappointing performance.

The monthly returns for January 1840 show that Portlock then had his staff of thirty-five deployed as follows: Oldham was in the field mapping geology; four men were collecting geological specimens and five were collecting natural history specimens; Du Noyer was executing a series of geological drawings; five men were cataloguing and arranging specimens in the Belfast office; nine men were conducting general field enquiries; and the remaining ten members of the staff were engaged in a multiplicity of different duties.[57] Despite their captain's tendency to aloofness, the team was evidently a happy one, but January 1840 was to be their last month together. Early the following month the work of the entire department came to a shuddering halt. It was 1828 all over again but this time there was no Carmichael Smyth to lend a judicial air to the descent of the axe. Economy had become the watchword of the day and on 10 February Portlock was instructed to cut back his staff and to focus his remaining resources upon bringing the geological survey of County Londonderry — his most advanced county — to an early completion. To facilitate him towards this end he was relieved of his duties in connection with the secondary triangulation of Ireland[58], and Lieutenant Henry James was instructed to join him in order to expedite the geological work.[59] But within only a few days another severe blow fell; Portlock was ordered to close the Belfast office and museum and to transfer all its contents to Mountjoy.[60] The reduction of the department started immediately. Thirteen of the staff were dismissed during the remaining weeks of February, and by September 1840 Portlock was reduced to a staff of only three.[61] The removal of the contents of the Belfast office to Dublin must have occasioned some difficulty because when a catalogue of the museum was made in 1843 the collection was found to consist of the following items[62]:

    44 specimens of mineral waters
   1900 minerals
   4824 fossils
    70 boxes of unnamed fossils, mostly duplicates
    16 boxes of soils
    22 cubes of building stones
  13,580 non-fossiliferous specimens, named and with
     their locations marked upon six-inch maps
   8356 zoological specimens
    19 volumes containing about 3000 botanical
     specimens
   1292 drawings of fossils
    35 stratigraphical sections
     5 plans of coalmines

 The sad fate of Portlock's department was closely related to the fate of the entire memoir project. Although it had been the subject of much acclaim, the Templemore volume had really been conceived upon too lavish a scale. It had been overfull with pernickety detail, and its production had proved far too expensive. Ever since the volume's appearance in November 1837, the future of the entire memoir project had hung in the balance. A final decision to terminate the scheme was not taken by Colby's superiors until July 1840, but by February of that year it was already clear which way the wind was blowing and it was in consequence of this that the axe fell upon Portlock's department. It is nevertheless remarkable that the geological baby should have been thrown out along with the memoir bathwater. Whatever the merits or demerits of the memoir project, it was generally agreed that Ireland was very much in need of a geological survey. Why, then, was the geological work terminated so precipitately?

 This question is not easily answered, but it seems probable that the explanation lies in the professional relationship existing between Portlock and Colby. There is no evidence of any bad blood between the two men, but there is evidence that they failed to understand each other. Portlock was a slow, meticulous worker determined to attain the highest standards of geological accuracy, but he became so engrossed in the minutiae of the task as to forget that he was supposed to be conducting a national geological survey – a survey which Colby and his superiors hoped to see completed within a period which was to be measured in years rather than in decades. For his part Colby had little comprehension of the finesse expected of the geologists belonging to the younger generation. In particular, being himself a mineralogical geologist of the older school, he was at a loss to understand what he believed to be Portlock's undue obsession with palaeontology. What Colby seems to have wanted was not a stratigraphical map based upon palaeontological principles and displaying an historical and academic approach to earth-history, but rather a severely practical document of the type which today would be categorised as an engineering geological map.[63] He believed that there should be mapped not stratigraphical units, but such rock properties as hardness, toughness, frangibility, cleavage, durability, elasticity, absorbency of water, specific gravity, and so on. On 18 February 1840, and with Portlock very much in mind, Colby wrote to Larcom:

> In truth geology has run too much into fossils – is too much made up of minute research – and has not yet a sufficiently obvious connection with reality.[64]

Perhaps Colby was right. Maybe what was needed *was* a series of maps of engineering geology. But he was too far in advance of his time. Not until 1974 was there published the earliest Irish map devoted to engineering geology.[65]

Another problem for Colby was that Portlock appeared to be displaying a trait which was anathema to Colby; his loyalty to the Ordnance Survey and to his regimental corps was increasingly suspect. Were all those papers which he read to the Geological Society of Dublin not designed to bring personal glory to Portlock rather than reflected credit to his corps? Was he not becoming increasingly embosomed with 'the brethren of the hammer' at the expense of his service allegiances? In short, was Portlock's infatuation with geology not causing him to forget his role as the holder of one of Her Majesty's commissions in the Royal Engineers? Perhaps all might have been forgiven had Portlock submitted for publication a steady stream of maps, no matter what their geological character, but in fact he had produced virtually nothing. Portlock, an engineer officer, seemed to have failed miserably where De La Beche, a 'mere' civilian, was succeeding so admirably. In February 1840, when Lieutenant James was attached to Portlock's department to hasten the geological work, Colby wrote to Larcom:

> . . . you must impress on James the importance of vigorous exertions to re-establish the character of the Survey officers.[66]

Thus by the beginning of 1840 Colby probably harboured grave doubts as to Portlock's suitability for the post of director of the geological survey. Since all activities connected with the memoir project were being terminated, it perhaps seemed expedient to include Portlock's geological work in the general shut-down so that, after a decent interval, the geological survey could be given a fresh start under a more suitable officer. As we will see, Colby already had an idea as to who that officer might be.

Portlock's Londonderry Report

The 1840 decision to abandon the memoir project aroused widespread protest among informed Irishmen. Sir Denham Norreys and a deputation of the Irish Members of Parliament waited upon the Master-General to urge a reprieve for at least the geological survey of Ireland, but all that they secured was firstly an agreement that Portlock would be allowed to publish a geological memoir of County Londonderry, and secondly a promise that Lieutenant James would be sent over to England to examine the organisation of De La Beche's survey, the implication being that he might learn some lessons which could then be applied in the Irish context.[67]

When the Master-General gave his consent for the completion and publication of Portlock's study of County Londonderry, both he and Colby understood that the work was already in an advanced stage and that publication would follow almost immediately. In this they were mistaken. Portlock's striving for perfection caused him to examine and re-examine his ground time and again, while in order to resolve problems of Londonderry geology he deemed it necessary to extend his surveying deep into the nearby counties of Fermanagh and Tyrone. But even Portlock's most ardent sympathisers would be hard-pressed to explain why he found it necessary to keep two of his sadly depleted staff collecting around Carrickfergus, County Antrim, for a total of 43½ man days during April 1840.[68] Perhaps this work in Antrim was just one sign of that refusal to be hustled which now became so evident and which was to drive both Colby and Larcom to despair.[69] As early as 16 July 1840 Colby wrote to Larcom:

> This semi-revival of the memoir will I fear do little good. There can be no doubt but our friend Portlock has a great deal of Geological knowledge of Ireland – But I much fear he will not give it in a condensed form.[70]

As 1840 dragged on into 1841, and 1841 on into 1842, the memoir was still unfinished and Colby regretted that he had ever allowed Portlock to undertake the completion of the work.

Repeatedly Portlock was asked for estimates of the time he needed to finish his studies; equally repeatedly he overran both his own deadlines and those set for him by the Survey. Time and time again Colby protested that Portlock was losing himself in a maze of academic issues which were really of no concern to a Survey officer, and in August 1842 Colby suggested to Larcom that Portlock was 'afflicted with a diseased desire for temporary petty fame'.[71] On a number of occasions Colby tried to wipe his hands of the whole affair because, as Portlock's superior, he was himself receiving blame for the continual procrastination. The dénouement came on 12 December 1842 when Colby wrote to Portlock as follows:

> It is with very great regret that I have to apprize you that the repeated failures of your Estimates and delays of your Publication which was ordered by Lord Vivian to be ready before the Session of Parliament for 1841; have put it out of my power to request the Honble Board to rescind their order of the 29th March 1842, which directs that you shall be relieved from the duties of the Survey. I have repeatedly requested to be relieved from all responsibility for the Geological Survey of Ireland; that you might take on yourself the full responsibility for your own Estimates and Progress to the Inspector General, and to the Honble Board. But these requests have been denied to me, and I have this day informed the Inspector General that the Honble Boards order of the 29th March, which relieves you from the duties of the Survey, will be carried into effect on or before 31st January next, by which time your present publication ought certainly to be completed.[72]

So Portlock's devotion to geology earned him dismissal from the Ordnance Survey and forced his reversion to the normal duties of his corps. Early in 1843 he was posted to Corfu, then a part of the British Empire, but he accepted his banishment without rancour and immediately applied himself to studying the rocks of his new island home.[73]

Portlock's final return upon his geological work for the Ordnance Survey — a return for January 1843 — bears at its foot the following inscription:

> By very great and unremitting Labour my Report was finally completed and given to the Public.[74]

The long-awaited volume was published in February 1843 when 1500 copies were printed and put on sale at 24 shillings apiece.[75] It is a bulky octavo tome containing eight hundred closely printed pages together with a map, landscape sketches, figures of fossils, and coloured geological sections. In format the volume is clearly modelled upon the memoir on southwestern England which De La Beche had completed for the Ordnance Survey four years earlier[76] and, whatever may have been Colby's views upon the nature of modern geology, he had, through his Survey, given birth to two classics of regional geology.

The fold-out map in Portlock's report occupies a sheet with a size of about 89 x 74 centimetres, and the map is dated at Dublin in November 1842 (Fig. 4.5). It represents the whole of County Londonderry, together with large portions of counties Fermanagh and Tyrone, and the base is a half-inch (1:126,720) map derived from the index sheets which accompanied the six-inch survey. Indeed, the six-inch sheet lines and numbers are still present upon the geological sheet. The geology was reduced from the six-inch field-sheets to the half-inch scale by Thomas Oldham, and the various basic stratigraphical categories (Old Red Sandstone, Carboniferous Limestone, coal series, etc.) are represented in the usual manner by colour washes set within engraved boundary lines. Following the tradition of his Templemore map, Portlock has added a variety of symbols to indicate localities where rock was actually exposed, and to accompany the symbols he introduces a complex system of alphabetical and numerical notation designed to show the precise type and stratigraphical horizon of the rocks present in any exposure. Thus 'A' represents sandstone, 'C' shale and clay, 'L' limestone, 'S' slate, and 'T' basalt, while the figure '1' indicates the Silurian (his lowest fossiliferous formation), '2' the Old Red Sandstone, '3' the Carboniferous, and so on up to '7' the Tertiary. Beyond this, lower-case letters and expon-

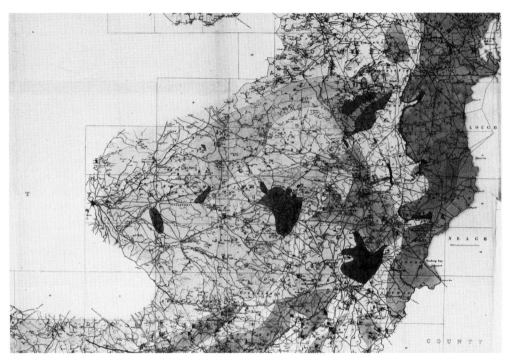

Fig. 4.5. A portion of the map accompanying Portlock's Report of 1843. This is the country lying between Six Mile Cross, Dungannon, and Aughnacloy, all three towns being located deep in County Tyrone. The county which Portlock was supposed to be mapping was County Londonderry.

ential numerals are added to differentiate the various formations so that 'Gh' is hornblendic-granite, 'Smq' is mica-schist passing into quartz-schist, '2Ax' is the Old Red conglomerate, and '3Cs^1' is the Lower Carboniferous shale. It may have looked confusing, but Portlock was satisfied that his system minimised the loss of detail arising from the reduction of a six-inch field-survey to the half-inch scale. Certainly the many thousands of symbols and notations upon the published map offer a wealth of detail testifying to the thoroughness with which Portlock and his assistants had done their work. At the time of its appearance the map was unquestionably by far the most accurate representation of the geology of that particular portion of Ulster.

Excellent as it is, Portlock's report never became a best-seller; only 106 copies had been sold by November 1843.[77] In far-off Corfu Portlock probably cared little about such matters, but his connection with Ireland was by no means at an end. Between 1849 and 1851 he commanded the Royal Engineers in the Cork District, in 1852 he was president of Section C at the British Association meeting in Belfast, and in 1862, having retired from the army with the honorary rank of major-general, he returned to Ireland to spend the evening of his life. He died at his home, Lota in The Cross Avenue, Blackrock, County Dublin, on 14 February 1864.

Captain James takes over

Twice — in 1825 and 1830 — Colby had inaugurated a geological survey of Ireland, and twice — in 1828 and 1840 — that survey had come to an abrupt end. Now, in 1842, Colby resolved to try yet again. Portlock was still in Ireland, but he was in such bad grace with the military authorities that Colby felt quite free to disregard his presence. The man now chosen by Colby to carry the mantle, once worn by Pringle and then by Portlock, was Captain Henry James (he had been promoted to captain's rank in June 1842) who for the next three years was to stand close to the centre of the vortex which developed around the vexed question of the Ordnance Survey's involvement with Irish geology. It was a confused period of personal jealousies, administrative intrigue, and governmental indecision.[78] It was also a period during which the Ordnance Survey mapped scarcely so much as one square kilometre of Ireland's geology.

James was a Cornishman, born near St Agnes in 1803.[79] He passed through the Royal Military Academy at Woolwich, and in 1826 he became a lieutenant in the Royal Engineers. In 1828 he transferred to the Ordnance Survey and he was sent to Ireland where he spent the next fifteen years surveying his way across the country from County Down into County Cork. It was during those long and perhaps rather tedious years that he found his interest in geology, and some of his earliest recorded field-work was carried out on the shores of Carlingford Lough in company with a Major Patrickson during the summer of 1834.[80] His geological leanings attracted Colby's attention, and in February 1840 James was instructed to join Portlock in order to lend assistance with the preparation of the Londonderry material for publication. James does not feature among the many individuals to whom Portlock paid generous tribute in the preface to the Londonderry Report, so presumably his assistance was minimal, but in 1841 James was given the onerous task of rearranging the Survey's collections at Mountjoy after their transference from Belfast. This museum duty he did not complete until 1843, partly because it was additional to his normal Survey responsibilities, and partly because in the latter part of 1842 he was temporarily transferred to England.

As we have seen (p. 104), the 1840 abandonment of Portlock's geological survey had raised some Irish hackles, and as a sop to the critics it had been agreed that James should visit England in order that he might examine the methods of geological survey being employed there by Sir Henry De La Beche (he was knighted in 1842). The visit was delayed for two years but at length, on 10 August 1842, Colby wrote to De La Beche saying that he was sending James over to join the British survey at Ross-on-Wye. For the next four months James worked as a temporary member of De La Beche's staff, mapping in Hereford and Gloucester. By 27 December 1842 James was staying with friends at Davenham Hall in Cheshire, his apprenticeship with De La Beche over, and he seems to have given his chief every satisfaction because, from Davenham, James wrote to De La Beche thanking him for the flattering terms he had recently employed in a letter to Colby recommending James as a thoroughly competent geological surveyor.[81] Only four short years later De La Beche's correspondence was to contain a very different assessment of James's geological abilities.

After Portlock's fall from grace, Colby had clearly earmarked James as the officer best fitted to take charge of any revived geological surveying in Ireland, and in December 1842 Colby decided that the time had come to put his plan into operation. On 19 December he wrote to De La Beche inviting him to assume responsibility for a new Irish geological survey with James serving as his Local Director in Dublin. De La Beche agreed, and on 20 January 1843 Colby placed the following proposals before the Inspector-General of Fortifications[82]:

1. The geological survey of Ireland should be resumed and placed under the control of De La Beche as part of an enlarged geological survey of the entire United Kingdom.

2. James should be the Local Director for Ireland with full responsibility for the day to day affairs of the Irish branch.

3. John Phillips (1800-1874), who in April 1841 had joined the Geological Survey of England and Wales to take charge of the palaeontological work, should now assume responsibility for the Survey's palaeontology throughout the United Kingdom.

4. The Irish geological survey should receive an annual subvention of £1,500, as compared with the annual sum of £2,200 which was being paid to the Geological Survey of England and Wales.

The authorities offered no immediate reply to Colby's proposals, and, when James completed the reorganisation of the museum at Mountjoy in May 1843, he was posted to Athlone, there to resume normal Ordnance Survey duties and to bide his time until his geological fate was decided. But at this point the whole subject of a geological survey of Ireland became caught up in a very much larger issue — the question of a revival of the old parish memoir project.

Far from being dissipated with the passage of time, the Irish sense of grievance engendered by the 1840 decision to abandon the memoirs had actually intensified and gathered momentum. On 19 January 1843 the Royal Irish Academy prepared for submission to the Lord Lieutenant an address in which the Members urged the revival of the memoir project *in toto*[83] and in London on 19 June a meeting was convened at the home of the Marquis of Downshire further to consider the memoir question, with the Marquis himself in the chair and Lord Adare acting as secretary.[84] Those present included Irish peers, Irish Members of Parliament, and the proprietors of Irish estates, and, on the proposal of the Earl of Clare and the O'Connor Don, the meeting passed a motion pressing for the publication of a complete coverage of parish memoirs on the pattern of the Templemore volume. A deputation drawn from those present at the meeting later waited upon the Prime Minister, Sir Robert Peel, to present him with the resolution. Back in Ireland the agitation continued unabated and interest in the memoir project was such that a Dublin publisher thought it worth his while to issue a volume consisting merely of press reprints relating to the subject.[85]

Faced with such pressure of opinion Peel took the obvious step. On 30 June 1843 he set up a commission, consisting of three members, to examine the entire memoir issue.[86] The three gentlemen concerned were John Young the M.P. for Cavan, Captain Henry George Boldero the Member for Chippenham, and Lord Adare who could hardly be counted as an impartial party because not only had he taken a prominent part in the affair at the Marquis of Downshire's house a few days earlier, but he was also a close friend of Larcom's. The commission was directed to examine the scientific, historical, and statistical work of the Ordnance Survey in Ireland, and in particular it was asked to resolve three problems:

1. Should any future geological survey of Ireland be linked with parallel historical and statistical investigations?

2. Was there any justification for the Ordnance Survey becoming involved in the mapping of Irish geology?

3. Should there be produced any further memoirs on the Templemore pattern?

The commission began its hearings on 14 July 1843 and its enquiries revealed that the Ordnance Survey had collected geological samples in fifteen Irish counties (Antrim, Armagh, Cavan, Down, Fermanagh, Kildare, Londonderry, Louth, Mayo, Meath, Monaghan, Sligo, Tyrone, Waterford, and Wexford), but it also emerged that, apart from the area covered by Portlock's memoir, the work had been so patchy and imperfect that there existed neither a complete geological map nor a complete geological collection for so much as a single Irish parish. The commission soon reached the conclusion that little of the old work would be of any value in a renewed geological survey, and they therefore sought De La Beche's advice over the expense likely to be incurred by a geological survey of Ireland beginning *de novo*. De La Beche replied on 12 August; he estimated that it would take about ten years to complete a fresh geological survey of the whole of Ireland and that the annual cost would be £1,500.

The members of the commission signed their report in November 1843, the geological recommendations being as follows:

1. That geology was of sufficient national importance to justify the making of a new and complete geological survey of Ireland.

2. That the survey should be entirely independent of any antiquarian, historical, statistical or other surveys that might be undertaken.

3. That the new survey should be under the control of De La Beche with James as the Local Director in Ireland, and that the survey should receive an annual grant of £1,500 during the period of about ten years which, it was understood, would be required to complete the project.
4. That all the fossils collected in Ireland should be examined by John Phillips.
5. That in addition to the Museum of Economic Geology at Craig's Court in London, founded in 1837, there should also be local geological museums in Dublin and Edinburgh. Irish specimens collected by the survey but not required at any of these three museums should be distributed among any other local museums which might submit requests.

The commission had produced a report pleasing to public opinion in Ireland, but the recommendations left the government itself quite unmoved. There was no flurry of implemental activity in the corridors of power, and the memoir project was left to rot in its grave as the geological survey of Ireland became the victim of yet further vacillation. While the government procrastinated, there developed a behind the scenes power struggle over who should control any new geological survey of Ireland which might eventually be established. By a majority of two to one the commission had named James as the man best fitted to fill the proposed new Local Directorship in Dublin, but one member of the commission — Captain Boldero — was of the belief that the new post should go not to James but to De La Beche's palaeontological assistant John Phillips.[87] Two opposing battle lines were now drawn up. On the one side there was James supported by Lord Adare and John Young, and urged strongly by Colby and Larcom, both of whom were determined that the new venture should be military in character and firmly under the control of the Ordnance Survey. On the other side there stood Boldero, De La Beche, and Phillips, all equally determined that any new Irish survey should be entirely civilian in character and under the direction of a geologist of high scientific standing, namely under the direction of De La Beche himself.

Colby and Larcom were resolute that Irish geology was a matter for the army, and let it be quite clear that they meant not just a survey directed by Royal Engineer officers and staffed with civilians as Portlock's department had been, but rather a survey staffed throughout by military personnel. On 20 April 1844 Colby wrote to Larcom:

> There can be no doubt, but that a military organisation would execute the Geological Survey of Ireland with much more rapidity, and uniformity than the proposed civil arrangement.[88]

It is difficult to see exactly how Colby could reconcile such a belief with the facts of recent Ordnance Survey history — captains Pringle and Portlock had both obviously failed in Ireland whereas the civilian De La Beche was equally obviously a success in England — but two of the reasons which caused Colby to prefer a military survey do deserve mention. Firstly, it was a period of prolonged peace in Europe — the Crimean episode was still a decade away — and Colby felt that if the Royal Engineers were being denied the opportunity of glory upon the field of battle, then surely they should be allowed the opportunity of distinction in the field of science. It is as though he had a mental picture of some regimental colour already emblazoned with the names of such victories as 'Ciudad Rodrigo 1812', 'Vittoria 1813', and 'Waterloo 1815', and to these he was determined to see added 'The Geology of Ireland 1845-1855'. Repeatedly in his letters he expressed the hope that his corps may win credit and honour through its geological exertions. Perhaps the fact that he himself had never seen any active service merely intensified this desire to launch his men upon a new and victorious geological campaign following upon the virtual completion of the Irish topographical survey. His point is not to be dismissed lightly; it is a fact that the staff of the Geological Survey itself, from the days of De La Beche in the 1840s down to Sir Edward Bailey in 1952, repeatedly referred to their mapping programme in military terms. Here, for instance, is part of a report sent by 'Lieutenant-General John Palaeorlynchus' (John Phillips) to his 'commanding officer' (De La Beche) from 'Camp' Standish, Gloucestershire, on 31 December 1842:

> Having found a part of the country endeavouring to practise a dilusion, by wearing false colours as Lias, the fact being that they are really of the Keuper formation, he directed that no quarter should be given under such circumstances. In consequence five square miles of valuable territory ... has been wholly reduced to submission ... A small body of the Lias Infantry under the direction of General Gryphite having taken up positions amidst floods & fens, it is intended to surround & capture them tomorrow.[89]

The male of the species seemingly revels in a little military make-believe! Secondly, Colby held that a decision to constitute the new geological survey of Ireland as a purely civilian enterprise must result in its being plagued with constant staff problems. Newly-trained civilian recruits would soon discover that geologists were in great demand in the colonies, and there would follow an unacceptably high rate of staff-turnover as men left to seek their fortunes overseas. Military surveyors, on the other hand, would be less free to desert their posts in this manner. Events were to show that here Colby did have an argument of some validity.

These were two military reasons why Colby the Colonel was determined to command any new geological survey of Ireland, but there were two other still more basic reasons for Colby's stance — reasons which concerned not Colby the soldier so much as Colby the man. Firstly, like most competent administrators, Colby had proclivities towards empire-building. His devising of the memoir project back in the 1820s was proof enough of that fact. Now, in the early 1840s, he was merely 'king' of the topographical survey — a survey which showed signs of shrinking as the six-inch work in Ireland was completed. How much better it would be to bestride both topography and geology as an 'emperor'.

Secondly, there was the problem — the serious problem — of De La Beche. A few years earlier, in 1835, Colby had been delighted to enlarge his empire by employing De La Beche as the Ordnance Survey's geologist, but since then De La Beche had revealed himself as a scheming and successful empire-builder in his own right. Already De La Beche had founded a Museum of Economic Geology outside the limits of Colby's suzerainty, he had curried favour in high places, not least with the Prime Minister himself, Sir Robert Peel, and in 1842 he had been rewarded with a knighthood. Never was Colby to receive his sovereign's accolade — his sole honour was a Danish title — and he must have been increasingly jealous of the status being accorded to a man who, in theory at least, was his own subordinate. Colby doubtless viewed De La Beche as a pushful civilian upstart — a cuckoo in the Ordnance Survey nest. The knowledge that the military career to which De La Beche had once aspired had been terminated in 1811 by his expulsion from the Royal Military College at Marlow[90] can hardly have improved De La Beche's image in Colby's soldierly eyes. Why, Colby must surely have wondered, should De La Beche achieve such kudos for producing geological versions of a mere handful of British one-inch maps? Surely he, Colby, was far more deserving of public acclaim. Not only had he been conducting the topographical survey of the whole of England and Wales, but in Ireland he had brilliantly carried through the unprecedented task of a nationwide survey on the six-inch scale. Were his two thousand six-inch sheets of Ireland not vastly more significant than De La Beche's twenty or so one-inch geological maps of England and Wales? For Colby this must have been a particularly painful issue because not only were the geological maps small in number, but a few of De La Beche's geological peers had recently expressed serious doubts as to the validity of some of the geological interpretations presented in the maps.[91] If these were truly Colby's sentiments, then they must be adjudged as entirely reasonable. The modern critic can only wonder at the vagaries, both contemporary and posthumous, which make one individual the recipient of public acclaim while another individual is left to languish in a forgotten limbo.

The situation must have seemed clear to Colby. If geology was a path to public esteem, then he and nobody else was going to be in charge of an Irish geological survey. Further than this, Colby must have read De La Beche's intentions. The ultimate objective of De La Beche's imperialism could only be the severance of the existing link between the Ordnance Survey and De La Beche's geological department, thus bringing into being an independent geological survey with De La Beche at its head. This dismemberment of his empire Colby had to stop. The best way to check De La Beche was to ensure that the geological surveying of Ireland was an Ordnance Survey affair, conducted by soldiers acting under Colby's orders. At the time of writing his memorandum of 20 January 1843, Colby was prepared to allow De La Beche control of the proposed Irish geological survey, with a soldier – James – as his Local Director in Dublin, but a few months later Colby evidently changed his mind. If possible De La Beche was now to be firmly excluded from any involvement in Irish geology; James was to operate directly under Colby's control. If this exclusion could be achieved, Colby must have reasoned, then with Irish geology securely in his own hands, there was far less chance of De La Beche being able to prize British geology out of the Ordnance Survey empire. Control of the proposed Irish geological survey thus became crucial in the power struggle between Colby and his subordinate.

We turn now to De La Beche himself but, before considering his motives in the Irish affair, a word of introduction is necessary. Historians have always treated De La Beche with reverence as one of the heroic figures of British nineteenth-century geology. His accomplishments were indeed remarkable in that, quite apart from his direct contributions to geological science, he made indirect contributions through the foundation of no less than four geological institutions.[92] He was, of course, devoted to his chosen science, but he was also an ambitious man determined to use geology as a vehicle for his own self-interest. This latter facet of his character has been inadequately appreciated. In society he could be charming; this description of him was written by a young lady in 1836:

> ... Henry De La Beche came to luncheon. The last named is a very entertaining person, his manners rather French, his conversation spirited and full of illustrative anecdote. He looks about forty, a handsome but care-worn face, brown eyes and hair, and gold spectacles.[93]

His geological colleagues, however, saw another side of his character. Murchison wrote to Sedgwick observing:

> De La B is a dirty dog, there is plain English & there is no mincing the matter. I knew him to be a thorough jobber & great intriguer....[94]

Andrew Crombie Ramsay noted in his diary that De La Beche pretended to be open, frank, and cordial, but was really 'an artful dodger, for ever working for his own interest, heedless of that of others'.[95] Joseph Beete Jukes was another who knew De La Beche well, and a few years after De La Beche's death Jukes wrote:

> Poor Sir Henry started the survey very much for his own personal honour & glory....[96]

The events connected with Irish geology in the period between 1843 and 1845 can only be understood if the real De La Beche is substituted for the De La Beche of the hagiologies.

De La Beche sought status, both within the scientific community and in society at large. Around 1843 his short-term objective was to remain within the Ordnance Survey framework and to secure control of the proposed geological survey of Ireland, thus thwarting Colby's scheme for running the new survey as a military organisation headed by Colby's puppet in the form of Captain James. De La Beche's long-term objective was to place himself at the head of an independent geological survey entirely outside the ambit of the Ordnance Survey and with its sphere of operations extended to encompass the entire British Isles. If he failed to attain the first of these objectives, then the second was automatically removed from his grasp. If James

Fig. 4.6. Sir Henry Thomas De La Beche, Director of the Geological Survey of Great Britain and Ireland from its foundation until his death on 13 April 1855. His spectacle frames were especially made for him in gold by Dollands of London so that there would be no danger of magnetic interference while he was taking compass bearings in the field. In 1818 he married Letitia Whyte of Loughbrickland, County Down, but the marriage proved unhappy and the couple separated in 1826. Reproduced from an illustration in Archibald Geikie's Life of Sir Roderick I. Murchison.

became the Local Director in Ireland and recruited a staff drawn from the Royal Sappers and Miners, as Colby intended, then De La Bouche was firmly tied to the Ordnance Survey's coat-tails for years to come. It is worth remembering that those coat-tails were cut to a military pattern in a very real sense because, although civilians, De La Beche and his officers were expected to wear a tight-fitting dark-blue uniform adorned with brass buttons bearing a crown and crossed-hammers emblem and the letters O.G.S. (Ordnance Geological Survey).

The question of whether it was Colby or De La Beche who from a scientific standpoint was the better qualified to conduct the geological survey of Ireland seems never to have been raised; the matter was simply one of administrative power politics. In the game both men had their pawns. Colby's pawn was his candidate for the new Dublin Local Directorship, Captain James. James was an ambitious man. In Irish geology he espied a small empire which he could carve out for himself within Colby's larger domain; he wanted the independence and authority associated with the new command. But his interest in geology itself was really only that of a dilettante. He was a careerist in the science, and although he had professed a concern for geology ever since the early 1830s, only twice did he present papers to the Geological Society of Dublin, then the chief forum for the discussion of Irish matters geological.[97] It was a poor showing as compared with the twenty-one papers which Portlock had presented to the same society between 1832 and 1841.

De La Beche's pawn was a man of utterly different scientific calibre — John Phillips, a nephew of William Smith, a former professor of geology at King's College London, and the author of various geological works ranging from elementary texts to his monumental *Illustrations of the geology of Yorkshire* first published in 1829. His administrative qualities had also been thoroughly tested because since 1832 he had been the assistant general secretary of the British Association, in which capacity he had played a major part in the organisation of the Association's annual meetings. If De La Beche could secure control of the new Irish survey, then it was Phillips whom he wanted as his Local Director in Dublin. And a splendid choice he would have been. Among the quartet chiefly involved in the affair of Irish geology Phillips was unique. There is no evidence that with him personal ambition was a motivating factor of any significance, and throughout the affair he remained upon the friendliest of terms with his rival, James. Phillips was just a devoted geologist. Frank, open, and guileless, he perhaps never appreciated that he was being used by De La Beche as a tactical element in a grand strategical scheme.

Initially, as this struggle developed, De La Beche was convinced that he would win. In the *De La Beche Papers* in the National Museum of Wales there is a series of letters dating from the first half of 1844 making it clear that De La Beche and Phillips both regarded it as a foregone conclusion that De La Beche would be taking charge of the new survey, as Colby had suggested in his memorandum of 30 January 1843, and that with a little effort Phillips could be eased into the Irish Local Directorship in place of James. In view of his anticipated future Phillips, with De La Beche's encouragement, accepted appointment to the newly-created chair of geology in Trinity College Dublin, a post offered to him by a decision of the College's Board taken on 30 December 1843.[98] The appointment was to date from 25 March 1844, with Phillips continuing to perform his Geological Survey duties in Britain by working there during the university vacations. De La Beche presumably believed that when the time came there would be no objection to Phillips retaining his chair but laying aside his British survey duties and substituting the Dublin Local Directorship. In the mid-nineteenth century it was certainly common enough for geologists to hold academic and Survey posts concurrently, Thomas Oldham, Andrew Crombie Ramsay, and Archibald Geikie being just three of those who did so. In Dublin it was generally understood that Phillips's arrival heralded the inauguration of a De La Beche-style civilian geological survey, and on 16 March 1844 the *Dublin Evening Post* commented sarcastically that Phillips possessed both the attributes necessary to secure advancement in Irish official circles, he 'being neither an Irishman nor a Papist'.[99] James, however, viewed Phillips's appointment to the Dublin chair in a very different light. The Board of Trinity College received Phillips's acceptance of the post on 6 January 1844 and only four days later we find James in Athlone writing to Larcom with some jubilation claiming that victory must now be theirs because plurality of office was hardly likely to find official favour, and by accepting the chair Phillips must surely have debarred himself from the Dublin Local Directorship.[100] But at that time James's own fate hung in the balance because in that very same month he received a posting to Bermuda and it needed all the influence that Colby and Larcom could muster in order to have James detained in England pending a decision about the control of the new Irish survey.[101]

So certain was Phillips of his own future in Ireland that in February 1844 he began to discuss with De La Beche the details of the new survey's organisation. He already knew something of Irish geology because he had been present at the meetings of the British Association in Dublin in 1835 and in Cork in 1843, and on 3 February 1844 he wrote to De La Beche discussing the financial provisions of the new survey and observing:

> The Field Geology of Ireland presents to my mind nothing of difficulty beyond what must belong to a new Survey. The assistants should in my opinion be all *Irish*. Even a knowledge of Erse is necessary in some of the wild districts of 'Ould Ireland'.[102]

On 13 and 14 May, after arriving in Ireland to assume his academic duties, Phillips wrote to De La Beche about the rooms on the upper floor of the Dublin Custom House which were being made available to house the new survey.[103] The accommodation offered was not to Phillips's liking and he suggested that instead a city house should be acquired for the survey; he understood that suitable premises might be obtained in Rutland Square (now Parnell Square) for between £2,000 and £3,000. By April 1844 even James and Colby were convinced that the new post must go to Phillips[104], and on 27 May De La Beche wrote to Phillips:

> I have every expectation that the Irish matter will go straight for you — and God grant that this may be so, but at the same time it is but right to say that it requires much watching and care — this it shall have from
>
> Ever yours
> H.T. De La Beche[105]

It was nevertheless during May 1844 that we discern the earliest signs of a growing loss of confidence within the De La Beche camp. In that month we find De La Beche enquiring of Phillips exactly how much time his academic duties would leave available for geological survey work. Somebody somewhere was clearly urging that Phillips could not possibly be allowed to hold two posts simultaneously. Phillips replied from his rooms in 35 Trinity College on 18 May 1844 protesting that his academic duties were light and that there was not so much as a single day when he was not free for at least a part of the time to attend to geological survey matters.[106] The university demanded from him a total of only thirty-six lectures spread through the three terms of the year and this, he emphasised, was no labour to one who had been teaching geology for eighteen years. For seven and a half months of each year the university would have no claim whatsoever on his time, and he laid stress upon the valuable cross-fertilisation of ideas which must result from the combination of Survey with academic duties. But such argument was to no avail, and during June 1844 De La Beche and Phillips both recognised that the Dublin Local Directorship was fast slipping from their grasp. In July Phillips learned that his tenure of the Trinity chair was likely to preclude him not only from the new Local Directorship, but also from retention of that vacation post which he already held within the British Survey.[107] The struggle was clearly intensifying. Colby laid a trap for De La Beche. On 21 July 1844 Colby wrote to Larcom as follows:

> I have written a letter to De La Beche calling for returns for information for the Master General and Board, which can hardly be given in such a manner as not to show his *unfitness* to direct and organise the Geo. Survey of Ireland.[108]

Perhaps the stratagem succeeded. Certainly on 23 July 1844 the contest ended in Colby's overwhelming victory. On that day there was completed a Treasury minute authorising the recommencement of the geological surveying of Ireland.[109] Colby was to be in overall control of the new department with James as its executive head. De La Beche, it was observed, was far too heavily committed to the mapping of British geology to allow any extension of his powers onto the opposite side of the Irish Sea. He and Phillips were to be firmly excluded from Irish geology. The new department was to be military in character and James was instructed to recruit his staff from the Royal Sappers and Miners, but he was permitted to draw upon assistance from two non-military sources. Firstly, it was agreed that Robert Kane (1809-1890), the Irish chemist, should perform all the survey's chemical analyses and, secondly, it was understood that Edward Forbes (1815-1854), who had just joined the Geological Survey of England and Wales as De La Beche's full-time palaeontologist, should visit Ireland periodically in order to examine the Irish fossil collections. The involvement of Forbes particularly pleased Colby because he had no wish to see James himself following Portlock into what Colby regarded as the morass of stratigraphical palaeontology. In one other respect the minute was less pleasing to Colby: no money was granted for the use of the new survey and it was to be financed from

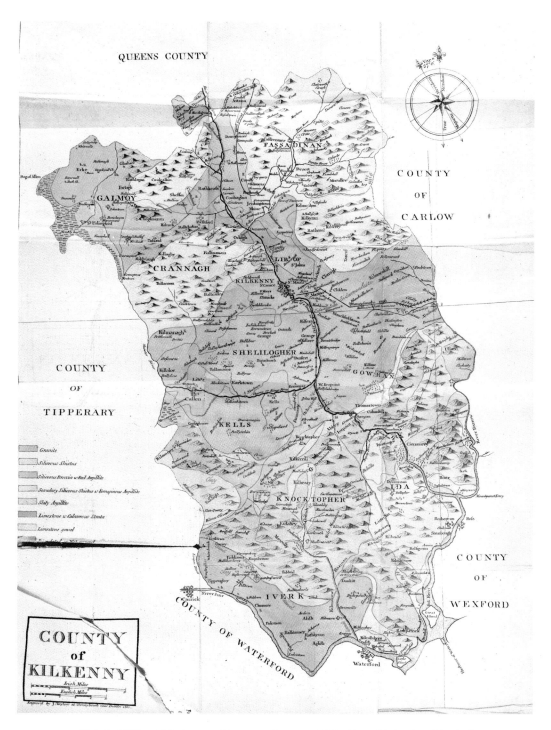

Plate 1. William Tighe's map of County Kilkenny published in 1802. (Reduced in size.)

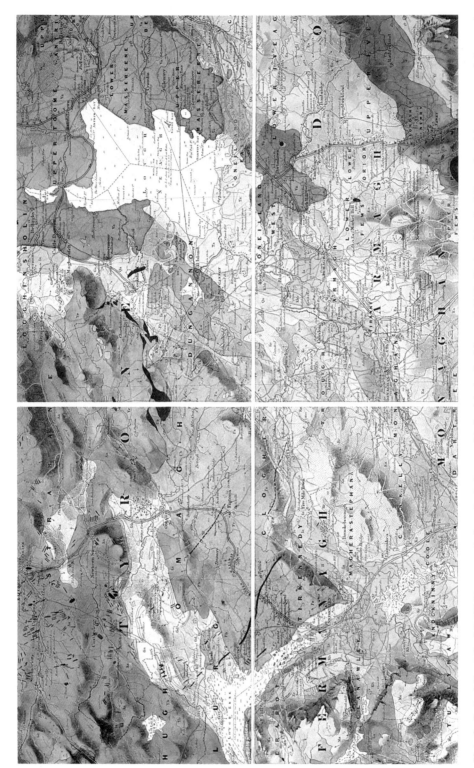

Plate 2. Part of Sheet 2 of Griffith's quarter-inch geological map of Ireland in its final form of 1855. (Reduced in size.) Reproduced from a copy of the map in the author's collection and formerly in the library of Lord Talbot de Malahide (1805-1883) who was president of the Geological Society of Dublin from 1855 until 1858.

Sheets of Many Colours

regular Ordnance Survey funds.

De La Beche must have been annoyed and disappointed, but it is surely with Phillips that our sympathies must lie. The Treasury minute is dated 23 July 1844 and the astute De La Beche must surely have heard the news soon thereafter. But in his moment of defeat he seems to have spared no thought for Phillips; he left Phillips to hear the news for himself, an event which took place on 13 August 1844, Phillips's informant being none other than the jubilant James himself.[110] Phillips could feel trebly aggrieved. De La Beche had abandoned him and he had been excluded from the Dublin Local Directorship, but the authorities were also administering what was tantamount to a personal slight. They were proposing to import Forbes to do precisely that type of palaeontological work which the Dublin-resident Phillips was himself so well qualified to perform. Even Larcom admitted that Phillips had been 'ill-used'.[111] It is small wonder that Phillips wrote to Adam Sedgwick in Cambridge complaining that the Irish geological survey was 'a mass of intrigue and personal contests'.[112] He took an early opportunity of shaking the Irish mud from his feet; after only one year in office, he resigned his chair in March 1845 in order to return to full-time employment with the Survey in England. Colby was left to devise his third set of plans for the geological conquest of Ireland. It was to be a very brief campaign — a campaign which ended not in victory but in Colby's total surrender. In July 1844 Colby thought he had won a war; in reality he had won nothing more than a battle.

Griffith creates a diversion

The struggle for control of the geological survey of Ireland was essentially a two-cornered affair involving Englishmen, but all the while, hovering in the background, there was a third and Irish figure who had nursed ambitions in Irish geology for far longer than either Colby or De La Beche. That figure was of course Richard Griffith. Modern readers of all the reports and correspondence relating to the geological surveying of Ireland, and dating from the period 1839 to 1845, could be forgiven for temporarily forgetting that since 1839 there had existed a geological map of the entire country on the quarter-inch scale. In reality Ireland was far less of a *terra incognita* than either Colby or De La Beche cared to admit. As we have seen, from 1825 onwards Griffith must have viewed the Ordnance Survey's geological activities with both interest and pain, because any official map was bound to supersede his own beloved work. At last, in the spring of 1844, with the whole future of Ordnance Survey geology back in the melting-pot, Griffith decided that the moment was apposite for him to enter the arena in order to see what he could salvage for himself. In a two-cornered and equally matched struggle such as that in progress between Colby and De La Beche was there not every possibility that a third party arriving late upon the scene might easily be allowed to carry off the prize as a kind of face-saving compromise? It was the type of situation which Griffith relished because, as we have already discovered, he was a master at turning events towards his own ends. Therefore, on 25 March 1844, he penned a long letter to the Chancellor of the Exchequer (Edward Goulburn) drawing attention to his own geological activities over a period of more than thirty years and implying that any new geological survey of Ireland was a totally unnecessary squandering of public money.[113]

In the letter he emphasised that his geological work had always been characterised by the closest attention to detail — that wherever possible his survey was based upon six-inch field-mapping — and he continued:

> I have now nearly completed a geological map of Ireland on a large scale which I hope to be able to publish in about a year.

From the context of the letter it would seem that he was here referring to the publication of the geologically-coloured county index maps (mostly on the half-inch scale) which Patrick Ganly had been constructing, but sadly this claim must be dismissed as another of those optimistic boasts to which Griffith was so prone. Certainly he never published any map of Ireland upon a scale larger than the quarter-inch. He continued his letter:

> I may say that the field work of the Geology of Ireland is now nearly completed, and little remains to be done beyond the marking in of small igneous protrusions in certain localities which hitherto I have not had leisure to complete. I have had much assistance in tracing the boundaries of the rock districts of Ireland, from the persons employed under my direction on the boundary department of the Ordnance Survey, & also on the Valuation. Having given them sufficient instruction in Geology, several of the boundary surveyors, & the valuators devoted their evenings (after the official business of the day had terminated) in tracing rock boundaries & in taking the dips of the strata. By means of such assistance and by my own labour within the last thirty years, I have accumulated a mass of accurate information which could not have been collected by a public establishment without much delay & a very considerable expenditure.

Griffith now proposed to lay before the government all this cartographic information together with his collection of fossils as arranged by Frederick McCoy, and he clearly believed that once the authorities saw the scope and quality of his work, then they must immediately dismiss from their minds all thought of the necessity for any new survey. He offered to allow somebody else to prepare all the work for publication should that be the official wish, but very naturally he emphasised that he would himself like to be afforded the opportunity of undertaking the task. In support of his own candidature he offered the following persuasive argument which, when stripped of its icing of mock generosity, amounts to Griffith saying 'Only I could perform the work adequately':

> I should naturally wish to have an opportunity of putting the whole together myself as of course a vast quantity of the information I possess which will be requisite in comparing the rocks and strata in one part of the Country with another is in my head & cannot be communicated until it has been committed to paper in a regular manner but still if it be wished by the Government to transfer the work to other hands, I shall not hesitate to hand over all my Geological maps & documents to the person who may be named. But in this case I hope that means may be taken that I shall get credit for the data contributed by me.

He concluded his letter with the following words which he must have hoped would ring pleasantly in the Chancellor's ear:

> What we require here is a good fossilist, & if such were appointed I think I could undertake the Geological department and complete it at a comparatively trifling cost to the public.

Griffith was really in an impossibly difficult position. He was clearly eager to conduct the new geological survey himself — and let it be said there was nobody better qualified for the task — but an effective presentation of his case was difficult without divulging the full extent to which he had committed his boundary and valuation surveyors to the investigation of purely geological problems. He had been running an unofficial geological survey ever since 1830, but so discreet had he been that his activities had passed unnoticed, while the Ordnance Survey's involvement with geology had twice been dragged out into the public gaze and axed on both occasions on the ground that geology was interfering with the Survey's more legitimate duties. Now, in 1844, he needed to draw attention to the enormous volume of geological information at his disposal without at the same time raising embarrassing questions about the precise relevance of geology to land valuation or about the quite remarkable acumen his men had displayed in making so many geological discoveries, supposedly exclusively during their off-duty hours. Once such matters were brought into the open, it could only be a matter of time before the

Valuation Office felt the Chancellor's pruning knife putting an end to Griffith's continuing geological studies. Equally, he had to avoid all suspicion of a misappropriation of public funds for geological purposes because clearly such suspicions must debar him from being entrusted with the new survey. Griffith had to sail as best he could between the Scylla of ill-advised revelation and the Charybdis of undue modesty.

Griffith's appearance upon the scene took Colby by surprise. He must have heard of Griffith's intentions before that letter was written to the Chancellor of the Exchequer on 25 March 1844 because on 23 March Colby wrote to Larcom:

Griffith certainly is an unexpected element in the game — but it will do us no harm.[114]

When the details of Griffith's letter became available Colby, Larcom and James all realised that they had to take Griffith's proposal far more seriously than they had initially thought necessary. Hitherto they had regarded Griffith as the author of a very satisfactory quarter-inch geological map of Ireland; they thought he had produced a good outline of Irish geology just as Greenough had in his map produced a good outline of the geology of England and Wales. But Colby had believed that, because of the limitations imposed by the quarter-inch scale, Griffith's map could be dismissed as little more than a cursory first approximation in Irish geology. He believed that his own men, working at the six-inch scale, would be refining all Griffith's lines and filling in a wealth of detail which Griffith, perforce, would have overlooked. This is what De La Beche was doing in England and Wales; he was employing a one-inch survey to correct and amplify what Greenough had recorded upon the scale of one inch to six miles. Now, however, in his letter to the Chancellor, Griffith was maintaining that far from being his basic map, his six quarter-inch sheets comprising the map of Ireland were derived documents compiled by a generalisation of detailed field surveys conducted in many cases upon the six-inch scale. Griffith was thus claiming that he and his surveyors had been working upon a far larger scale than even De La Beche was employing in England and Wales. Clearly if Griffith was telling the truth — and we now know that he was — then Colby's project for a new geological survey of Ireland was indeed rendered superfluous. The Colby camp was thrown into a flutter as an attempt was made to discover whether Griffith really did have in his possession a large number of geological maps carrying far more detail than that depicted upon the quarter-inch map of 1839.[115]

What Griffith needed at this juncture was geological friends — friends who would support his claims and verify the accuracy of his detailed surveys. But friends were what Griffith lacked. As we saw earlier (p. 81), he seems to have antagonised many of those Irish geologists who might now have been expected to rally to his support. Even the Geological Society of Dublin, of which he had long been a prominent member, seems to have turned against him; on 15 May 1844 the society's council drew up, for submission to the Lords of the Treasury, a memorandum urging the extension into Ireland of the geological work of the Ordnance Survey.[116] It may be significant that the society's president at the time was Charles William Hamilton with whom Griffith had recently clashed over their differing interpretations of the structure of the Killarney region. Griffith's sole supporter in Dublin seems to have been the ingenuous and but recently-arrived Phillips. He was amazed at the high quality of Griffith's work. He wrote to Sedgwick observing that Griffith had shown him Ordnance Survey six-inch and county index sheets 'coloured with a minuteness that impressed me exceedingly'. From Dublin Phillips made excursions expressly to check the accuracy of Griffith's work and he reported to Sedgwick as follows:

Upon the whole it appeared to me that with Griffith's maps in hand, & Griffith's hearty concurrence, the business of surveying for a new map would be easy and rapid: rather a re-examination & inspection, than an entirely new survey such as we make in England and Wales. . . . I think that to throw away and discredit so noble a map, & and such admirable sections, and to discard the advantages arising from the collections of fossils made by Griffith would be an unaccountable & lamentable error.[117]

Phillips, let it be remembered, was a geological cartographer of high competence thoroughly versed in the techniques of De La Beche's school of geological surveying. Griffith could hardly have found a better qualified adjudicator, yet even Phillips had to admit that in Ireland Griffith's work was dismissed as neither credible nor acceptable.[118]

In pursuit of his claim to be allowed to head the new geological survey, Griffith in April 1844 took a copy of his quarter-inch map of Ireland over to London to lay it before the Chancellor. As a result, the Chancellor referred Griffith's proposals to the Master-General and Board of Ordnance, and they, ironically at De La Beche's suggestion, invited Griffith to lay all his work before a committee of assessment just as in 1835 De La Beche himself had laid his maps of the West Country before a committee prior to the formal establishment of his Geological Survey of England and Wales.[119] In Griffith's case the committee consisted of Adam Sedgwick from Cambridge, William Buckland from Oxford, Henry Warburton the president of the Geological Society of London, and C.W. Hamilton the president of the Geological Society of Dublin. For the committee's consideration Griffith sent over a copy of his quarter-inch map of Ireland, some of his geologically coloured six-inch sheets, some geological sections, and a selection of the geologically coloured county index maps.

The committee convened during the 1844 meeting of the British Association held in York between 26 September and 2 October. Buckland was absent because of a family bereavement and a last minute effort to secure Roderick Murchison as a substitute failed because he was travelling on the continent. Griffith, De La Beche, and James were all present in York to keep a watchful eye upon their respective interests, and in the privacy of quiet corners in meeting rooms De La Beche and James were doubtless ready to offer forthright comments upon the feasibility of Griffith's proposals. The committee held two meetings and by the evening of 27 September the decision had been made.[120] Sedgwick voted in favour of Griffith's proposal, but Hamilton and Warburton both recommended that the new geological survey of Ireland should be the responsibility of the Ordnance Survey. Hamilton later explained to a meeting of the Geological Society of Dublin (Griffith was in the audience) that he had voted against Griffith's proposals because they would have deprived Ireland of the 'inestimable value' of a fresh and more detailed survey.[121] In any case, as Griffith must have known, the committee's deliberations at York were something of a farce because two months previously the Treasury had authorised a new geological survey of Ireland and placed its administration squarely into the hands of Colby and James. Griffith had to bow off as best he could but before leaving him we must indulge in a fascinating historical speculation. Suppose back in 1825 Griffith had found himself running Colby's original geological survey of Ireland rather than at the head of the Boundary Survey, then might the later Geological Survey of England and Wales have been an offspring of the Geological Survey of Ireland? Might Griffith rather than De La Beche be lauded today as the founder of the Geological Survey of Great Britain and Ireland?

Colby quits the field

Colby considered himself to have defeated both De La Beche and Griffith. Now, during the summer of 1844, he and James set about the organisation of their new geological survey. It was very like 1826 all over again because we find James purchasing books and collections of specimens for the use of his men during their training. So far as field operations were concerned there was initially some uncertainty as to where a start should be made. At one time Fermanagh and Leitrim were the favoured counties, partly because of the presence there of the Connaught coalfield, and partly because of the hope that it would prove possible to extend the geological lines already plotted by Portlock in the neighbouring territory to the northeast.[122] James himself was happy enough with this suggestion and he proposed to make a start on the shores of Lough Allen somewhere near the Arigna coalmines, although he later changed his mind and decided that it would be more convenient for the survey to start not in Connaught but in County Dublin.[123] Colby had yet other ideas; he was determined that James should begin in County Donegal. His reasons were simple enough. Firstly, he insisted that it was logical to start at one of Ireland's extremities. Secondly, he maintained that contours were essential for successful geological mapping and, while such lines had not hitherto been a feature of the six-inch map, they were in the autumn of 1844 being surveyed in County Donegal. Finally, the Survey's budget for 1844-1845 had not included any money to cover the cost of bringing Forbes over to Ireland, and Colby therefore thought it prudent to commence operations in a county where unfossiliferous rocks predominate.[124] It was naturally Colby's wish that prevailed, and on 26 November 1844 James was ordered to proceed to County Donegal taking with him six Sappers and Miners to be trained as geologists.[125] Little though they accomplished, surely no-one will begrudge those six men a brief individual mention on the pages of history; their names were Henry Brown, Joseph Longland, Richard J. Loveday, James A. Lyons, Daniel McInnes, and John Shearer.[126] The returns for December 1844 and January 1845 show James then to have been based at Lifford and Strabane (he took his Christmas dinner for 1844 with his brother officers in their mess at Londonderry) and during the first two months of 1845 his party was at work mapping the geology of the County Donegal six-inch sheets numbers 61, 62, 70, 71 and 79.[127] James's idea was to map the lowlands during the winter and the highlands during the summer but, lowlands or highlands, there was no escaping the fact that they were at grips with the Irish Dalradians — some of the most complex of Ireland's rocks. That Colby and James sent untrained sappers off to do battle with such strata is revealing either of the character of the map to which they aspired or of the superficiality of their geological understanding. If all they hoped for was a map of the petrographic type, showing an outcrop of granite here, of schist there, and of limestone yonder, then presumably this was within the capabilities of their unskilled team. Such a map, however, must have been viewed as anachronistic by the geological world at large. If, on the other hand, they intended to produce a true geological map of the type which had become standard by the 1840s — a map containing in-built historical and structural interpretations — then we can only express amazement that Colby and James should have been so naive as to imagine that sappers would be capable of such sophisticated interpretative studies. In his address to the Geological Society of Dublin in February 1845, C.W. Hamilton highlighted the danger of expecting sappers to produce high-quality geological research:

> ... it was treating Geology as if it were a work which could be done by the day or by the piece; it was throwing experience, and that tact which experience gives, out of the question, and acting upon the supposition that a clinometer could be put into the hands of a sapper, to be used as mechanically as a theodolite, wholly forgetful of this, that the theodolite itself reveals the errors of the observer, while in geological observations there is no such proof. The report of the sapper was to be taken as authority upon the most abstruse instances of relation; and the head, under whose superintendence all such observations were to be reduced and regulated, was liable, however zealous, to be led into error, without any means of correcting it.[128]

From the surviving records it is impossible to discover either the type of map that Colby and James hoped to produce or the manner in which the sappers were to perform their task. Indeed, during the summer and autumn of 1844 the whole picture of the Ordnance Survey's geological mapping of Ireland seems to have been clouded with confusion.

Perhaps from the geological standpoint Colby lacked that attribute which Napoleon regarded as an essential ingredient in the character of any successful general — luck. Certainly now for the third time in less than twenty years things began to go wrong in Colby's Irish geological domain. Firstly, there was James. Events since July 1844 had raised serious doubts in

Colby's mind as to James's ability to administer the geological department. Having secured the post to which he had aspired, James was now spending too much of his time not in the making of geological maps, but in being self-important, in public posturing, and in preening his new dignity. On 20 September 1844 Colby wrote to Larcom:

> I do not anticipate any difficulties about the Irish geology, if we can really get our friend James to give his earnest attention to it in a business like manner.[129]

By 19 November things had got worse and Colby wrote to Larcom that James 'is not supporting an Ordnance Geology by any strenuous efforts'.[130] Secondly, there was Kane. He was supposed to be assisting the new survey by conducting its chemical analyses but we find Colby complaining that Kane was less interested in a geological survey of Ireland than in some kind of Irish resource survey in extension of the studies he had just published in his now classic *Industrial resources of Ireland*.[131] Thirdly, there was finance. Colby was aggrieved at the Treasury's failure to make any financial provision for the new survey, and during the autumn of 1844 he waited upon the Master-General to discuss yet again the whole future of Ordnance Survey geology in Ireland. Colby's Irish geological empire was coming under renewed attack; Larcom was on the alert. He wrote to a fellow officer on 25 November 1844:

> Be assured I will keep and am keeping every engine in motion that may prevent my favourite scheme of a military geology being defeated.[132]

The story recounted in this chapter is about to reach its dénouement but it has to be a climax devoid of high drama. The final fate of the Ordnance Survey's involvement in Irish geology was decided without any internal inquiry such as that of 1828, without any public collection of evidence such as that undertaken by the commission of 1843, and without any adjudicating body such as that summoned to York in 1844. All was now to be decided in private and behind closed doors — the doors in all probability being those of Drayton Manor in Staffordshire, the family seat of the Prime Minister, Sir Robert Peel. That the matter was finally resolved with so little public ado — and resolved almost entirely to his own satisfaction — reflects both the determined self-interest and the powers of intrigue possessed by one man: De La Beche. De La Beche had refused to accept his defeat of July 1844 as final and clearly he had been lobbying his influential friends in an effort to ensure that the entire question of the Ordnance Survey's involvement in geology was re-opened at the highest government level. By the autumn of 1844 news of these activities had reached Colby's ears and relations between Colby and De La Beche now became decidedly strained. De La Beche played upon the fact that the Master-General did not really approve of officers such as James being seconded from their corps for special scientific duties[133], but this was now a situation where allies were more important than the actual weapons of debate. And De La Beche had secured for himself the most powerful ally possible — Sir Robert Peel. That remarkable figure was no stranger to the world of science. For several years he had been intimately associated with the scientific circle centred upon the Geological Society of London; scientists were regular members of house parties of Drayton Manor; geologists such as De La Beche, Richard Owen, Roderick Murchison and William Buckland were preferred for public honours during Peel's ministry of 1841-1846; and in 1848 there was to be a move to elect Peel to the presidency of the Royal Society. Peel once confessed to some distaste of De La Beche's 'roundabout way of doing a job' in order to secure his ends[134], but the two men do nevertheless seem to have become friends. This relationship was now to prove decisive. It allowed De La Beche to broach with the Prime Minister the vexed question of the geological survey of Ireland, and further than this it allowed De La Beche the opportunity of urging Peel's thoughts along a particular path. Indeed, when late in 1844 Peel crystallised his thinking into an official memorandum, his arguments were so nicely orientated towards De La Beche's objective of an independent and civilian survey that it is tempting to imagine De La Beche as actually having been present in Peel's study as the document was prepared, peering over the Prime Minister's shoulder as he wrote, offering a felicitous suggestion here and a mental nudge there.[135]

In his memorandum Peel observed firstly that the union of the topographical and geological surveys had resulted in unnecessary inconvenience, and he therefore wondered whether it might not be advisable to release the Ordnance Survey from all its geological obligations so that a completely fresh civilian department could be established and charged with responsibility for the investigation of geology throughout the United Kingdom. Secondly, he recollected that there was in Britain one government-financed geological enterprise which was already quite outside the jurisdiction of the Ordnance Survey — namely the Museum of Economic Geology which since 1835 had stood in Craig's Court, London. The museum was under De La Beche's direction, and it housed the collections made by the Ordnance Survey geologists, yet by a strange accident of history it was not under the ultimate control of the Master-General and Board of Ordnance, but rather under the aegis of the First Commissioner of Her Majesty's Woods, Forests, Land Revenues, Works and Building. Since the First Commissioner already had the museum under his wing, Peel mused, why should he not also assume responsibility for a refurbished geological survey of the United Kingdom? This idea became the burden of a Treasury minute drawn up on 27 December 1844[136]; the Ordnance Survey would lose its responsibility for geology, a new geological survey of the United Kingdom would be established, and the new survey would be answerable to the First Commissioner. The proposals were put to the First Commissioner (Lord Lincoln, later the Duke of Newcastle) on 1 January 1845 and they met with his instant approval. Indeed, Lord Lincoln seems to have been almost enthusiastic about the idea of enlarging his already extensive empire because on 13 January he completed a comprehensive memoir explaining exactly how the proposed new department might operate.[137] The salient features of this important memorandum are as follows:

1. Both the existing Geological Survey of England and Wales and the new survey in Ireland should be transferred from the Ordnance Survey to the office of the First Commissioner, where both surveys would henceforth be under the immediate control of De La Beche as General Director.

2. De La Beche should divide his time between the British and Irish surveys but, to assist him in the performance of his duties, there should be appointed a Local Director on either side of the Irish Sea. The Local Director in England and Wales should be Andrew Crombie Ramsay (he had joined De La Beche's staff in 1841) and the Local Director for Ireland should be James who would be seconded from the Royal Engineers in order to fill the office.

3. An annual grant of £1500 should be made available for the purposes of the Irish survey, it being understood that the entire island would be surveyed within a period of about ten years.

4. A Museum of Economic Geology should be set up in Dublin with Robert Kane as its Director, his annual salary to be £300. The museum should be responsible for conducting all the survey's chemical analyses, and in return it should become the repository for all the Ordnance Survey's museum material in store at Mountjoy and for all future specimens collected by the new survey. The Treasury had intimated its willingness to finance the erection of a new building for the museum, but the First Commissioner suggested that it would prove more economic to purchase an existing house and then convert it to museum use.

The Treasury accepted all these proposals in a letter to Lord Lincoln dated 31 January 1845[138] and only one small amendment was later introduced, this being to clarify Kane's status; it was decided that he should be responsible not to De La Beche, but directly to the First Commissioner himself.

While all these negotiations were in train Colby and his confrères remained strangely silent. Perhaps De La Beche had succeeded in keeping some of his intrigues secret; there is certainly evidence that Colby himself remained blissfully unaware of what was afoot until as late as February 1845.[139] By then it was too late for Colby to resist; his position was overrun, the battle was over, and the war was lost. Exactly why Colby was taken at such a disadvantage can only be a matter for speculation. Perhaps he had grown tired of the entire issue; maybe he felt let down by James; or possibly it was just the prospect of his impending retirement from the Ordnance Survey upon his forthcoming promotion to the rank of Major-General.[140] But for whatever reason he now became a passive and helpless spectator as his former empire was partitioned and as the geological territories were handed over to De La Beche. The partition took place on 1 April 1845, and on that date the Geological Survey of Ireland came into being as a component of the new Geological Survey of Great Britain and Ireland, the whole being under the control of the First Commissioner of Her Majesty's Woods, Forests, Land Revenues, Works and Buildings.[141] Ordnance Survey geology was dead and James had to disband that little force of six would-be sapper-geologists which had made so fleeting an appearance upon the Irish geological scene. For De La Beche it must have been a happy day. He had managed to secure virtually all that he had been playing for during the negotiations of the previous year. The only fly in his ointment was James; De La Beche would clearly have preferred to see Phillips sitting in the Local Director's chair in Dublin, but this was a matter of small moment for a man who had succeeded in severing the military link to place himself at the head of a completely independent and much expanded geological survey. For Colby, on the other hand, 1 April 1845 was the day when he finally saw dashed his long-cherished plans for a military geological survey of Ireland. The Ordnance Survey's involvement with Irish geological cartography had begun with such high hopes but now it ended in dismal failure with nothing to show for twenty years of work but those two maps by Portlock. When Colby had initiated the geological survey in 1824 it had been a true pioneering concept and, had all gone well, then Ireland must have been among the first of the world's nations to possess a national coverage of official geological maps. But all had not gone well, and the ten years' lead which the Irish survey had over its English counterpart had merely been frittered away. By the spring of 1845 De La Beche in England was in command of an efficient organisation with an excellent publication record, and he had gathered around him such figures as W.T. Aveline, H.W. Bristow, Edward Forbes, John Phillips, and A.C. Ramsay, all of them geologists of high repute. Ireland in contrast presented a sorry picture. James still had to find a civilian staff for his new survey and even his own geological qualifications were somewhat suspect. Ireland had, nevertheless, at last obtained a specialist geological survey, and from 1 April 1845 down to the present day that survey has had a continuous history of steady if uneven achievement. If it is any consolation to the Irish, they had secured for themselves a geological survey twenty-two years before the same privilege was accorded to their cousins the Scots.

REFERENCES AND NOTES FOR CHAPTER FOUR

1. DNB; Joseph Ellison Portlock, *Memoir of the life of Major-General Colby*, London, 1869, pp. xii + 316.
2. Portlock, *op. cit.*, 1869, 122, (ref. 1). See also OSL, Colby to Bryce 9 April 1832 for Colby's views on the publication of geological maps in Britain.
3. John Harwood Andrews, *A paper landscape: the Ordnance Survey in nineteenth-century Ireland*, Oxford, 1975, 144-179. See also Charles Frederick Arden-Close, *The early years of the Ordnance Survey*, Newton Abbot, 1969 (reprinted), *passim*.
4. Lieut Col. Colby's report to Sir Henry Hardinge K.C.B. on the present state and progress of the Irish Survey, a MS. document of 1826 in the Ordnance Survey, Dublin.
5. *Report from the Select Committee on the survey and valuation of Ireland*, H.C. 1824 (445), VIII, 51.
6. *Ibid*. Strangely, the sheet of symbols is nowhere referred to in either the report or the proceed-

ings of the Select Committee.
7. OSLR (0) 1031.
8. Portlock, *op. cit.*, 1869, 127, (ref. 1).
9. OSLR (0) 1727, 31 July 1828.
10. *Report on the Ordnance Survey of Ireland by Sir James Carmichael Smyth, with appendices, 1828*, Pringle to Colby 5 August 1828. Copies of this MS. report exist in the GSI and in the PROL (W.O. 44/115). There is also a microfilm copy in NLI (no. 6399).
11. OSLR (I) 5770.
12. *Report on the Ordnance Survey, op. cit.*, 1828, memorandum from Pringle 5 August 1828, (ref. 10).
13. The sole known surviving copy of the *Directions* is in the archives of GSI. I am grateful to Dr Jean Archer for recognising the significance of the document and for drawing it to my notice.
14. *Proc. Geol. Soc.* 1, 1826-33, 51-52.
15. *Report on the Ordnance Survey, op. cit.*, 1828, (ref. 10). Dr Jean Archer has pointed out to me that some of the books purchased still exist in the library of GSI and may be identified by their 'Ordnance Survey of Ireland' library stamps. The volumes include Kirwan's *Elements of mineralogy* (1794); Jameson's *Treatise on the external, chemical and physical characters of minerals* (1817) and *Manual of mineralogy* (1821); Conybeare and Phillips's *Outlines of the geology of England & Wales* (1822); and Macculloch's *Description of the Western Islands of Scotland* (1819). Many of these volumes are still in a near mint condition and they can therefore have received little study in the Ordnance Survey library.
16. OSLR (0) 1433, 13 November 1827.
17. *Report on the Ordnance Survey, op. cit.*, 1828, (ref. 10).
18. Copies of the panoramas survive in the Dublin offices of both the Ordnance and Geological surveys.
19. *Trans. Geol. Soc. Lond.* 3, 1816, plate 10.
20. For example, OSLR (0) 1409, 27 October 1827; 1446, 29 November 1827.
21. OSLR (0) 1651, 29 May 1828.
22. OSLR (0) 1603.
23. *Report on the Ordnance Survey, op. cit.*, 1828, (ref. 10); Andrews, *op. cit.*, 1975, chap. 2, (ref. 3).
24. OSLR (I) 7133.
25. *Report on the Ordnance Survey, op. cit.*, 1828, (ref. 10).
26. OSLR (0) 1808.
27. OSLR (I) 7209.
28. V.A. Eyles, John Macculloch, F.R.S., and his geological map: an account of the first geological survey of Scotland, *Ann. Sci.* 2, 1937, 114-129; D. Flinn, John Macculloch, M.D., F.R.S., and his geological map of Scotland: his years in the Ordnance. 1795-1826, *Notes Rec. R. Soc. Lond.* 36, 1981, 83-101; D.A. Cumming, Geological maps in preparation: John Macculloch on western islands, *Archs. Nat. Hist.* 10, 1981, 255-271; W.B. Hendrickson, Nineteenth-century state geological surveys: early government support of science, *Isis* 52, 1961, 357-371.
29. V.A. Eyles, The first national geological survey, *Geol. Mag.* 87, 1950, 373-382.
30. OSLR (0) 2449, 19 November 1829; 2502, 1 January 1830.
31. OSLR (0) 2528.
32. DNB; *Papers on subjects connected with the duties of the corps of Royal Engineers*, N.S. 13, 1864, ix-xxix; *The Times*, 16 February 1864, 12, and 17 February 1864, 12; LP 7515 contains a great deal relating to Portlock, especially a series of reminiscences by G.V. Du Noyer.
33. OSLR (I) 1565, 5 January 1826.
34. RGSIP, *Papers read before the Geological Society of Dublin 1831-1859*.
35. Joseph Ellison Portlock, *Report on the geology of the County of Londonderry, and parts of Tyrone and Fermanagh*, Dublin, 1843, iv.
36. The existence of this document was drawn to my attention by Norman Butcher who kindly allowed me to use the copy in his own collection.
37. The only copy of the document now known is in LP 7550.
38. *Proc. Geol. Soc.* 1, 1826-33, 380.
39. *Ibid.*, 447.
40. *Trans. Geol. Soc. Lond.*, second series, 5, 1840, 195-206.
41. *Ibid.*, 125-126.
42. OSL.
43. *Ibid.*
44. *Ibid.*, Board Minute, 15 June 1835.
45. Thomas Frederick Colby, *Ordnance Survey of the County of Londonderry*, Dublin, 1835, no pagination.
46. C. Babbage, Remarks on the statistics contained in the Ordnance Survey of the parish of Templemore, *Rep. Br. Ass. Advmt. Sci.*, Dublin, 1835, part 2, 118-119.
47. *Report of the commissioners appointed to inquire into the facts relating to the Ordnance memoir of Ireland*, H.C. 1844 (527), xxx, 19 and 71.
48. Thomas Frederick Colby, *Ordnance Survey of the County of Londonderry. Volume the first. Memoir of the city and north western liberties of Londonderry. Parish of Templemore*, Dublin, 1837, pp. 11 + 336 + 16.
49. F.J. North, Further chapters in the history of geology in South Wales, *Trans. Cardiff Nat. Soc.* 67, 1934, 31-103; Geology's debt to Henry Thomas De La Beche, *Endeavour* 3, 1944, 15-19; John Smith Flett, *The first hundred years of the Geological Survey of Great Britain*, London, 1937, pp. 280; Edward

Battersby Bailey, *Geological Survey of Great Britain*, London, 1952, pp. xii + 278.
50. Thomas Frederick Colby, *An address delivered at the sixth annual meeting of the Geological Society of Dublin*, Dublin, 1837, 10.
51. H.J. Seymour, The centenary of the first geological survey made in Ireland, *Econ. Proc. R. Dubl. Soc.* 3 (17), 1944, 227-248.
52. MMR, May 1838.
53. NLI MS 1441 is a field-diary of Du Noyer's covering the period 8 July 1839 to 9 November 1839.
54. OSLR (0) 7281.
55. J.B. Harley, The Ordnance Survey and the origins of official geological mapping in Devon and Cornwall, pp. 105-123 in Kenneth John Gregory and William Lionel Desmond Ravenhill (editors), *Exeter essays in geography in honour of Arthur Davies*, Exeter, 1971, pp. xviii + 258.
56. *Estimates of the office of Ordnance... for the year 1840-41*, H.C. 1840 (66), xxx, pp. 33.
57. MMR, January 1840.
58. OSLR (0) 9032, 9033, 10 February 1840; LP 7555, Colby to Larcom 22 February 1840.
59. LP 7555, Colby to Larcom 10 February 1840.
60. PROL W.O. 44/703. There is a microfilm of this set of documents in NLI (no. 6400).
61. MMR, February and September 1840.
62. *Report of the commissioners, op. cit.*, 1844, 72-73, (ref. 47). See also C.E. O'Riordan, Some notes on the dispersal of the Ordnance Survey of Ireland collections, *G.C.G.: the Geological Curator* 3 (2 & 3), 1981, 126-129.
63. Colby, *op cit.*, 1837, (ref. 50).
64. LP 7555.
65. Geological Survey of Northern Ireland, *Geology of Belfast and district: special engineering geology sheet, solid and drift*, Belfast, 1971, 1:21, 120.
66. LP 7555, Colby to Larcom 10 February 1840.
67. *Report of the commissioners, op. cit.*, 1844, 7, (ref. 47).
68. MMR, April 1840.
69. A great deal of material relating to Portlock's work between 1840 and 1843 is in LP 7555, and in PROL W.O. 44/703 (also NLI microfilm no. 6400).
70. LP 7555.
71. LP 7555, Colby to Larcom 2 August 1842.
72. PROL W.O. 44/703 (also NLI microfilm no. 6400).
73. J.E. Portlock, On the geology of Corfu, *Rep. Br. Ass. Advmt. Sci.*, Cork, 1843, part 2, 57. See also Some remarks on the white limestone of Corfu and Vido, *Q. Jl. Geol. Soc. Lond.* 1, 1845, 87-90.
74. MMR. January 1843.
75. Joseph Ellison Portlock, *Report on the geology of the County of Londonderry, and of parts of Tyrone and Fermanagh*, Dublin, 1843, pp. xxxi + 784 + 54 plates.
76. Henry Thomas De La Beche, *Report on the geology of Cornwall, Devon, and west Somerset*, London, 1839, pp. xxviii + 648. 1509 copies of this volume were printed and they were sold at 9s. 4d. each.
77. *Report of the commissioners, op. cit.*, 1844, viii, (ref. 47).
78. There is a great deal of correspondence related to this period in LP 7556.
79. DNB; *Proc. Geol. Soc.*, 1877-78, 34-35.
80. S. Patrickson, Description of a limestone district on the N.E. shore of Carlingford Bay, and of littoral deposits of shells and limestone, *J. Geol. Soc. Dubl.* 1(3), 1837, 180-182.
81. DLBP contain various letters relating to James's work in England. See also IGS 1/12, Colby to De La Beche 10 August 1842.
82. PROL W.O. 44/703 (also NLI microfilm no. 6400); IGS 1/12.
83. *Report of the commissioners, op. cit.*, 1844, 70, (ref. 47). PROL W.O. 44/703 (also NLI microfilm no. 6400).
84. *Report of the commissioners, op. cit.*, 1844, 17-18, 69-70, (ref. 47); LP 7553; PROL W.O. 44/703 (also NLI microfilm no. 6400).
85. *A collection of documents expressive of public opinion on the utility and importance of the Ordnance Memoir of Ireland*, Dublin, 1844, pp. 74. A copy is in LP 7554.
86. *Report of the commissioners, op. cit.*, 1844, viii, (ref. 47).
87. There is much material relating to the James versus Phillips affair in LP 7555, 7556, and 7557.
88. LP 7556. See also the introduction to LP 7557.
89. DLBP.
90. Paul Joseph McCartney, *Henry De La Beche: observations on an observer*, Cardiff, 1977, 4.
91. OSL, Colby to De La Beche 3 April 1837; De La Beche to Colby 8 April 1837; McCartney, *op. cit.*, 1977, 30-31, (ref. 90).
92. The Geological Survey (1835), the Museum of Economic Geology (1837), the Mining Records Office (1839), and the School of Mines (1851).
93. Caroline Fox, *Memories of old friends*, London, 1882, vol. 1, 5.
94. Quoted by J.B. Morrell in *British Journal for the History of Science* 9(2), 1976, 141-142.
95. RP 1/16, 14, 14 January 1851.
96. *Ibid.*, Jukes to Ramsay 22 November 1858.
97. RGSIP, *Papers read before the Geological Society of Dublin 1831-1859*.
98. G.L. [Herries] Davies, The University of Dublin and two pioneers of English geology: William Smith and John Phillips, *Hermathena* no. 109, 1969, 24-36.
99. The same words appear in *The Freeman's Journal* for 18 March 1844.
100. LP 7555.

101. LP 7555 contains various letters on this subject.
102. DLBP.
103. *Ibid.*
104. LP 7556, Colby to James 16 April 1844; Colby to Larcom 16 May 1844.
105. DLBP.
106. *Ibid.*
107. *Ibid.*, Phillips to De La Beche 24 July 1844.
108. LP 7556.
109. *Ibid.*, Colby to Larcom 14 August and 20 September 1844.
110. DLBP. Phillips to De La Beche 24 August 1844.
111. LP 7557, Larcom to Colby 10 February 1845.
112. SP, Phillips to Sedgwick 24 October 1844, IE. 97b.
113. VOLB, 16 December 1843-27 May 1845; PROL, W.O. 44/703 (also NLI microfilm no. 6400). Cambridge University Add. MS. 7652.
114. LP 7556.
115. *Ibid.*, James to Larcom 5 April and 23 April 1844; Colby to Larcom 25 April 1844.
116. RGSIP, *Minutes of meetings of council 1831-1863.*
117. SP, Phillips to Sedgwick 29 October 1844, IE 97c.
118. *Ibid.*
119. PROL, W.O. 44/703 (also NLI microfilm no. 6400).
120. Letters relating to the deliberations at York are in LP 7556. See also a letter by De La Beche written on 9 May 1844 in PROL, W.O. 44/703 (also NLI microfilm no. 6400). Cambridge University Add. MS. 7652, Griffith to Sedgwick 29 August 1844.
121. *J. Geol. Soc. Dubl.* 3 (3), 1846, 177.
122. LP 7556, Larcom to Colby 10 October 1844.
123. *Ibid.*, Larcom to Colby 11 October 1844.
124. *Ibid.*, Colby to Larcom 16 October and 26 November 1844.
125. *Ibid.* and MS of Colby's report to the Inspector-General, 15 January 1845, in the Ordnance Survey, Dublin.
126. LP 7556, Colby to Larcom 21 September 1844; MMR, December 1844.
127. MMR, January and February 1845; PC, James to Phillips 26 December 1844.
128. *J. Geol. Soc. Dubl.* 3 (2), 1845, 99-100. For Colby's comments upon Hamilton's address see LP 7557, Colby to Larcom, 16 April 1845.
129. LP 7556.
130. *Ibid.*
131. Robert Kane, *The industrial resources of Ireland*, Dublin, 1844, pp. xii + 417. The second edition (Dublin 1845) contains a geological map.
132. LP 7556.
133. *Ibid.*, Robinson to Larcom 21 October 1844.
134. McCartney, *op. cit.*, 1977, 37, (ref. 90).
135. *Geological Survey (Ireland)*, H.C. 1845 (238), XLV, pp. 7.
136. *Ibid.*
137. *Ibid.*
138. *Ibid.*
139. LP 7557, James to Larcom 7 February 1845; Larcom to James 10 February 1845.
140. He became a Major-General on 9 November 1846.
141. 8 & 9 Victoriae cap. 63. The act did not receive the Royal Assent until 31 July 1845.

5

A NEW SURVEY BREAKS GROUND

1845-1850

The first steam locomotive ever to be built in Ireland entered regular service with the Dublin and Kingstown Railway on 9 April 1841. She was a handsome 2-2-2 sporting plenty of burnished brass and bearing the name *Princess* in honour of Queen Victoria's first-born. (A proposal to call her *The Irish Teetotaller* in honour of Father Mathew's temperance campaign was rejected.) The constructional cost of the *Princess* was £1,050. 14s. 5d., or just about two thirds of the annual cost of the Geological Survey of Ireland which was inaugurated in April 1845. Never does Sir Henry De La Beche seem to have been asked to justify this additional public expenditure occasioned by the extension of his geological survey into Ireland. From Sir Robert Peel downwards, it was evidently just taken for granted that such a survey was essential in any nation aspiring to economic progress. In March 1832, when he had urged the importance of a geological survey of Devon and Cornwall, De La Beche had argued that such a project:

> ... would be of great practical utility to the Agriculturalist, the Miner, and those concerned in projecting and improving the Roads, Canals, and such other public works, undertaken for the benefit and improvement of the Country.[1]

Had he in 1845 been invited to justify the creation of the new Irish Survey, then he would doubtless have employed the same utilitarian logic — logic which he could have employed to good effect where Ireland was concerned because of the increasingly impoverished nature of the Irish economy. Such arguments invoking the public interest, however, would have represented far less than the whole truth. Nothing must be allowed to obscure the fact that De La Beche had his own purely personal reasons for wishing to see established the Geological Survey of Ireland. It represented for him the fulfilment of a long-cherished ambition — it was the prize in a power struggle with Colonel Colby. De La Beche had succeeded in extending the bounds of his geological empire, thus enhancing his personal status, and he had secured the opportunity of widening his geological experience at the tax-payer's expense through the making of regular visits to his new Irish domain. But although De La Beche was as yet oblivious to the fact, 1845 was hardly an auspicious year in which to commission a new Irish undertaking. It was the year when there arrived in Ireland the microscopic spores of the fungus *Phytophthora infestans* — the dreaded potato blight. The disease had been for some time ravaging potato crops in North America; it made its appearance in England in August 1845; and by the following month its presence in Ireland was being reported from counties Dublin, Wicklow, Wexford and Waterford. That Christmas there was little jubilation in many an Irish heart because the blight had spread rapidly and more than half of the nation's potato crop had been destroyed. For a large proportion of the Irish — and there were now more than eight million of them — the potato was the staff of life, and starvation threatened throughout the length and breadth of the land. Ireland had experienced many an earlier famine, but the catastrophe which overtook the country in 1845, and then held the nation in its grip for more than three years, was disaster upon such a scale that it has ever since been known simply as 'The Great Famine'. Those newly-appointed geological surveyors who now began to wend their rock-tapping way through the Irish countryside must have become all too familiar with the stench of field after field of rotting tubers. Sometimes they must have had to lay down their hammers and collecting bags to stand bareheaded as a famine funeral passed upon its mournful way to the local graveyard. All that nevertheless lay some months distant when, on 5 February 1845, Captain Henry James wrote to the Inspector-General of Fortifications intimating his willingness to be seconded from the Royal Engineers in order to become the Local Director of the new Geological Survey of Ireland.[2]

A team is chosen

As the Irish Local Director, James was of course accountable to De La Beche for all the activities of the Geological Survey of Ireland, and on 22 May 1845 De La Beche completed a 22-page set of instructions for his two Local Directors – James in Ireland and Ramsay in England and Wales – reminding them of the type of observations that the staff were to make in the field.[3] It was not until April 1846, however, that the First Commissioner drew up a set of regulations spelling out James's specific responsibilities in any detail. De La Beche forwarded a copy of the First Commissioner's document to Dublin on 29 April 1846 and from it James learned that each year he was expected to spend the eight months between 10 April and 10 December in the field assisting his staff, inspecting their work, and carrying out his own personal mapping programme. During the four winter months he was to be at his post in the Dublin office digesting the previous field-season's work, although if he wished he was entitled to spend one of these four months in London taking the opportunity to keep in touch with developments both at Survey headquarters and in the geological world generally. His annual salary was £300, but to this there was added an allowance of £50 to cover travelling expenses incurred within a fifteen-mile radius of his field-headquarters, wherever that might be. All his other travelling expenses were to be the subject of individual claims.

On assuming office one of James's first tasks was to recruit a staff and it seems that there was no shortage of candidates. So far as field-surveyors were concerned, the normal appointment was as a Temporary Assistant Geologist earning seven shillings per day for a six-day week. After successfully completing a six-month probationary period, such men were promoted to the rank of Assistant Geologist, in which grade they could earn up to ten shillings per day. Two other minor features of the terms of appointment deserve mention. Firstly, an Assistant Geologist was forbidden to keep a private collection of either fossils or minerals (one remembers that Portlock had seemingly had difficulty with surveyors who used their official collecting for private gain) and, secondly, no member of the Survey staff was allowed to make unauthorised geological communications to either private individuals or public bodies. Survey knowledge was to remain strictly Survey property.

As his first field-surveyor James recruited a man who has already made a fleeting appearance in these pages – Richard Griffith's palaeontologist Frederick McCoy.[4] McCoy was the son of a noted Dublin physician who between 1849 and 1873 was the first professor of materia medica in the newly-founded Queen's College Galway. McCoy junior early developed a passion for natural history, and if the date widely quoted for his birth – 1823 – be correct, then it follows that he was only sixteen years of age when he joined the Geological Society of Dublin and when he began to give courses of public lectures in natural history. Such was his reputation that about 1840 he was invited to undertake the examination and arrangement of Griffith's rapidly growing fossil collection (see p. 66), and collaboration between the two men resulted in the synopses of the Irish Carboniferous and Silurian fossils which were published in the 1840s.[5] McCoy also worked upon various other Dublin collections, including those of the Royal Dublin Society and the Geological Society of Dublin[6], but in the case of the latter society he was evidently adjudged to have neglected his curatorial duties and his engagement was terminated in February 1842.[7] Whatever his failing may have been, there is no doubt but that by 1845 he was one of Ireland's most experienced palaeontologists. Initially James hoped to employ him as the Irish Survey's palaeontologist responsible solely for the examination of the fossils collected by the Irish staff, but De La Beche made it clear that Irish organic remains were to be the responsibility of the Survey's London-based palaeontologist and McCoy was therefore appointed to the Survey as an ordinary field-surveyor. In view of his palaeontological expertise, and despite his lack of any training as a field-geologist, he was appointed as a Temporary Assistant Geologist at the full rate of pay of ten shillings per day. He was offered his post on 22 May 1845 and he accepted two days later. His sturdy frame, reddish hair, florid complexion, and warm heart gave him many of the attributes of the stock Irishman, and some

Fig. 5.1. Henry James poses for the photographer soon after his receipt of the accolade in 1860. He was then a colonel and Director of the Ordnance Survey, and it was about this time (in 1863) that his services to cartography earned him appointment to the Spanish Order of Isabella the Catholic. Reproduced from the original photograph in the archives of the Geological Survey of Ireland, by kind permission of the Director.

might regard his blunt determination and pepperiness when thwarted as two further characteristics that he owed to the land of his birth. It is odd that McCoy, the first surveyor appointed to the staff, should have been destined to enjoy a scientific career far more distinguished than that vouchsafed to any of the men who followed him into the ranks of the Irish Survey. He and James were the only serving or former members of the Irish staff ever to receive the accolade of knighthood, and it is equally odd that these two future knights should both have departed from the Survey under circumstances which at the time seemed to augur ill for their future careers.

The second surveyor appointed was Walter L. Willson (died in 1878) who was offered a post as Temporary Assistant Geologist on 18 June 1845. He had previously worked for five years as a civilian employee of the Ordnance Survey, and he accepted James's offer on 23 June. The third surveyor was Lewis Edwards who arrived in Ireland on 8 September 1845 having been brought over from Britain to assist with the mapping of the Castlecomer and Slieveardagh coalfields. He was supposed to have had previous experience of this type of work and he certainly came well recommended, but he was soon found to be incompetent and he was dismissed as from 18 October 1845. The last field-surveyor to be appointed during the inaugural year was Andrew Wyley who before joining the Survey had taught for two years in the Rev. John Scott Porter's Classical, Commercial, and Mathematical School in College Square, Belfast. His name had been brought prominently before the new Survey because by chance one of his referees — a clergyman from Lisburn, County Antrim — had recently shared a coach with De La Beche on a journey from Waterford to Clonmel. His testimonials spoke highly of his personal qualities, although one is left wondering what was behind the enigmatic observation of an acquaintance some years later that Wyley was 'a cultured but very curious man'.[8] In his letter of application Wyley claimed to be proficient in mechanical philosophy, chemistry, mathematics, mensuration, surveying and draughtsmanship, but, while mathematics was his forte, geology was his pastime and he had already assembled a collection of Irish geological specimens — a collection which he presumably now had to dispose of. He was offered his Survey post on 27 September 1845 and he reported for duty at Arklow, County Wicklow, on 7 October.

In addition to the four surveyors, James appointed two Fossil Collectors and one General Assistant. The senior of the Fossil Collectors was James Flanagan (died in 1859) who was offered the post on 28 March 1845, although it was 24 May before he finally intimated his acceptance. He had worked earlier under Portlock, firstly on the Ordnance Survey triangulation of Ireland and then as a fossil collector on the survey of County Londonderry. Both Portlock and James had the highest opinion of his abilities[9] and as early as 15 April 1843 James had written to De La Beche observing that Flanagan was 'a very useful fellow, and a capital collector'.[10] In his new post he was paid four shillings and six pence per day. The second Fossil Collector was one Thomas Murphy, and the General Assistant was James Penny whose job it was to preside over affairs in the Survey's Dublin Office which, in 1845, was temporarily located in James Gandon's splendid Custom House on the north bank of the Liffey. For a short while this building became the hub of the Irish geological world because, in addition to the Geological Survey, it also housed the meeting room, library, and museum of the Geological Society of Dublin, and the embryonic museum of economic geology which Robert Kane was assembling in response to the First Commissioner's memorandum of 13 January 1845. But the Custom House was a geological hub that was soon to be split asunder because the Geological Survey and Kane's museum were shortly moved out to new premises, while the Geological Society of Dublin was evicted from its accommodation in September 1847 in order to provide more space for the increased establishment of the Poor Law Commissioners necessitated by the Famine.[11] It was nevertheless in the Custom House, on 28 October 1845, that James formally took over all the geological, botanical and zoological specimens which had been collected by the Ordnance Survey in Portlock's day. Colby deeply resented having to surrender all this material to De La Beche's new Survey[12], but that the

Fig. 5.2. James Flanagan, 'a very intelligent civil assistant' with the Ordnance Survey of Ireland, Portlock's fossil collector, and Fossil Collector with the Geological Survey of Ireland from 1845 until his death on 14 April 1859. He was, on 28 March 1845, the first, apart from James himself, to be offered a post with the new Geological Survey of Ireland. This sketch by George Victor Du Noyer was taken at Ferriters Cove, County Kerry, on 9 September 1856, twelve days after Murchison had left Dingle. The original drawing is in the archives of the Institute of Geological Sciences, London, and it is here reproduced by kind permission of Mr F.W. Dunning, Curator of the Institute's museum.

transfer did take place serves to emphasise the fact that the new Survey was in a very real sense the lineal descendant of the geological survey which Colby had started back in 1825.

On 21 May 1845 the First Commissioner wrote to De La Beche informing him that all his staff were to be regarded as belonging to the Geological Survey of Great Britain and Ireland, and he was instructed to make it clear to candidates that, if appointed, they were eligible for service either in Britain or in Ireland. But there is no evidence that such a clause was ever brought to the attention of candidates for places on the Irish Survey, and interchanges across the Irish Sea, in whatever direction, were very rare. One such transfer did, however, take place during the Survey's first field-season. On 21 June 1845 Trevor Evans James was sent to Ireland as an Assistant Geologist, having served since May 1840 as a member of De La Beche's staff in England and Wales. Presumably he was intended as a stiffening for the inexperienced Irish team, but his stay in Ireland was very brief. On 1 October 1845 James the Local Director wrote to James the Assistant Geologist to inform him that De La Beche required his services back in England and instructing him to proceed immediately to Derby. This was the official reason

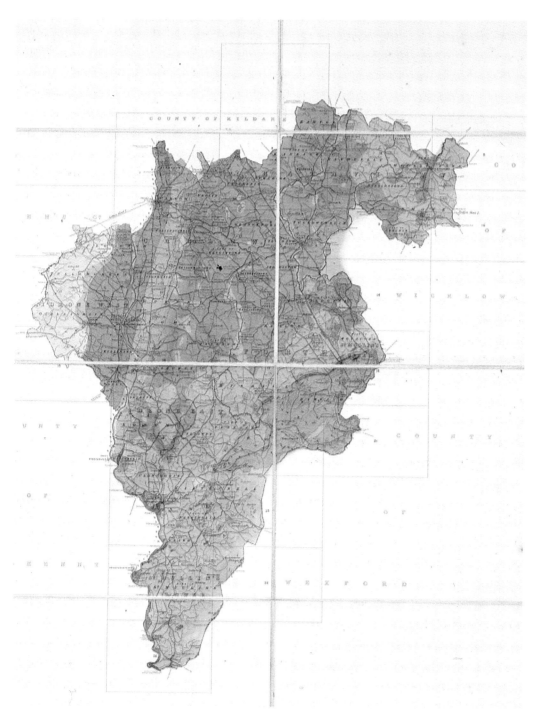

Plate 3. The coloured area of the Geological Survey's half-inch map of County Carlow published on 9 May 1849. (Reduced in size.) The title borne by the original is: Index to the Townland Survey of the County of Carlow/ Rule/Geologically Coloured. *The granite is red, the 'Silurian Slates &c' are purple (in the east centre), and the Carboniferous rocks are blue and grey. The widespread grey tint, best seen where it is washed over the granite, represents the 'Limestone drift (Tertiary)'.*

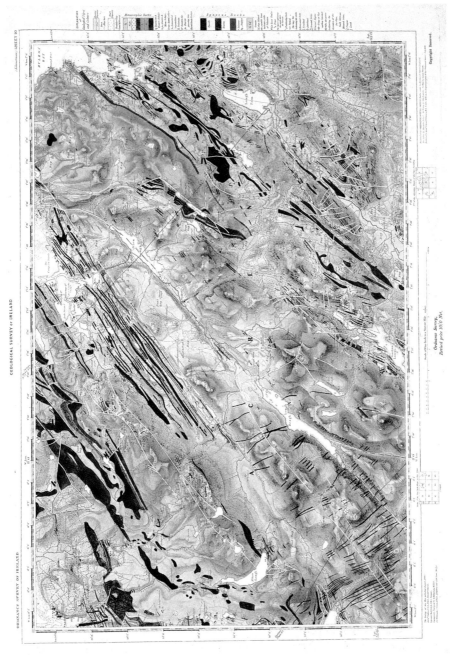

Plate 4. The final one-inch Geological Survey sheet to be published: Sheet 10 published on 21 November 1890. The surveyors responsible for mapping this complex area in northern County Donegal were Kilroe, Kinahan, McHenry, Nolan, and Wilkinson. The base map is hachured (cf. Fig. 6.7) and the two lines running across the sheet in a southeasterly direction are the lines of sections available on the Survey's sheets of longitudinal sections numbers 34 and 35. The map's earliest colourist was Mrs E. Williams of Camden Town, London. This, however, is not the sheet in its first published state. Originally upon the sheet an area in the southeast was classified as 'quartzite' and coloured yellow but some months later (perhaps after the microscopic examination of thin sections) it was decided to reclassify the rocks in question as 'quartzose grits' to be coloured in pink with yellow dots as here. Mrs Williams received the new pattern copy in February 1892 and the episode serves as a reminder that many of the one-inch sheets were modified from time to time without there being any reference to the fact in the map's marginal information.

for Trevor James's departure from Ireland; the real reason was that the local Director had taken a strong personal dislike to his Welsh namesake and had insisted upon his removal, although admittedly there had also arisen doubts about the Welshman's geological competence. Certainly De La Beche found it necessary to dismiss him from the Survey of Britain as from 30 September 1846.[13]

Although the permanent transfer of personnel across the Irish Sea was rare, three British-based members of the Survey did pay regular visits to Ireland during the new Survey's early years. Firstly, there was De La Beche himself. He was present in Ireland during the summer of 1845 and he returned regularly thereafter, often staying for several months on end. He seems to have taken a far deeper interest in the work of the Irish branch than did any of the subsequent holders of his office, and his visits became very much more then mere tours of inspection. Examine the field-areas being worked by his staff he certainly did, but he also liked to engage himself in some actual surveying, and during the Survey's early years he spent much time running long geological sections across the mountains of southeastern Ireland and down to the shores of the St George's Channel.

The second regular visitor was Edward Forbes, the Survey's distinguished Palaeontologist and a man of great personal charm and magnetism.[14] His premature death, only a few months after leaving the Survey in order to succeed Robert Jameson in the Edinburgh chair of natural history, came as a profound shock to British science at large. In Ireland he was responsible both for naming the Survey's fossils and for arranging them in the museum, but he too liked to savour the Irish countryside and he was frequently to be seen at work in the country's more fossiliferous localities. Some of his habits must have seemed a trifle odd to those of his Irish hosts who were not inclined to scientific pursuits. When he was preparing his work on the British mollusca, for example, he always carried in his pocket a tin-box containing half a dozen of the slugs he was in the process of describing. While staying at the old vicarage in Monkstown, County Cork, he horrified the maid by baiting her damp kitchen with turnip slices and other delicacies suitable to the appetites of his slimy friends, seeking in this way to increase the variety of species present in his pocket menagerie.[15]

The final London-based visitor was (Sir) Warington Wilkinson Smyth (1817-1890), a Cambridge graduate who had displayed such energy behind an oar upon the Cam that he had earned himself both the sobriquet 'the steam engine' and a place in the winning university boat of 1839. After widespread experience upon the continent, he joined the United Kingdom Survey as Mining Geologist in June 1845[16], and in Ireland between 1845 and 1851 he undertook special studies of the mines of County Wicklow[17], and of the Castlecomer and Slieveardagh coalfields.

The Survey staff of course received reimbursement for all their travelling expenses incurred in the course of duty, and equally the Survey provided all the instruments necessary for successful work in the field. Some indication of the range of equipment provided for a surveyor comes from a note written by James on 3 November 1845 in which he records that Trevor James had drawn the following items from the Survey stores and was now liable for their return as a result of his departure from Ireland: 2 sketching cases, 1 50-foot measuring chain, 1 set of arrows for use with the chain, 1 pair of parallel rulers, 1 clinometer, 2 hand-magnifiers, 2 protractors, 1 chisel, 2 hammers, 1 3-foot rule, 1 moleskin bag and strap, 1 portfolio, 2 notebooks, 1 acid bottle, and 1 box with key, the box presumably being intended to contain all the equipment during the officer's removal from one field-station to another.[18] Now that the link with the Ordnance Survey was broken, members of the Geological Survey were no longer issued with military-style uniforms, although the officer in charge of the Geological Survey of Ireland did attend levees of the Vice-Regal Court resplendent in the dress of a civil servant fourth class, consisting of a blue, gold-embroidered coat, white kerseymere breeches and white silk stockings, the tout ensemble set off by a black beaver cocked hat with ostrich feather and a black-scabbarded sword. The military link might have gone, but for some decades in Britain there did persist a tradition which served to remind the surveyors of their former military connection. Upon joining the Survey, it seems, each officer was issued with a set of brass Survey buttons bearing the crown and crossed-hammers emblem, and these buttons were worn at those dinners when the self-styled 'Royal Hammerers' gathered in London to celebrate the end of each field-season with feasting, songs, and general merriment.[19] Whether sets of Survey buttons were ever issued to the staff of the Irish branch is uncertain, but one fact is very clear; although representatives of the Irish staff might occasionally join the London conviviality of the Royal Hammerers, the Irish Survey itself was throughout the nineteenth century sadly deficient in all that joie de vivre, bonhommie, and esprit de corps which, if we can believe the historians, were so marked a feature of Survey life in Britain. Since the Irish staff seemingly never dined together, one must conclude that a set of Survey buttons but rarely graced an Irish waistcoat.

Mapping begins

Ireland has an area of 32,000 square miles (83,000 square kilometres), and its solid rocks range in age from the Pre-Cambrian schists of Connemara and Donegal to the Tertiary basalts and granites of Antrim and Down. Where in this geological patchwork was the new Survey to commence its field operations? Portlock had mapped a large area in mid-Ulster, and in 1844 James had started work in County Donegal, but instead of ordering a revision and extension of these earlier surveys, De La Beche very wisely resolved to make a completely fresh start. On 22 May 1845 he wrote to James directing him to open his campaign in southeastern Ireland in an area comprising counties Dublin, Wicklow, Kilkenny, Carlow, Waterford and Wexford, and James himself was instructed to proceed to a field-headquarters located in either County Waterford or County Wexford. De La Beche decided to tackle this portion of Leinster for two reasons. Firstly, his British surveyors were now at work just across the St George's Channel in Wales, and he hoped that a mapping of the coastal regions on either side of the sea would enable his two survey teams to elucidate each other's geological problems. Secondly, southeastern Ireland contained such known mineral resources as the copper lodes at Avoca and the coal-seams at Castlecomer, and from the outset De La Beche was anxious to demonstrate the economic potential of the Survey's work.

All the Irish geologists from Griffith and Ganly to Pringle and Portlock had grown accustomed to using the Ordnance Survey's six-inch (1:10,560) maps as the basis for their field-observations, and admirably had they served their purpose. The maps contain such a wealth of topographical detail that the location of almost any outcrop is immediately obvious without recourse to the use of a compass or a chain, and the sheets themselves are so spacious that an abundance of detailed geological observation can be recorded directly upon the map itself. Now it was the turn of James's men to try their hand at the art of large-scale field-mapping. By the August of 1845 six-inch Ordnance Survey sheets had been published for the whole of Ireland save for County Kerry (the Kerry sheets were published in November 1846) and the existence of the six-inch maps gave the Irish Survey a distinct advantage over its British counterpart because, as we have seen (p. 102), in the absence of a six-inch coverage of England and Wales, De La Beche's men had been working there within the cramped confines of the Ordnance Survey's one-inch map.

James placed his first requisition for maps with the Ordnance Survey on 5 June 1845 when he ordered the following items from Mountjoy:

1. Three sets of wet impressions and one set of dry impressions of each of the 141 six-inch sheets covering counties Waterford, Wexford and Wicklow.
2. Six copies of each of the half-inch county index sheets for counties Waterford, Wexford and Wicklow.
3. Three copies of the quarter-inch map of Ireland which Larcom had constructed in

1836-37 for the Railway Commissioners and which was the base for Griffith's geological map of 1839.

It was a straightforward order; James clearly intended to do his field-work upon the six-inch maps and then reduce the geological lines to the half-inch and quarter-inch scales. The reference to wet and dry impressions of the six-inch maps, however, does need a word of explanation. The six-inch maps were printed from copper plates and during the printing process the sheets of paper were wetted in order to improve their ink-absorbency. But as the paper dried out, some contraction occurred and there was a resultant slight loss of accuracy in the scale of the sheet. For certain purposes this scale distortion was held to be unacceptable, and in such cases dry impressions were taken from the copper plates. The inking upon the dry paper was much inferior to that upon the wet, but such dry impressions could be used for the plotting of compass-bearings and for the representation of any distances which it had been felt necessary to chain out on the ground.

There was one other map for which James felt a need. On 7 June 1845 he wrote to Griffith asking for two copies of the quarter-inch geological map of Ireland, one copy to be handed over to De La Beche for use in the Survey's London headquarters and the other to be placed in the branch office in Dublin. In response to the request Griffith produced two specially up-dated versions of the map, both of which have survived to the present; one is still in the office of the Geological Survey of Ireland in Dublin and the other now rests in the Institute of Geological Sciences in London. The London map is dated in manuscript 'Dublin, May 1846', but the Dublin copy carries that misleading and controversial printed inscription 'Dublin March 28th 1839' (see p. 75) although manuscript dates in the map's margin make it clear that at least two of its six sheets were printed in January 1845. Griffith has introduced into the two maps a welter of geological refinements and they became the prototypes for the new edition of Griffith's map published in the summer of 1852 (see p. 75). So many were the modifications introduced into the two maps for the Survey that Griffith had to provide the maps with fresh, manuscript keys containing forty-two colour-boxes as compared with only twenty-six colour-boxes in the original edition of 1839. There is another feature of the two manuscript keys that is also of interest. In all the early copies of Griffith's map the key is upside down in the sense that the oldest formations appear at the top, and he perhaps felt a little embarrassed that his map failed to conform to the now established convention that stratigraphical columns should young upwards. Certainly in preparing the manuscript keys for the two maps destined for the Survey he took good care to see that his stratigraphical column was now placed in its more traditional attitude. James and his young upstarts were never going to be given the opportunity of poking fun at the doyen of Irish geology.

By the time the maps requested of the Ordnance Survey were delivered at the Custom House on 23 June, both McCoy and Willson were securely in harness and the time had come for that exciting step which in Survey parlance was always known as 'breaking ground'. James and his men were embarking upon a task which they imagined would be completed within little more than ten years. They were seriously mistaken, The century was to be almost at its close before the work was completed, and of the men who broke ground during the summer of 1845, only McCoy seems to have been left alive by the time the project was concluded — and by then he had long been resident in far distant Australia. During the first field-season the main weight of the Survey's effort was thrown against the slates and limestones of southern County Wexford, and James and his wife took up residence in Wexford town in a furnished house rented on a weekly basis. From there, on 30 July, James wrote to John Phillips reporting that a period of excellent weather had allowed them to enjoy long days in the field.[20] Willson, James observed, was 'a young hand, not very promising', but his senior Fossil Collector, 'Old Flanagan', was 'an honest chap and a very good collector'. Another letter written that summer — a letter from Forbes to Ramsay — affords one of those rare glimpses into the life of the surveyors at their field-station. The letter was written on 14 September in the little one-street town of Fethard lying on the south Wexford coast (Fig. 5.3):

> I am here in a little village near Hook Point, in the midst of Mountain Limestone fossils, examining their distribution — all very interesting. The Captain [James], a very nice fellow named Willson, who is of his staff, and that thorough Welshman, little J [T.E. James], peppery uncomfortable, and marvellously stupid and uninformed (as I find on close quarters), are my companions. We make a very merry mess, however, and the Welsh squire's absurdities — for he is in misery in Ireland — make us laugh. Sir Henry [De La Beche] was with us till two days ago, working like a trooper, and when not at work telling funny stories. In a few days I leave this and go with the Captain (who sports a ferocious pair of egg-brown moustaches) to look at the Pleistocene beds in Wexford. Thence I go with Sir Henry to Dublin.[21]

Meanwhile, Smyth was further to the north taking a look at the Castlecomer coalfield. Thirty years earlier Griffith had been sanguine of its potential but Smyth was far from impressed with what he saw. On 30 September he wrote to Ramsay from Carlow:

> We are going to get in the Kilkenny coal which proves to be a very poor shallow concern, just the leavings of a tearing denudation, the dregs at the bottom of a tay-cup, and yet poor Paddy thinks no small beer of his coalfield![22]

The Fossil Collectors too were busily engaged. Flanagan was set to work upon the (Ordovician) slate exposures along the pleasant eastern shores of Waterford Harbour around Arthurstown and Duncannon, while Murphy was given the less attractive task of grubbing around for Pliocene shells in the so-called Wexford 'manure gravels' which are really glacial tills containing an ice-dredged fauna.[23]

James was well pleased with his first field-season; on 4 October 1845 he wrote to Phillips reporting the conclusion of 'a short, but brilliant campaign this summer'.[24] Here again we see employed the military analogy, but James and his men saw themselves not as battling merely to lay bare the secrets of regional geology; they saw themselves as battling to liberate a region from the geological hegemony represented by Griffith's map. McCoy in particular took great satisfaction in reporting the discovery of what he took to be errors in Griffith's map and in August and September 1845 he made comments such as these in his letters to James from Enniscorthy in County Wexford[25]:

> I am sure you will be surprised to hear that the three great masses of Quartz Rock south and east of the town represented with their intervening slates on the Railway Map are not to be found.
>
> The boundary of the granite and metamorphic rocks is nearly two miles wrong on the Railway Map.
>
> I found a hill of granite . . . in the middle of what is slate on the Railway Map.

By some chance McCoy found himself being driven around by the same carman that Patrick Ganly had employed during a visit three years earlier. From the carman McCoy discovered that Ganly had spent six weeks exploring the local geology, whereas Ganly had told McCoy that he had devoted a mere four days to the region. In view of the true length of Ganly's stay, McCoy was sharply critical of what Ganly had achieved:

> Could he have lost his senses without anyone knowing it? For it is scarcely possible he should not know Quartz Rock from Greenstone or Slate from Granite.[26]

Until only a few months earlier, let it be remembered, McCoy had himself been deeply involved in the palaeontological studies underpinning Griffith's map, but clearly McCoy's allegiances now lay elsewhere. For the moment there were two rival geological surveys at work in Ireland — the official survey under De La Beche and the unofficial survey under Griffith —

Sheets of Many Colours

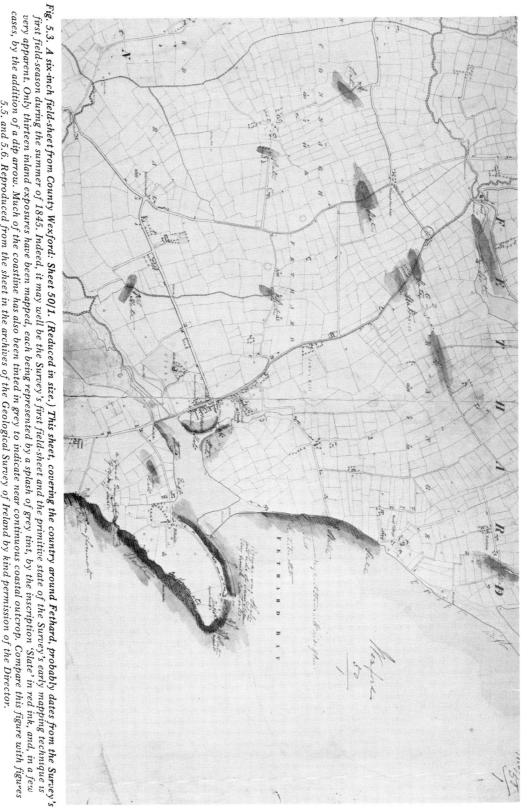

Fig. 5.3. *A six-inch field-sheet from County Wexford: Sheet 50/1. (Reduced in size.) This sheet, covering the country around Fethard, probably dates from the Survey's first field-season during the summer of 1845. Indeed, it may well be the Survey's first field-sheet and the primitive state of the Survey's early mapping technique is very apparent. Only thirteen inland exposures have been mapped, each being represented by a splash of grey tint, by the inscription 'Slate' in red ink, and, in a few cases, by the addition of a dip arrow. Much of the coastline has also been tinted in grey to indicate near continuous coastal outcrop. Compare this figure with figures 5.5, and 5.6. Reproduced from the sheet in the archives of the Geological Survey of Ireland by kind permission of the Director.*

and both surveys were at work in southern Leinster. It was like 1827 all over again. Then Griffith had taken himself off to Ulster to map the geology of regions that were already being investigated by the Ordnance Survey (see p. 48); now, almost twenty years later, he sent Ganly down to Leinster to re-investigate the geology of areas where De La Beche's men were at work. James wrote to Phillips on 4 October 1845 observing that Griffith:

> ... is so conscious of the inaccuracy of his work that he has actually had one of his men (Ganley [sic]) dodging our lads from Dublin & Enniscorthy, New Ross, Waterford, & Bunmahon. We have seen him at work at all these points in *our* district![27]

A month earlier James had written to McCoy commenting upon sightings of Ganly within the area being mapped by the Survey and continuing:

> ... it is therefore evident that Mr Griffith is getting up the geology of the country and of the district in which we are now engaged, a somewhat singular proceeding — however my principal object in writing now is to guard you against assisting Mr Griffith in any way, it is clear that two Surveys are now in progress and you can have no connection whatever with the other.[28]

The first annual report of the Geological Survey of Ireland was despatched to De La Beche on 2 February 1846, and in it James informed his chief that the survey of County Wexford was finished, that maps, sections and illustrations of the county were in active preparation, that large portions of counties Carlow, Kilkenny and Waterford had been mapped, and that a great deal of information had been gathered about the mines in both counties Waterford and Wicklow. This was an impressive record for only ten months of work. According to the report 5 per cent of the country was already mapped, and it really did begin to look as though ten years of work might well bring a new geological map of Ireland to successful completion. There is nothing to suggest that De La Beche was surprised at the progress claimed by the youthful Survey, but if he did entertain any doubts about the quality of such rapid mapping, then those doubts were soon to prove fully justified. The contents of that first annual report owed far more to James's vivid imagination than to the field activities of his inexperienced staff.

As the first year in the life of the Survey drew to its close, James very naturally began to wonder about the scale upon which the Survey's findings should be published. Immediately he was confronted by a base-map problem of precisely the same type as that which had plagued Portlock during the previous decade. The six-inch was too large a scale for publication, while the county index maps were drawn on too small a scale. There was still no intermediate Irish map corresponding to the one-inch (1:63,360) series which was the base for all De La Beche's published maps in England and Wales. The existence of the six-inch sheets gave the Irish surveyors an enormous field advantage over their British counterparts, but the absence of any Irish one-inch map placed the Irish branch at a grave disadvantage when it came to publication. The Ordnance Survey had for long been talking about the compilation of a one-inch map of Ireland, and from the outset De La Beche had understood that an Irish one-inch base-map would be available by the time the geologists had their material ready for publication. But by 1846 the Ordnance Survey had still taken no final decision on the matter and James therefore set Penny to work upon the task of reducing the six-inch sheets of County Wexford to the one-inch scale. It was almost like Griffith trying to triangulate his own base-map back in 1819, and on 20 February 1846 James wrote to De La Beche:

> We have reduced the six inch map of Wexford to the one inch scale, but this is a most laborious operation, and I shall be glad if you will write a note to Col. Colby, merely a private one, to know when we may expect to have outline impressions of the Counties of Waterford, Carlow, Kilkenny, and Wicklow on the inch scale. If we are not to get them from the Ordnance then we must continue our labours.[29]

But on this occasion not even De La Beche's subtle brand of diplomacy could secure the necessary decision, and throughout the first ten years of its existence the Irish Survey continued to be dogged by the old familiar base-map problem.

Exit James

In the spring of 1846 James was confronted by a problem more immediate than that posed by the absence of a suitable base-map — finance. On 12 May he wrote to De La Beche venting his belief that it was impossible to maintain the then establishment of the Irish branch upon the very limited annual budget of only £1500. As an economy measure he suggested that Wyley, his most recent recruit, should be transferred to England just as soon as he had completed his current Irish assignment. De La Beche's own remedy was more drastic; he wanted to see Wyley struck from the Survey's payroll altogether. On 16 May De La Beche wrote to Ramsay:

> In the reductions to be made on the Irish Survey young Wyley must be removed from it — My impression is that he is a very inferior man for our purposes to Henfrey — I may be mistaken, but it strikes me as true.[30]

The Henfrey here referred to is George Henfrey who had joined De La Beche's British staff in 1845, but in the event it was Henfrey who quitted the Survey in 1846 while Wyley remained a member of the Irish staff until 1855. So far as Ireland was concerned it was Murphy the junior Fossil Collector who was dismissed. He had given entirely satisfactory service but the winds of economy blew cold and his appointment was terminated in May 1846.

In the same month as Murphy's dismissal, James himself left the Survey on four weeks' leave of absence. The Commissioners of the Board of Works had asked him to survey the sites of a number of harbour works which it was proposed to construct around the Irish coast as a famine-relief measure and, in view of the national emergency, it was eventually agreed he might undertake the task, although De La Beche considered that the youthful Survey could ill afford the loss of its Local Director at the beginning of a new field-season. Perhaps this journey around the Irish coast gave James the opportunity of pondering upon the Survey's financial problems and upon the geological course that his personal career was now taking. Certainly when he returned to Dublin he brought with him a bombshell. On 20 June he wrote to De La Beche asking to be relieved of his duties as Local Director. For De La Beche this was a complete surprise; nothing in their relationship hitherto had suggested that James was in any way dissatisfied with his lot, and even now his resignation was supported by a minimum of explanation. In his letter he merely observed:

> [I am] finding that my position on the Geological Survey of Ireland is not such as I had reason to believe it would be when I undertook the duties of Local Director, and that the powers and means which have been entrusted to me are insufficient to enable me to carry on the duties in a satisfactory manner either to the public or to myself . . . .

In his official correspondence over the next few weeks James was a little more forthcoming about the reasons for his action and he attributed his discontent to four facts. Firstly, he claimed that, when he accepted the post, De La Beche had given him to understand that one of his chief tasks would be the collection of specimens for Kane's museum of economic geology, but now De La Beche had informed him that the assembly of museum material must not be allowed to interfere with the Survey's primary function of field-mapping. Secondly, James protested that he was allowed no control over the Irish activities of either Forbes or Smyth. Thirdly, and with some justice, he pointed out that both of the supposedly experienced Assistant Geologists sent over from Britain to augment his staff had proved to be incompetent and had been dismissed. Finally — and here we come closer to the nub of his resignation — he complained of what he termed the drudgery of geological field-work.

It is difficult to take most of James's complaints very seriously; they look like trumped-up excuses offered to justify a resignation which was being made for other, more fundamental reasons that James did not care to divulge. If the truth were known, his departure was probably occasioned by two discoveries. Firstly, conceited though he undoubtedly was, by June 1846 he must have recognised his personal geological limitations. It was suggested earlier that he was merely a shallow careerist in the science (p. 112), and during the 1845 field-season his close association with such real, full-blooded geologists as De La Beche, Forbes, and Smyth must surely have convinced him that he was not really of their genre. They were inspired with a missionary zeal to shed light upon the darkest secrets of earth-history, and their field-mapping was an exciting and almost spiritual exercise. But, as James freely admitted, geological fieldwork was for him merely drudgery and he must have been disturbed to discover from the regulations promulgated in April 1846 that this boring activity was supposed to engage his attention for no less than eight months of each year. Secondly, there was the question of status. He was a highly ambitious man and by 1846 he must have recognised that the Local Directorship was not going to yield that independence and self-importance which he so desired. He was now a captain of four years standing and he probably felt that a command subordinate to the civilian De La Beche and consisting of only three Assistant Geologists, one Fossil Collector, and one General Assistant was hardly fit duty for an officer of his experience. Colby had never been happy about James's secondment to serve in such a civilian organisation[31] and perhaps by 1846 James himself had come around to a similar point of view. Maybe, too, James's action was tinged with an element of pique arising from the knighthood which had been conferred upon his museum colleague Kane on 16 February 1846. James can hardly have overlooked the fact that in age Kane was no less than six years his junior.

Surprised though he may have been, De La Beche was far from being despondent at the turn of events. James had never been his first choice for the Dublin post, and a closer acquaintanceship had evidently convinced De La Beche of his Local Director's limitations. Certainly De La Beche now felt that he had been relieved of an incubus. On 9 July 1846 he wrote from Dublin to Ramsay:

> It had been a lucky thing that the Captain resigned, otherwise I really don't know the amount of mess into which our Irish Survey might have got. His notion seems to have been that he was to do little or nothing personally in the field — and altogether he seems to have considered the affair as mere employment — until he could step into something else — That something else has turned up in the shape of Engineer to the Portsmouth Dock Yard — and right well ought we to be pleased therewith.[32]

James left his post in Dublin on 4 July 1846 but this was by no means the end of his career in the public service. From 1846 he was superintendent of the constructional works in Portsmouth dockyard and he achieved sufficient scientific eminence to ensure his election to Fellowship of the Royal Society in 1848. Two years later he returned to the Ordnance Survey as superintendent of the Edinburgh office and from there he was in 1854 translated to the survey's Southampton headquarters as Director-General. Thus the man who had failed as Local Director in Dublin found himself only eight years later at the head of an organisation far larger than the Geological Survey of Ireland was ever to become, and, strange to relate, his regime at Southampton proved to be an outstanding success. He received his knighthood in 1860 and he rose to the rank of Lieutenant-General before ill-health forced his retirement in 1875.

De La Beche's new broom

James's successor as Local Director was a Dubliner — Thomas Oldham.[33] He was born on 4 May 1816, the eldest son of Thomas Oldham and his wife Margaret Boyd, his father being a broker with the Grand Canal Company. Oldham junior attended school in the city and then entered Trinity College Dublin where he received his B.A. in the spring of 1836. At that time the University possessed no school of engineering, but it was towards the industrial arts that Oldham now turned his attention, and he left for Edinburgh, there to study the subjects appropriate to his chosen profession of civil engineer. In the Scottish capital he attended university classes (including the lectures of the same Robert Jameson who had been Griffith's teacher thirty years earlier) and he gained professional experience by working upon civil engineering projects in and around the city. When he returned to Ireland in March 1838, however, it was to assume not a post in engineering, but an office with the Ordnance Survey as Portlock's chief geological assistant. In that capacity he played a major role in the field-mapping of mid-Ulster, and it was to him that there fell the task of reducing all the six-inch work down to the half-inch scale upon which Portlock's map was eventually published. Portlock himself was generous in his praise of Oldham whom he described as 'possessed of the highest intelligence and the most unbounded zeal'.[34] Oldham remained with the Ordnance Survey until the final relics of Portlock's department were disbanded in January 1843[35], but he soon found himself another geological post because in June of the same year he became curator of the museum of the Geological Society of Dublin. This was the office from which McCoy had recently been removed, but Oldham filled the post to everybody's satisfaction and it became

Fig. 5.4. Thomas Oldham. When he left Ireland for India he was aged thirty-four and this illustration is probably a likeness taken later in life. Reproduced from a portrait in Archibald Geikie's Memoir of Sir Andrew Crombie Ramsay.

his stepping-stone into the academic world. In November 1844 he was appointed assistant to Sir John MacNeill, the first professor of civil engineering in Trinity College, and then, in April 1845, he succeeded Phillips to become the University's second professor of geology.

Although Oldham ceased to be curator to the Geological Society of Dublin in January 1845, he remained one of the society's staunchest supporters, and between November 1843 and January 1851 he presented to the society no less than twenty-six papers.[36] It was a record that not even Portlock could match, and over the same period, in addition to his curatorship, Oldham varyingly served the society as president (1848-1850), as secretary, as assistant secretary, and as a member of council. His scientific attainments were recognised by his election to Membership of the Royal Irish Academy in 1844 and to Fellowship of the Royal Society (on the same day as James) in 1848. His most famous geological discovery was that fossils existed in the ancient Bray Group rocks of County Wicklow, a discovery which he made in 1840 but which was not brought to the notice of the Geological Society of Dublin until 12 June 1844.[37] These were among the oldest fossils then known, and on 15 November 1848 Edward Forbes announced to the same society that he proposed to name the new forms *Oldhamia* in honour of their discoverer.[38] Oldham was unquestionably in a geological class far superior to that of James and he was to prove an excellent Local Director. His formal letter to De La Beche accepting the office was written in his chambers at 5 Trinity College on 27 June 1846, and he became Local Director as from 4 July, the University evidently being happy that he should hold both his new office and his chair concurrently. Thus the combination of academic and Survey duties, which had been denied to Phillips in 1844, was permitted to Oldham only two years later.

The terms of Oldham's contract were identical to those under which James had operated, but in one significant respect their appointments were totally different. James had in large measure been foisted upon a reluctant De La Beche; Oldham was De La Beche's personal choice for the post. On 9 July De La Beche wrote to Ramsay from Dublin:

> The more I see of Oldham the more I am disposed to consider he will be successful – I like him much – besides he will agree so much better with Sir Robert Kane – the Captain rather roughed up the latter.... I will go with Oldham into the country on Monday for about 3 weeks.[39]

Those three weeks served to strengthen De La Beche's conviction still further and on 26 July he again wrote to Ramsay, this time from Bunclody, County Wexford:

> Oldham and self continue to get on famously and I am right well contented with him. Our captain was a failure – the more I see of the work *supposed* to be done the more I see that somehow he seems to have supposed that this said work was to be done as mechanically as that of sappers on a Trigl. Survey and his mind seems to have been engaged in making up the supposed good work of others and not in getting a grasp of it himself. Oldham appears to me to have a philosophical mind – quite ready to go ahead in the school we have been forming – As a colleague you will find him worth 50 of the other.[40]

The Survey's other senior staff were also quick to recognise that in Oldham they had found a kindred spirit who was ready to be brought into the innermost conclaves of their geological brotherhood. Soon after Oldham's appointment Forbes wrote to De La Beche:

> I am glad to hear all things go well in Ireland. The opinion you entertain of Oldham's talents is what I expected. I have always been impressed with the notion, that his abilities are of a far higher order than those of any other Irish geologist.[41]

Similarly, on 8 August Ramsay wrote to his chief from a Welsh field-station at Dolgelley observing of Oldham:

> I like his notes — They are hearty and cordial.... I feel as if I should like him better and better the more I know him. I am certain we are much better adapted for each other than the grandiose Captain, and we shall pull together in your triumphal car like fun.[42]

Amidst all these pleasantries James introduced the only jarring note. On 2 July, in one of his final acts as Local Director, he wrote the following paragraph to De La Beche:

> I do not know whether or not you are aware of the fact, that there has for some time past existed a sort of feud between Oldham and McCoy, and that at the very last meeting of the Geological Society here there were read some angry letters between them arising out of observations made by Oldham on McCoy's work on the Carboniferous fossils at some former meetings. I was in the country at the time, and have not seen the letters, but it is clear that Oldham's appointment as Local Director, makes McCoy's position peculiarly unfortunate, and I should think it would be adviseable to remove him to England.[43]

At this distance in time it is impossible to decide upon the rights and wrongs in this altercation between Oldham and McCoy, but it evidently generated some heat because McCoy wrote a paper critical of Oldham which the Council of the Geological Society of Dublin rejected at its meeting on 17 June 1846 with the observation 'that Council consider it desirable that all personal allusions should be avoided in controversial papers of a scientific character'.[44] De La Beche nevertheless decided not to take the matter very seriously and he rejected James's advice that McCoy should be transferred to England. Perhaps this was a mistake because the Oldham-McCoy affair continued to smoulder until the autumn of 1846 when it was doubtless a contributory factor in McCoy's decision to resign his Survey appointment.

Oldham sets to work

Within a few days of assuming his new office, Oldham arrived at an unpleasant conclusion: James had left the affairs of the Survey in chaos. Oldham could find no adequate record of the Survey's activities during his predecessor's tenure of the Local Directorship, and he soon realised that in his annual report of February 1846 James had grossly exaggerated the extent of the ground covered during the first field-season. In particular, James's claim to have completed the mapping of County Wexford now proved to be quite false. On 6 August 1846 Oldham wrote to De La Beche:

> Of the fair plans and sections alluded to by Capt. James in February last I find none in this office, with the exception of three sections, so prepared that I could not recommend them for publication.

There was even a suspicion that James was guilty of misappropriating public property. In July 1845 the Stationery Office had specially bound all the fossil drawings prepared by Du Noyer for Portlock's 1843 memoir, but Oldham could now find no trace of the volume. It later transpired that James had taken it with him to Portsmouth, and when taxed with the matter, the culprit meekly protested that he had only been trying to restore the volume to Portlock. But Portlock was in Corfu not Portsmouth, and one surmises that other items may well have disappeared from Dublin in James's train because when Oldham checked through his inventories in February 1847 he discovered that a number of other articles which had been handed over to James in 1845 were now no longer present in the Survey's collection. Happily the volume of drawings for Portlock's memoir did find its way back to Dublin where today it is one of the treasures of the Geological Survey.

One of Oldham's first tasks as Local Director was to inspect the work of his staff in the field. Unfortunately it did not always meet with his approval. Wyley's mapping was, of course, already known to be suspect, and Oldham was certainly far from satisfied with what the former Belfast teacher had to show him. Under Oldham's guidance, however, there was some improvement and on 22 October 1846 he wrote to De La Beche from Ashford, County Wicklow, remarking upon Wyley's progress and recommending that his pay should in consequence be increased to 8/6 per day. To this proposal De La Beche speedily assented, but Oldham was never entirely happy about Wyley. As late as 3 October 1849 we find him writing to De La Beche from Wexford complaining that Wyley was never going to make a good field-man, that he was 'extraordinarily erratic', and that all his mapping had to be checked very carefully.[45]

It was nevertheless for McCoy that Oldham reserved his most severe strictures. On 29 August 1846 he wrote from Bunclody, County Wexford, to McCoy at Borris, County Carlow, complaining in no uncertain terms about the quality of McCoy's work in the Bunclody region. Oldham claimed that alluvial deposits had been completely disregarded; that geological locations had been misplaced on the six-inch field-sheets by as much as half a mile; that rock outcrops had been marked where there were none and vice versa; and that many important localities had clearly never been examined at all. He demanded an explanation of McCoy before he reported the matter to De La Beche. In his reply, dated 1 September, McCoy protested that Oldham's complaints were insufficiently detailed to allow of an adequate rejoinder, but this exchange must have a bearing upon the fact that McCoy decided to resign from the Survey as from 30 September 1846.

We cannot know how far Oldham's strictures arose from a genuine concern for the quality of the Survey's work and how far there was an element of personal hostility towards McCoy stemming from those previous clashes to which James had drawn attention. It nevertheless does seem probable that Oldham's complaint of slipshod work was grounded in fact, and it should be remembered that four years earlier a similar complaint had earned McCoy virtual dismissal from his curatorship at the Geological Society of Dublin. Oldham certainly continued to complain bitterly about the quality of McCoy's work for months after the offender had left the Irish scene. On 7 April 1847, for example, he wrote to De La Beche about the mapping in eastern County Wexford:

> Error after error turning up near Courtown. Limestones marked where not existing — and omitted where they are — and similar mistakes of other kinds — The whole of that area, being a critical one, will have to be gone over carefully — Is not this too bad. Some of McCoy's work.[46]

On 7 May Oldham at Courtown Harbour wrote again to De La Beche in the same vein:

> I have been here for the last ten days — making the most of all the fine weather we had, which was not much — a district so badly examined I never saw — but this is all past now.[47]

For McCoy this contretemps was but a temporary set-back in an otherwise outstanding career. From Ireland he went to a post in the Woodwardian Museum at Cambridge where Sedgwick was soon enthusing over the young Irishman's talents[48], and in 1849 McCoy re-crossed the Irish Sea to assume the chair of mineralogy and geology at Queen's College Belfast. Five years later there came another move, this time to the antipodes where he became professor of natural science in the new university of Melbourne. There he speedily acquired renown because of his scientific attainments, his pugnacious character, and a sartorial perfection that made him one of the dandies of Victoria. He received his knighthood in 1891 and when he died in 1899 he was the grand old man of Australian science. In 1935 the Melbourne students founded a McCoy Society in his honour.[49]

One of McCoy's problems in 1846 may have been that he was inadequately briefed as to the duties of a field-geologist. De La Bech's *Instructions* of May 1845 had been singularly unhelpful in this respect. They were replete with advice about the kind of phenomena to be observed, but nowhere do they answer the basic question of exactly how a geological map was to be made. In what manner were the surveyors to explore their ground? Were they, for example, to be content with running traverses or were they to map methodically outcrop by outcrop? Presumably James was expected to train his staff in Survey technique by drawing upon the experience he had himself gained while working as a temporary member of De La Beche's staff in England back in 1842, but was James a satisfactory mentor? In any case James's English experience had been with a one-inch survey and not with the six-inch sheets now being used in Ireland. In the archives of the Geological Survey of Ireland there do survive some six-inch field-sheets from counties Carlow and Wexford dating from about 1846 and these certainly indicate that the youthful Survey's mapping did leave much to be desired. On the sheets (they are mostly quartered Ordnance Survey sheets) the surveyors have merely followed the principal roads and a few of the streams, marking in crudely and vaguely the general localities where different types of rock had been observed (Figs 5.5 and 5.6). There is little or no attempt to record dips, to draw geological boundaries, or to depict the character of the terrain. In short, they are field-sheets of an extremely rudimentary type and they are certainly far inferior to the field-sheets being produced for Griffith by Patrick Ganly a few years earlier. A final point about these crude documents deserves mention: at least one of them (Carlow 19/3) is evidently the work of the much castigated McCoy.

The poor quality of the early mapping, and the consequent necessity for a re-examination of the ground, was one reason for the Survey's very slow progress between 1846 and 1848, but there were two other major contributory factors. The first of these was the disarray into which the Survey was thrown in 1846 when it had to remove itself from the Dublin Custom House to the newly acquired building which was to be its more permanent home. Secondly — and much more seriously — there was increasing delay caused as a result of the Survey being burdened with the task of collecting large numbers of geological specimens for museum use. This time-consuming work can hardly have been a popular duty with the staff in the field and it was a responsibility which was to drive both Oldham and his successor to despair.

In compliance with the First Commissioner's letter of 13 January 1845, number 51 St Stephen's Green East had been acquired to house both the Survey and Kane's museum of economic geology, and on 7 October 1846 Kane informed De La Beche that he was now ready for the Survey to move into the new premises. Some reconstruction of the building was necessary in order to make it suitable for its new function and this work was not finally completed until 1850, but by July 1847 the Survey was already in occupation of its new home. The building is a Georgian structure and it is interesting to note that in 1846 Kane issued Oldham with a caveat: would he please refrain from placing heavy specimens in his upper rooms because the fabric might not be able to withstand their weight! Kane's fears for the building's safety hardly seem to have been justified because it still stands a century and a quarter later,[50] having so far escaped the attention of the developers who have done so much to destroy the character of a square which has long been known to Dubliners simply as 'The Green' (Fig. 5.7). Kane was of course responsible for any chemical analyses requested by the Survey, and it was intended that the geologists should hand over to the museum all rock, mineral, and fossil specimens for which they had no further use. Sadly, the transfer to the Green occasioned some friction between Kane and Oldham as a result of a dispute over the ownership of the property which had hitherto been located in the Custom House. Kane's demands were certainly comprehensive and at one stage he was even claiming as museum property all the Survey's hammers, collecting-bags and instruments! Eventually it was agreed that everything should be handed over to Kane except for two categories of article: firstly, those items necessary for the Survey's day to day activities, and secondly, the fossils, plants and books which had once been the property of Portlock's department in the Ordnance Survey. The apportionment of the material was declared complete on 9 July 1847, but the dispute caused De La Beche to regret bitterly that the Dublin museum, unlike the Craig's Court museum in London, was not under Survey control, Kane being directly responsible to the First Commissioner himself. On 26 December 1846 De La Beche wrote to Ramsay:

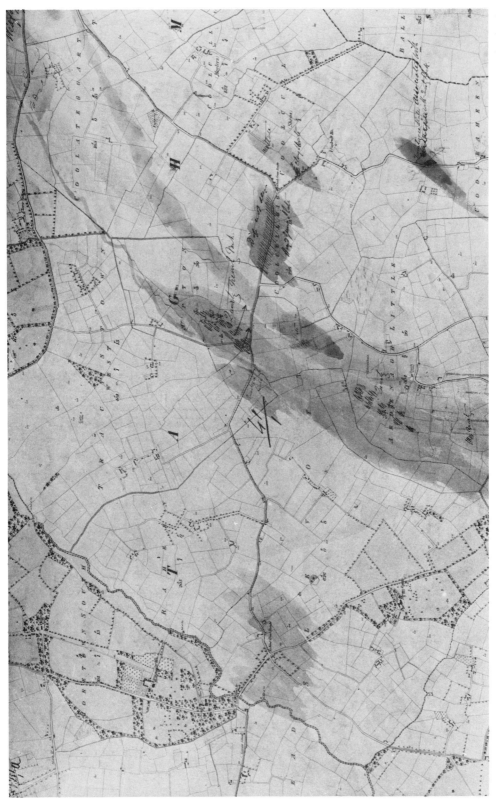

Fig. 5.5. An early six-inch field-sheet: County Wexford, Sheet 41/1 covering the country west of Taghmon. (Reduced in size.) This very rudimentary mapping (perhaps by Frederick McCoy) probably dates from 1845 or 1846. Compare this figure with Figure 5.6. Reproduced from the map in the archives of the Geological Survey of Ireland, by kind permission of the Director.

Sheets of Many Colours

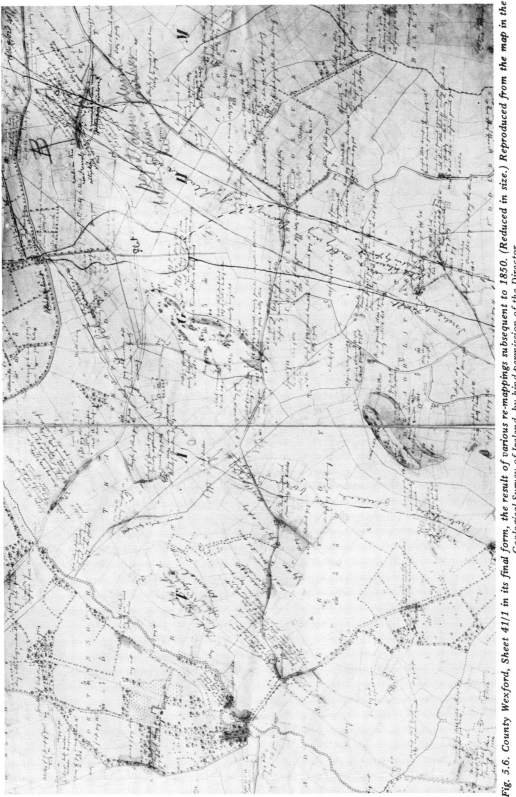

Fig. 5.6. County Wexford, Sheet 41/1 in its final form, the result of various re-mappings subsequent to 1850. (Reduced in size.) Reproduced from the map in the Geological Survey of Ireland, by kind permission of the Director.

> Touching *Muzzy* [museum] matters, it would be better not in any way to allude to the Museum in Dublin – the arrangements connected with which have been a blunder – or mistake, whichever you like.[51]

The fiery Kane was certainly very much a law unto himself, and De La Beche was powerless when in 1848 Kane extended the scope of the museum to include manufactured goods. It was known henceforth as the Museum of Irish Industry. For almost a quarter of a century the Survey and the museum were cohabitants of 51 St Stephen's Green and throughout the period Kane was to remain a permanent thorn in the Survey's flesh, but since the museum was never a part of the Geological Survey of Ireland, its history finds no place in the present study.[52]

The upheaval occasioned by the removal from the Custom House was soon over, but the delay caused to the mapping programme by the collection of museum samples dragged on from year to year. It had always been understood that the Survey would collect rocks, minerals and fossils for the museums in St Stephen's Green and at Craig's Court, but in the summer of 1846 Oldham received instruction that henceforth he was in addition to collect on behalf of the museums in the newly-established Queen's colleges in Belfast, Cork, and Galway. This was just about bearable; it merely meant that his men had to find five samples of each type of specimen instead of the two which had sufficed hitherto. The real problem was a request made by Kane – it was he who added the straw which all but broke the camel's back. On 16 April 1846 he wrote to James asking the Survey to collect soil samples, what he wanted being outlined as follows:

> A collection of specimens, showing every kind of soil and subsoil which is met with in each county – the locality of each being marked and the area over which it extends being defined as well as circumstances allow – some note of the characteristic vegetation of each kind of soil should be attached.

Fig. 5.7. Number 51 St Stephen's Green East, the headquarters of the Geological Survey of Ireland from October 1846 until March 1870. The two wings were added to the building between the autumn of 1848 and the spring of 1850 in order to accommodate Sir Robert Kane's Museum of Irish Industry. The Survey had its own premises on four floors of the southern wing (on the right of the illustration). Today the building houses the Office of Public Works, but the forty panels of Irish marble erected by Kane in the entrance hall in 1850 are still to be seen there. From a photograph by Terence Dunne.

Some of the samples were intended for display in the museum, while others were to go into Kane's laboratory for chemical examination. It was a tall order. Kane was asking the Survey to undertake nothing less than a national soil survey, and on 24 September 1846 the First Commissioner formally instructed the Survey to begin the collection of soil samples. Initially De La Beche was anxious to co-operate because he hoped that analysis of the soils would yield some indication of their agricultural potential, thus allowing the Survey to be of direct assistance to the Irish economy. In theory it was a splendid idea, pedologically it was premature by decades, and in practice the project was a total failure. Not until 1969 was there published the first map of Ireland's soils.[53]

In his first annual report, dated 5 March 1847, Oldham frankly admitted that the Survey had been making but slow progress and he went on to remind De La Beche of the difficulty he was facing in trying to run a survey in famine-stricken Ireland:

> The very distressed state of this country & the consequent demand for all persons at all competent to superintend the execution of numerous works intended for the relief of the poorer classes, has further rendered it extremely difficult to procure such aid as we require, so that our staff of assistants has been smaller than we wished.

As a result, the post left vacant by McCoy's resignation four months earlier was still unfilled; Willson and the erratic Wyley now represented Oldham's entire field-force. Nothing had yet been published, but Oldham reported that 1500 square miles (3900 square kilometres) of southeastern Ireland had now been adequately mapped including 23 of the 47 six-inch sheets of County Wicklow, 22 of the 54 sheets of County Wexford, 9 sheets of County Carlow, 9 of County Kilkenny, and 2 of County Waterford. In addition he recorded that Smyth had completed his examinations of the Castlecomer coalfield, the mineral district of County Wicklow, and the copper mines at Knockmahon in County Waterford. Finally, large numbers of specimens had been collected both for the Dublin museum and for the museums of the three Queen's colleges.

Oldham probably felt uneasy at having to admit that his two-year-old Survey still had nothing to show as a publishable earnest of its labours. History illustrated only too clearly what happened to a survey which failed to bear sufficient fruit, and Oldham must have been uncomfortably aware of the fate that had overtaken the director of an earlier geological survey of Ireland — indeed, his own former chief — who was now living out his banishment in Corfu. Oldham therefore made a small propitiatory offering to the Earl of Bessborough, the Lord Lieutenant. This took the form of three sheets of horizontal sections based upon the Survey's work in southeastern Ireland, and covering the mining districts of County Carlow, Queen's County, and County Wicklow. The three sheets were delivered to Dublin Castle on or just before 7 April 1847. A week later, on 14 April, two of the same sheets were introduced to a wider audience when Oldham exhibited them at the monthly meeting of the Geological Society of Dublin.[54] The precise nature of these sheets of sections which were offered for public display in April 1847 is nowhere made clear but they were most probably manuscript versions of some sheets of horizontal sections which the Survey was to publish in the summer of the following year.

During the field-season of 1847 Oldham concentrated his resources — such as they were — on completing the survey of County Wicklow, partly because of its mineral resources, and partly because it afforded a convenient base-line for the geology of the whole of Leinster. In the June of that year Willson was stationed at Dunlavin in the slate country on the western side of the Wicklow Mountains, and Wyley was based at Roundwood amidst the slates to the east of the mountains, although for some reason Flanagan was collecting at Camolin down in County Wexford. Forbes was in Ireland twice in 1847, noting that it was now 'a land of misery and tears'[55], and in March the Famine actually caused a minor interruption in the work of the Survey because on the 20th of that month Oldham circulated the following letter to all his staff:

> Wednesday next the 24th inst. being appointed by Her Majesty & Council, as a day to be set apart for general humiliation in consequence of the awful distress prevalent in this country, attendance to their duties will not be required from the assistants on that day.

Another and much more serious hinderance was the inaccessibility of a good deal of County Wicklow's mountainous interior. On 18 June 1847, for example, we find Willson writing from Dunlavin to report that he had failed to obtain convenient lodging from which to map six-inch Sheet 16 and a part of Sheet 18, two sheets which include the mountains around Glenbride and the head of the Kings River. He therefore sought permission to hire a horse and car by the day to facilitate him in reaching the more remote parts of his area. This Oldham sanctioned immediately. Despite all these difficulties, the mapping of County Wicklow was completed in the autumn of 1847, and by the time he wrote his second annual report, on 8 March 1848, Oldham was able to inform De La Beche that in addition half of County Kildare had also been mapped together with large portions of counties Carlow and Dublin.

The Survey's systematic sheet by sheet mapping of course had to have priority, but in April 1847 Oldham was granted authority to undertake an investigation of a slightly different type. The sections laid bare by the rapidly expanding railway system of England and Wales had presented geologists with unique opportunities for study, and at its Glasgow meeting in 1840 the British Association had set up a committee charged with the responsibility of placing upon permanent record as much as possible of the geological information revealed during railway construction.[56] De La Beche was a member of that committee, and the Geological Survey was certainly paying close attention to the 'railway geology' of England and Wales as early as 1841. Now it was decided to extend this type of work into Ireland, and the new task was assigned to Portlock's former artist George Victor Du Noyer[57] (1817-1869) who more recently had been earning his living as a drawing master at the newly-established College of St Columba at Stackallen House, County Meath. In May 1847 he was appointed as a temporary Survey officer and instructed to prepare geological sections of the strata exposed along all the lines of railway radiating from Dublin. It was just over sixty years since the Dublin Society had appointed Donald Stewart to perform a similar task along the banks of the Grand Canal (see p. 7), and it is also interesting to note that with Du Noyer's appointment exactly half of the Survey's staff — Oldham, Du Noyer, and Flanagan — were now ex-members of the team which had worked with Portlock ten years previously. Du Noyer's duty carried with it an element of risk because a geologist engrossed in his studies at the track-side can so easily fall victim to an unnoticed train. C.T. Clough of the Geological Survey and H.E. Strickland of the University of Oxford are just two geologists who lost their lives in this manner, and very few devotees of the science can have been in a position to follow the example of Edward Greenly who worked the railway sections of Anglesey in safety by placing his wife in some position of vantage so that she could convey early warning of the approach of a train.[58]

Du Noyer started work by following the line of the Dublin and Drogheda Railway which had been opened in 1844, and in the November and December of 1847 — with De La Beche's permission of course — he brought the results of his study to the notice of the Geological Society of Dublin.[59] By the March of the following year he had drawn detailed sections not only along the Dublin and Drogheda Railway, but also along the Midland Great Western which by then had pushed its head almost as far west as Mullingar. Difficulties arose, however, when he transferred his attention to the line of the Great Southern and Western Railway because the company refused him access to its cuttings. This was evidently the first time that a Survey officer anywhere in the British Isles had been refused such a facility and De La Beche was annoyed. The Parliamentary act of 1845 empowered a Survey officer to traverse private land, as the act rather quaintly put it, in order:

> ... to break up the Surface of any Part of such Land for the Purpose of ascertaining the Rocks, Strata, or Minerals within or under the same.

In June 1848 De La Beche used these powers when he served the directors of the railway company with legal notice of his intention to send Du Noyer onto their property. Reluctantly they acquiesced, and after completing his work on the Great Southern and Western in the late summer of 1848, Du Noyer moved on to examine the line of the Dublin and Belfast Junction Railway northward of the Boyne.[60] Throughout all this railway work Du Noyer proved himself to be a most competent Survey officer, and on 30 September 1848 he was transferred from his special assignment to the regular staff of the Survey as a belated replacement for McCoy. To his annoyance, however, he was informed that his sixteen months of special service would not exempt him from the six-month probationary period normal for an Assistant Geologist, and it was 14 April 1849 before De La Beche sanctioned his appointment to a permanent post. Du Noyer's railway sections were never published but they still exist in the Survey's Dublin office where they occupy rolled strips of squared paper each many metres in length. The sections themselves are beautifully drawn in Du Noyer's artistic hand at a vertical and horizontal scale of one inch to forty feet, and they afford a valuable record of geological exposures now lost because the walls of most of the railway cuttings are today overgrown.

The Survey's first publications

As the mapping of County Wicklow drew to its completion in the autumn of 1847, the base-map problem became yet again a matter of acute concern. De La Beche and Oldham were both anxious to publish the Survey's Irish work upon a one-inch map comparable to that being used in Britain but, despite De La Beche's entreaties[61], the Ordnance Survey had still not resolved upon the compilation of an Irish one-inch series. The efforts of the geologists to produce their own one-inch map had evidently proved less than successful and, since the Survey must publish, De La Beche and Oldham decided that there was no alternative but to call into service county index maps of the same series as had furnished Portlock with his base-map in 1843. This was a decision they took only with the greatest reluctance because, like Portlock, they recognised the absurdity of mapping upon the six-inch scale and of then eliminating a good deal of the carefully plotted detail during the course of reduction to county index sheets mostly drawn to a scale of only half an inch to the mile (1:126,720). In any case the half-inch scale was only twice that of Griffith's existing geological map of Ireland. The very fact that the index maps were county maps was in itself unfortunate because, as the Dublin Society had recognised in 1809, a county is an arbitrary and quite inappropriate unit for geological study. De La Beche himself now pointed out that an important structural entity such as the Leinster coalfield extends into three different counties and its representation upon three county maps must destroy its unitary character. But any attempt to fit the county maps together was foredoomed to failure because most of the counties are drawn upon their own individual graticule. The only available alternative to the county indexes was the six-inch map itself, and while the publication of six-inch geological sheets might be called for in certain complex mining districts, the production of a nationwide geological coverage upon that scale was entirely out of the question, not least because it would have involved the issue of some two thousand different sheets. In October 1847, therefore, De La Beche asked the Ordnance Survey to prepare a slightly modified version of the County Wicklow index to serve as the basis for the forthcoming geological map of the county.

The geological map of County Wicklow was published on 26 July 1848[62] with the title *Index to the Townland Survey of the County of Wicklow geologically coloured* and Oldham proudly put the map on display at the meeting of the British Association held that year in Swansea between 9 and 16 August.[63] The base-map is derived from the ordinary county index at a scale of half an inch to the mile (1:126,720) and the geological map still displays the six-inch sheet lines. The map has nevertheless been modified in three ways. Firstly, engraved lines have been added to represent both the geological boundaries and the course of the various horizontal sections that the Survey had prepared for publication. Secondly, part of County

Dublin has been added to the map so as to represent the northern terminus of the Leinster batholith along the shores of Dublin Bay between Booterstown and Dalkey. Finally, in the southeast, part of northern County Wexford has been added in order to avoid the abrupt ending of the Croghan igneous complex at the county boundary. The geology is represented by hand-applied colour-washes and all told the map affords an accurate enough representation of the county's solid geology, although Oldham's men had clearly experienced some understandable difficulty in differentiating between the 'Cambrian' (Bray Group) rocks and those belonging in the adjacent 'Lower Silurian' (Ordovician) division. Cartographically the map is open to criticism because of its attempt to be a dual purpose sheet representing both the solid and the drift geology. A grey tint superimposed over a very fine stipple has been introduced to show 'Tertiary clays and gravels ("glacial drifts")' and, while locally the result is legible and even pleasing, there are other areas where the addition of the tint and stipple has resulted in a confusion of the entire geological picture. But whatever its failings, the map is certainly a very attractive example of geological cartography and Oldham could feel justly proud of the sheet as the Irish Survey's first publication. The map is now rare because only two hundred copies were printed and coloured, these being sold to the public at seven shillings apiece as compared with a price of two shillings and sixpence charged by the Ordnance Survey for the standard county index sheets. Incidentally, those 'Tertiary clays and gravels', which Oldham was at such pains to depict, he thought were the residue of a recent marine inundation. Eight years had elapsed since Louis Agassiz had identified the relics of former Irish glaciers[64], but Oldham still preferred to regard stadial moraines in Glenmalur, County Wicklow, as some form of shingle bar formed by the waves of a transgressive sea.[65]

The county indexes may have been on too small a scale to satisfy the Survey, but their use as a geological base-map initially did seem to offer one slight advantage. Even Oldham himself admitted as much during a meeting of the Royal Irish Academy on 13 November 1848.[66] The advantage stemmed from the fact that Kane was then preparing a series of county maps showing land classification and using as a base-map the same county indexes as the Survey was now employing for its geological sheets. Kane's classification recognised the five categories of 'waste lands', 'soils of inferior value', 'soils of medium value', 'soils of superior value', and 'soils of factitious value', the categories being delineated partly upon the basis of the land values calculated by Griffith's valuators, and partly upon the results of a chemical examination of the soil-samples being collected by the Geological Survey. His map of County Wicklow had already been completed by the autumn of 1848 and it was hoped that a comparison of Kane's maps — they were known as agrological maps — with the Geological Survey's maps drawn to exactly the same scale would establish the existence of important and hitherto unrecognised relationships between geology and agriculture. In this way the activities of the Survey might be given a direct relevance to the country's agricultural improvement, but it all turned out to be something of a chimaera because after a few years Kane abandoned the construction of agrological maps and none of them was ever published.[67]

At the same time as the County Wicklow map made its appearance, the Survey also published four sheets of horizontal geological sections across the county, drawn to the standard Survey scale of six inches to one mile (1:10,560).[68] The sections cover a total length of some one hundred and thirty miles (210 kilometres) and they had been surveyed on the ground by De La Beche, Oldham, and Smyth. It was evidently copies of some of these sections which Oldham had both presented to the Lord Lieutenant and shown to the Geological Society of Dublin in April of the previous year, but their publication was delayed until 1848 because they were intended to accompany and illustrate the geological map of County Wicklow. It is interesting to note that in constructing the sections all the topographical detail seems to have been taken directly off the six-inch Ordnance map, and here the Irish Survey had a decided advantage over its British counterpart. In the absence of a six-inch coverage in Britain, the geologists were having to survey their own section lines with chains and theodolites, and this was to prove a formidable undertaking in the mountainous districts of North Wales where De La Beche's men had just started work. The 1848 list of publications contained one other item; Smyth's plan of the Avoca mines drawn to a scale of one inch to 300 feet (1:3,600) and entitled:

> Plan and sections of the Ovoca Mines County Wicklow. Comprising the copper and iron-pyrites mines of Ballymurtagh, Ballygahan, Tigroney, Cronebane, and Connary.

The plan's exact date of publication is uncertain but one copy was presented to the Royal Irish Academy on 9 November 1848[69], and another copy was seemingly presented to the Geological Society of Dublin at its monthly meeting five days later.[70]

On 9 May 1849 the Survey published its second county map, this one being for County Carlow, and four weeks later (on 13 June) there appeared the companion map for County Kildare (Plate 3).[71] These two maps are in exactly the same style as the earlier County Wicklow sheet, and again the drift deposits ('Limestone drift (Tertiary)' in Carlow and 'Limestone gravels and clays' in Kildare) are rather unsatisfactorily shown by means of a tint washed over a finely stippled background. Along with the Carlow and Kildare sheets there should have been published a similar map of County Wexford, that of course being the county where the Survey had first broken ground, but in his annual report for 1848 Oldham explained that he had delayed publication of the Wexford sheet because he had found so much of the Survey's early mapping to be inaccurate. Perhaps he had McCoy's work chiefly in mind, and he certainly sent his men back to County Wexford during the 1848 field-season to conduct an extensive programme of revision and re-mapping.

As was then the practice, many of the newly-published county maps went as complimentary copies both to individuals and to institutions, but the public at large was evidently not very enthusiastic about the three sheets. As late as 3 April 1855 Messrs Hodges and Smith, now of Grafton Street, Dublin, the sole Irish agents for the Survey's maps, had taken only 73 copies of the Wicklow sheet, 15 of the Carlow sheet, and 20 of the Kildare sheet.[72] This was hardly big business, and Oldham himself was far from happy about the maps. In his annual report for 1848 he complained that the county index maps were 'exceedingly ill adapted for the proper exhibition of the features of Geological structure', and in his report for the following year he returned to the subject at some length. He emphasised again that the index maps were on far too small a scale for the Survey's purpose, and he now pointed out that while the index sheets used for the maps of Wicklow, Carlow, and Kildare had all possessed a common scale of half an inch to one mile (1:126,720), there was no such uniformity of scale among the index sheets for the remaining twenty-nine counties. The forthcoming geological map of County Dublin, for example, was based upon an index sheet with a scale of one inch to one and a half miles (1:95,040). He concluded by entering a strong plea that the Ordnance Survey should again be pressed to prepare a one-inch map for the Survey's use.

In the autumn of 1848 the Survey's involvement in the collection of soil samples for the museum engendered serious friction between Oldham and Kane. The Survey's complaint was that the collection of the soils was seriously hindering the task of mapping the solid geology, and certainly the surveyors must have spent an appreciable proportion of their time digging holes and carrying soil-bags. Between 20 February 1847 and 16 November 1848 no less than 330 soil samples were despatched to Kane from counties Carlow, Dublin, Kildare, Kilkenny, Wexford, and Wicklow, and from Queen's County, each sample being accompanied by a form upon which the collector had to answer fifteen questions about both the soil and its environment. For his part Kane complained incessantly that the individual samples he received were too small for analysis — he wanted at least two pounds (900 grams) of soil in each sample — that the samples were being inadequately documented, and that they were from sites too widely scattered to afford a satisfactory picture of the areal distribution of soil-types. The whole sad subject seems to have come to a head on 13 November 1848 at a meeting of the Royal Irish Academy when an angry Kane taunted Oldham by reminding him that the Survey was not collecting soils merely of its own volition as a gesture of goodwill towards the museum. Rather had the Survey received specific instruction to undertake the work from no lesser individual than the First Commissioner himself acting 'upon the application of Sir Robert Kane'.[73] On the following day Kane addressed a long and forceful letter to De La Beche putting the museum's case and stressing in particular that the Survey had provided him with the woefully inadequate total of only 48 soil samples from the whole of County Wicklow. On 27 November De La Beche forwarded the letter to the First Commissioner together with a covering statement of his own presenting the Survey's attitude. In his memorandum De La Beche observed that if the Survey was to become seriously involved in the assessment of agricultural potential, then it should have upon its staff an agricultural geologist qualified in soil chemistry and trained to consider soils within the full complex of their environmental setting. This was a strikingly modern-sounding pedological proposal and De La Beche suggested that such an officer should be appointed for a trial three-year period at a cost of £250 per annum. De La Beche's efforts at empire building had the happy knack of succeeding, but this particular proposal never came to anything and the storm over soil samples gradually passed away. As early as 27 November Oldham wrote to his staff reminding them of their continuing responsibilities in respect of the collection of soil samples, and this duty they by no means shirked because between 6 December 1848 and 24 September 1850 there arrived at the museum a total of 451 soil samples collected in counties Carlow, Dublin, Kilkenny, Waterford, Wexford, and Wicklow. Relations between Oldham and Kane nevertheless remained cool as is shown by the following letter from Oldham to De La Beche dated at 51 St Stephen's Green on 4 June 1849:

> Sir R. Kane had the Lord Lieutenant here on Saturday and never said one word to me about it, though I saw him that forenoon, and never once brought him near our part of the house. This is the *friendly* way he acts towards us.[74]

Exit Oldham

In 1850 the staff of the Irish Survey consisted of Oldham, three Assistant Geologists (Du Noyer, Willson, and Wyley), one General Assistant (J.G. Medlicott who had replaced Penny in October 1846), and Flanagan the Fossil Collector. Of these six men only Du Noyer and Flanagan remained in Ireland by 1857, and it looked like a realisation of Colby's prophesy that a civilian survey would prove to have an unacceptably high rate of staff-turnover. The first to leave Ireland was Oldham himself. In March 1850 the Board of the East India Company offered him the newly-created post of Superintendent of the Geological Survey of India at a salary of £800 per annum. For Oldham, who was eager to marry a certain Miss Dixon, this offer was a sore temptation, but he nevertheless responded to the Indian overtures with a negative. De La Beche was informed of what had happened, but he did nothing to make it easier for Oldham to remain in Ireland and all the Irishman's hesitation was overcome when the East India Company's offer was renewed later the same year at a salary now more than three times the £300 he was earning as Local Director in Dublin. Oldham submitted his resignation to De La Beche on 14 November 1850, the resignation to take effect from the end of the same month, and in a letter to the First Commissioner he explained the reasons for his having felt less than happy about the Irish post. Firstly, he pointed out that the status accorded to the Local Director in Dublin left much to be desired. He was regarded as equivalent to the Local Director for England and Wales, but this was unjust because for most of the year the latter officer had his Director close at hand, whereas the Local Director in Dublin was much more isolated and he therefore carried a far greater burden of personal responsibility. Secondly, the high cost of living in Ireland was making it difficult for him to exist upon his Dublin salary. He pleaded the inadequacy of the £50 paid him in lieu of travelling expenses incurred within a 15-mile radius of his various field-headquarters, and he protested that over the previous four years these local travelling expenses had on average amounted to £73 per annum with the additional £23 coming out of his own pocket. Similarly, he observed that the cost of his accommodation while on tours of inspection was proving to be very high, and again he frequently had to dig into his own resources because hotel-charges were commonly well beyond the limit of the reimbursement which he could claim from the Survey. All this rings true — certainly far truer than those excuses which James had offered in explanation of his own resignation four years earlier.

Oldham remained in Dublin until January 1851 (his last appearance at a meeting of the Geological Society of Dublin was on 8 January[75]), and soon thereafter he and the new Mrs Oldham left for the East where Oldham assumed his fresh geological duties in the following March. In India he found that the establishment of the geological survey consisted of one peon and one writer; there were no European assistants, there was no provision for field-work, and the survey's few records were kept in a box in the Surveyor-General's office.[76] It was a far cry from 51 St Stephen's Green, and one of Oldham's first acts in India was to write back to Dublin inviting Medlicott to go to his assistance in Calcutta. Medlicott was perhaps relieved to have the opportunity of shaking the dust of Ireland from his feet; as a Protestant he had just effected a clandestine marriage with a Catholic and was experiencing the opprobrium of his enraged father, a Church of Ireland clergyman. Medlicott resigned from the Irish Survey in September 1851 and he joined Oldham in December, thus becoming the first of five members of the Irish Survey who were drawn off to India by Oldham's attractive offers. It was a bleeding of geological expertise which the Irish Survey could ill-afford, although, of course, no blame attaches to Oldham for triggering the emigration. In Ireland he had proved to be an excellent Local Director who had lifted the Survey out of the Jamesian slough and then gone on to establish a reputation for work of high quality. In India he was just as successful and he remained Superintendent there until ill-health forced his resignation in March 1876. He died at Rugby, in England, on 17 July 1878.

REFERENCES AND NOTES FOR CHAPTER FIVE

1. OSL, De La Beche to the Board of Ordnance 28 March 1832.
2. In this chapter all documents not given a specific location will be found in the archives of the Geological Survey of Ireland. For two general histories of the Geological Survey see John Smith Flett, *The first hundred years of the Geological Survey of Great Britain*, London, 1937, pp. 280, and Edward Battersby Bailey, *Geological Survey of Great Britain*, London, 1952, pp. xii + 278.
3. *Instructions for the Local Directors of the Geological Surveys of Great Britain and Ireland.* A copy is in the archives of the Geological Survey of Ireland.
4. ADB; DNB; *Q. Jl. Geol. Soc. Lond.* 56, 1900, lix-lx; *Geol. Mag.* N.S. dec. 4, 6, 1899, 283-287; *Nature, Lond.* 60, 1899, 83; *Victorian Nat.* 16 (2), 1899, 19.
5. Richard Griffith and Frederick McCoy, *A synopsis of the characters of the Carboniferous Limestone fossils of Ireland*, Dublin, 1844, pp. viii + 207 + 29 plates; *A synopsis of the Silurian fossils of Ireland*, Dublin, 1846, pp. 72 + 5 plates.
6. *Catalogue of organic remains, now exhibiting at the Long Room, in the Rotundo, collected by the late Henry Charles Sirr, Esq. formerly Town Major of the city of Dublin*, Dublin, 1841, pp. 9; *Catalogue of recent shells, now exhibiting at the Long Room, in the Rotunda collected by the late Henry Charles Sirr, Esq.*, Dublin, 1841, pp. 52; *A catalogue of the museum of the Geological Society of Dublin*, Dublin, 1841, pp. iv + 28.
7. RGSIP, *Minutes of meetings of council 1831-1863*.
8. A.W. Rogers, The pioneers in South African geology and their work, *Trans. Geol. Soc. S. Afr.*, annexure to volume 39 (1936), 50-62.
9. Joseph Ellison Portlock, *Report on the geology of the county of Londonderry, and parts of Tyrone and Fermanagh*, Dublin, 1843, ix; *Memoir of the life of Major-General Colby*, London, 1869, 291.
10. DLBP.
11. G.L. [Herries] Davies, The Geological Society of Dublin and the Royal Geological Society of Ireland 1831-1890, *Hermathena* no. 100, 1965, 66-76.
12. LP 7557, Colby to Larcom 22 February and 6 March 1845.
13. DLBP, De La Beche to T.E. James 23 April 1846.
14. DNB; DSB; *Q. Jl. Geol. Soc. Lond.* 11, 1855, xxvii-xxxvi; George Wilson and Archibald Geikie,

Memoir of Edward Forbes, F.R.S., Cambridge and London, 1861, pp. x + 589.
15. Wilson and Geikie, *op. cit.*, 1861, 490, (ref. 14).
16. DNB; *Geol. Mag.* N.S. dec. 3, 8, 1890, 383-384; *Nature, Lond.* 42, 1890, 205; *Q. Jl. Geol. Soc. Lond.* 47, 1891, proceedings 51-54.
17. W.W. Smyth, On the mines of Wicklow and Wexford, *Records of the School of Mines and of Science Applied to the Arts* 1(3), 1853, 349-412.
18. RP.
19. Archibald Geikie, *Memoir of Sir Andrew Crombie Ramsay*, London, 1895, 44-45, 142; Edward Greenly, *A hand through time*, London, 1938, vol. 1, 120.
20. PC.
21. Geikie, *op. cit.*, 1895, 72, (ref. 19).
22. RP.
23. DLBP, James to De La Beche 9 November 1845. See also H. James, Note on the Tertiary deposits of the Co. Wexford, *J. Geol. Soc. Dubl.* 3(3), 1846, 195-197.
24. PC.
25. *McCoy Papers* (Melbourne), McCoy to James 27 August, 7 September, 25 September 1845.
26. *Ibid.*, McCoy to James 25 September 1845.
27. PC.
28. *McCoy Correspondence* (Sydney).
29. DLBP. See also DLBP, James to De La Beche 4 February 1846. History repeats itself. Today (1983) the lack of a promised 1:50,000 map has left the Geological Survey of Ireland with an identical base-map problem.
30. RP. See also RP, De La Beche to Ramsay 20 May 1846.
31. LP 7557, Larcom to Colby 10 February 1845.
32. DLBP.
33. DNB; DSB; *Q. Jl. Geol. Soc. Lond.* 35, 1879, proceedings 46-48; *Geol. Mag.* N.S. dec. 2, 5, 1878, 382-384; *Proc. R. Ir. Acad.*, minutes 15 March 1879, 68-69; *J. R. Geol. Soc. Irel.* 5, 1879-80, 132-133.
34. Portlock, *op. cit.*, 1843, ix, (ref. 9).
35. MMR, January 1843.
36. RGSIP, *Papers read before the Geological Society of Dublin 1831-1859*; *J. Geol. Soc. Dubl.* 5(3), 1853, i-xxv.
37. T. Oldham, On the rocks at Bray Head, *J. Geol. Soc. Dubl.*, 3(1), 1844, 60-61. See also *The Geologist* 2, 1859, 371.
38. E. Forbes, On Oldhamia, a new genus of Silurian fossils, *J. Geol. Soc. Dubl.* 4(1), 1848, 20.
39. DLBP.
40. RP. Part of this letter is in Geikie, *op. cit.*, 1895, 84, (ref. 19).
41. DLBP.
42. DLBP.
43. DLBP. Also James to McCoy 3 July 1846 in *McCoy Correspondence* (Sydney).
44. RGSIP, *Minutes of meetings of council 1831-1863*.
45. DLBP.
46. DLBP.
47. DLBP.
48. Adam Sedgwick and Frederick McCoy, *A synopsis of the classification of the British Palaeozoic rocks, with a systematic description of the British Palaeozoic fossils in the Geological Museum of the University of Cambridge*, London and Cambridge, 1855, v-xvii.
49. Ernest Scott, *A history of the University of Melbourne*, Melbourne, 1936, 34-35.
50. The building today (1983) houses the Office of Public Works, but in the hall there are still to be seen the panels of various Irish marbles erected by Kane in March 1850.
51. RP.
52. A full gallery by gallery account of the museum is in *Dublin University Magazine* 42, 1853, 230-244. See also Thomas Sherlock Wheeler *et al. The natural resources of Ireland*, Dublin, 1944, 18-27.
53. *Ireland: general soil map*, Dublin, 1969, National Soil Survey, Soils Division, An Foras Talúntais, scale 1:575,000. See also M.J. Gardiner, The National Soil Survey, *Ir. Geogr.* 4(6), 1963, 442-453.
54. RGSIP, *Minutes of general meetings 1831-1875*; *J. Geol. Soc. Dubl.*, 3(4), 1847, 242, 3(4), 1848, 276 (where the date of presentation is incorrectly stated).
55. Wilson and Geikie, *op. cit.*, 1861, 414, (ref. 14).
56. *Rep. Br. Ass. Advmt. Sci.*, Glasgow, 1840, part 1, xxviii.
57. *Geol. Mag.* 6, 1869, 93-95; *Proc. R. Ir. Acad.* 10, 1870, 413-414; M.J.P. Scannell and C.I. Houston, George V. Du Noyer (1817-1869). A catalogue of plant paintings at the National Botanic Gardens, Glasnevin, *J. Life Sci. R. Dubl. Soc.* 2, 1980, 1-13.
58. Greenly, *op. cit.*, 1938, vol. 1, 277, (ref. 19).
59. G.V. Du Noyer, Remarks on the geological sections exposed by the cuttings on the Dublin and Drogheda Railway, *J. Geol. Soc. Dubl.* 3(4), 1847, 252, 255-260.
60. G.V. Du Noyer, On the cuttings exposed on the line of the Dublin and Belfast Junction Railway, *ibid.*, 4(1), 1848, 31-35.
61. See, for example, *Report from the Select Committee on Ordnance Survey (Ireland)*, H.C. 1846 (664), XV, 48-49; GSI. Letter Books, De La Beche to Earl of Bessborough 12 August 1846.
62. Many of the geologically-coloured county index maps have a MS. date and monogram in the bottom right-hand corner. These inscriptions were added at the Ordnance Survey by a senior engraver before the maps went to be coloured and the dates thus do not indicate the date of publication of any particular example of the

map.
63. *Rep. Br. Ass. Advmt. Sci.*, Swansea 1848, part 2, 71-72. See also T. Oldham, On the maps and sections of the County of Wicklow, published by the Geological Survey, *J. Geol. Soc. Dubl.* 4(1), 1848, 20.
64. Gordon Leslie [Herries] Davies, *The earth in decay: a history of British geomorphology 1578-1878*, London, 1969, 281.
65. T. Oldham, On the more recent geological deposits in Ireland, *J. Geol. Soc. Dubl.* 3(1), 1844, 61-71; Some further remarks on the more recent geological deposits in Ireland, *ibid.*, 130-132; On the supposed existence of moraines in Glenmalur, Co. of Wicklow, *ibid.*, 197-199; On the 'drift' deposits of the County of Wicklow, *ibid.*, 302-303.
66. *Proc. R. Ir. Acad.* 4, 1847-1850, 230-235.
67. *Dublin University Magazine* 42, 1853, 240-242; *First report of the Department of Science and Art*, H.C. 1854 (1783), XXVIII, 460; R.C. Simington and T.S. Wheeler, Sir Robert Kane's soil survey of Ireland: the record of a failure, *Studies* 34, 1945, 539-551; *J. Dep. Agric. Repub. Ire.* 44, 1947, 15-28. See also *J. Geol. Soc. Dubl.* 4(1), 1848, 130-133.
68. Horizontal Sections numbers 1 to 4. They were later re-titled longitudinal sections (old series) numbers 1 to 4.
69. The plan is so inscribed in the library of the Royal Irish Academy.
70. RGSIP, *Minutes of general meetings 1831-1875*.
71. *J. Geol. Soc. Dubl.* 4(2), 1849, 146, 150.
72. *Geological Survey (Ireland) return relating to skeleton maps of the Geological Survey of Great Britain and Ireland*, H.C. 1854-55 (278), XLVII, 2.
73. *Proc. R. Ir. Acad.* 4, 1847-1850, 235.
74. DLBP.
75. RGSIP, *Minutes of general meetings 1831-1875*.
76. Clements Robert Markham, *A memoir on the Indian Surveys*, London, 1878, 216. See also Fifty years of geological surveying in India, *Nature, Lond.* 62, 1900, 105-106, and India's geologists, *ibid.*, 137, 1936, 410.

6

YEARS OF GREAT ACHIEVEMENT

1850-1869

Dublin is fortunate in the possession of a fine zoological garden. Located in the Phoenix Park, the zoo annually attracts hundreds of thousands of visitors but, while those visitors are very welcome to gaze at the creatures upon display, all attempts to establish any closer familiarity with the animals are firmly discouraged. It is surely in the interests of both the visitors and the inmates that such a rule should prevail, but in the middle years of the last century the officials at the zoo seem to have had very different ideas. In the 1850s and 1860s, for instance, there was one regular visitor to the zoo — a tall, distinguished-looking gentleman, bushy-bearded, and with a high, bald pate — who was permitted to take remarkable liberties with the animals. He seems to have possessed some mesmeric power over the beasts and even the largest of the great cats held no terrors for him. He scratched the two tigers, he took all manner of freedoms with the old lion, a lioness and her cubs, and he was on the best of terms with a normally bad-tempered hyena. It was with the leopardess, however, that he enjoyed his most remarkable relationship. She leaped with joy when she heard his booming voice from afar and she received him with every sign of doating affection. With her he played endlessly just as at home he played with his numerous domestic pets, he and his wife having converted their Dublin residence into something akin to a private menagerie. And who was this devoted animal lover? He was Thomas Oldham's successor as the Local Director of the Geological Survey of Ireland — Joseph Beete Jukes.

Joseph Beete Jukes

Born at Summerhill, near Bordesley in Birmingham, on 10 October 1811, Joseph Beete Jukes was the son of a local manufacturer, one of the lesser nouveau riche of the Industrial Revolution.[1] His interest in geology developed while he was still a boy, but the science seemed to offer little in the way of career opportunities. He therefore bowed to the wishes of his widowed mother who sought to enhance her family's respectability through the possession of a son in holy orders. With this in mind, Jukes in 1830 went up to St John's College Cambridge as an exhibitioner. There he found his pleasure in the countryside rather than in the study, and one wonders whether on his rambles around Teversham or Grantchester he ever encountered a student from Christ's College out hunting for beetles — the young Charles Darwin. One Cambridge experience Jukes and Darwin certainly did share: they both met and developed an admiration for Adam Sedgwick, the professor of geology. Darwin never attended Sedgwick's lectures but Jukes did; he joined the class against the advice of his tutor who held so dubious a subject as geology to be quite unsuitable a study for a prospective ordinand.[2] The professor's lectures nevertheless held Jukes enthralled. As he sat at Sedgwick's feet all thought of entering the Church melted away and he resolved instead to devote his life to geology. It was a moment of rebirth, and in consequence he in later years always addressed Sedgwick as 'my dear father'.[3]

Coming down from Cambridge in 1836, Jukes lived for almost three years in England as an itinerant geologist, walking the byways, hammer in hand, and covering his modest expenses by delivering geological lectures to such as would listen. Eventually he felt the call of lands more distant. From April 1839 until November 1840 he held the rigorous office of Geological Surveyor to the colony of Newfoundland[4] and, having failed to secure the chair of geology in University College London during the summer of 1841[5], he in April 1842 again left England, this time as naturalist aboard H.M.S. *Fly* (485 tons, 18 guns) bound for Australia and the East Indies. Interestingly her commanding officer had family connections with County Down: Captain Francis Price Blackwood, son of Sir Henry Blackwood who had been Nelson's 'Prince

Fig. 6.1. Joseph Beete Jukes at about the time of his accident at Kenmare, County Kerry, on 27 July 1864. Reproduced from the biography of Jukes published in 1871 and prepared by his devoted sister Cara Amelia Browne who was the mother of the geologist Alfred John Jukes-Browne (1851-1914). A.J. Browne, to the distress of certain members of the family, assumed the additional name 'Jukes' upon attaining his majority just three years after his uncle's death.

of frigate captains'. Jukes spent almost four years in the antipodes, visiting Singapore, Java, New Guinea, and Norfolk Island, in addition to Australia and the Great Barrier Reef[6] and, while his years in the *Fly* may have been less fruitful than those of Darwin in the *Beagle* or T.H. Huxley in the *Rattlesnake*, there can be no doubt but that Jukes was a widely experienced geologist by the time of his return to England in June 1846. His appetite for global travel now satiated, he on 14 August 1846 wrote to Sir Henry De La Beche soliciting a post with the Geological Survey.[7] An interview followed and the bronzed traveller clearly made the right sort of impact because on 18 August De La Beche wrote to Andrew Crombie Ramsay:

> I have had a very satisfactory interview with Jukes, who appears a very fine fellow, and to love knowledge for its own sake.[8]

Jukes was offered a post with the Survey at nine shillings per day and he was told to present himself at Bala, Merioneth, on 1 October for training under Ramsay and William Talbot Aveline.

Jukes had now entered his element. As a field-surveyor he was an outstanding success. Meticulous in his attention to detail, and possessed of a splendid eye for country, he became one of the finest field-geologists of his day. Within less than a year of Jukes's joining the Survey, Ramsay recorded in his diary that Jukes 'produces better work and understands it better than any man on the Survey'.[9] The four years that he now spent mapping in North Wales and the English Midlands were probably the happiest in Jukes's life. By day his bulky form proved to possess quite remarkable powers of endurance in even the roughest terrain; by night, with his loud ringing laughter, he was the life and soul of the party at the hostelries which commonly served as the Survey's field-headquarters. As one of his colleagues observed:

'A more joyous, generous, kindly spirit lived not among us'.[10] Innocent in the devious ways of the world, Jukes was a man without guile who existed solely for his chosen science. But those few idyllic years were soon to end.

On 22 September 1849 Jukes married Miss Georgina Augusta Meredith of Harborne near Birmingham, and shortly thereafter De La Beche offered him the post of Local Director of the Geological Survey of Ireland in succession to Oldham. Initially Jukes was reluctant to accept the offer; he dreaded the thought of severing those close personal ties which now bound him so intimately to his colleagues on the Survey in England and Wales. Two considerations nevertheless drew him towards Ireland. Firstly, there was finance. By the autumn of 1849 Jukes's salary was £300 per annum which was exactly the same as that of the Local Director in Dublin. (Not until 1854 was the Irish Local Director's salary raised to £400 per annum.[11]) De La Beche nevertheless assured Jukes that the cost of living in Ireland was far lower than that in Britain, so that, while the two salaries might be the same, he and his new wife would in Ireland be able to enjoy a level of affluence far higher than that to which they might aspire in Britain.[12] Here, surely, we have another striking example of De La Beche's duplicity. Thoroughly familiar as he was with conditions in Ireland, he must have known that in reality the cost of living there was high. Indeed, his inability to live on his salary as a married man had been one of the chief reasons for Oldham's quitting of the Dublin post. But the naive Jukes trusted his chief implicitly. The second reason for Jukes being tempted to go to Ireland was scientific. He was a geological perfectionist. He felt that hitherto all his field-mapping had fallen short of his ideal because it had merely been one-inch mapping — one-inch mapping which, because of the limitations of scale, had precluded all that attention to those minutiae of field-geology which so fascinated him and which he so ardently desired to record upon his field-sheets. Six-inch mapping, such as that being conducted in Ireland, seemed to offer a challenging opportunity of achieving the cartographic perfection he sought.[13] Faced with such considerations, his initial hesitation about Dublin disappeared. On 20 September 1850 he wrote to De La Beche from Llangollen accepting the Irish Local Directorship.[14] A few weeks later, on the morning of 1 December, his Survey colleagues Ramsay and Alfred Selwyn saw him aboard the Irish boat at Holyhead and that afternoon Jukes ascended the steps of 51 St Stephen's Green to assume the responsibilities of his new office.

Jukes had made a mistake. Almost immediately he began to discover that conditions in Ireland were not to his liking. The very day after his arrival in Dublin he wrote to Ramsay:

> It's very dark & dreary here & I can't say I admire Dublin. There's devil a decent place to be had for lodgings under £2 a week and there's no intermediate grade between the dirty & the magnificent.[15]

Perhaps as an Englishman he felt himself to be an unwelcome intruder upon the Irish scene. Certainly only a week before he crossed to Dublin he had written to Sir Roderick Murchison:

> Oldham tells me that in one of their leading papers there are, appended to a notice of his going to India, some sharp remarks on importing an Englishman to succeed him.[16]

It may be that nationality had something to do with Jukes's first major disappointment in Ireland. In October 1850 he had applied for the Trinity College chair of geology which Oldham had held concurrently with the Local Directorship, and Jukes could hardly have been more strongly supported; he submitted testimonials from Darwin, Murchison, Ramsay and Sedgwick. An interview was held for which Jukes shaved off his beard upon being advised that the College disliked bearded professors — but on 21 January 1851 the Board of the College decided by 5 votes to 3 to offer the chair not to Jukes but to the Carlow-born Rev. Samuel Haughton (1821-1897) who was already one of the Junior Fellows and Assistant to the Archbishop King's Lecturer in Divinity.[17] Thus Jukes was denied that professorial salary with which Oldham had supplemented his Survey income, and this was a serious matter because

Jukes very soon discovered the hollowness of De La Beche's promise that he and his wife would be able to live very economically in Ireland. Indeed, the high cost of living in Ireland is a constantly recurring theme in Jukes's early Irish letters. The two passages that follow are respectively from letters that Jukes wrote to Ramsay from Monkstown, County Cork, on 22 July 1851, and from Dublin on 6 March 1852:

> As to money, by Jove, sir, this country is awful! It may be cheap to live in if you can settle down anywhere; but for travellers like us, I assure you it is at least as expensive as England, without half the comfort. They all look upon me as a government officer and fair game, and they naturally, without previous concert, combine together to impose on us.[18]

> As to the expense of living in Ireland you must either make up your mind to it, or you must sink down into a state of dirt and discomfort which makes life miserable. There is no middle class here, & accordingly no accommodation for such a class. You can't have middle class lodgings nor middle class inns, there are no country lodgings to be got anywhere. There is no medium, you must either live like a gentleman of fortune, or you must altogether cease to mingle in gentleman's society.[19]

For whatever reasons, Jukes seems to have been quite unable to adapt himself to conditions in Ireland and from almost the very moment of his arrival his character underwent a strange transformation. Gone was the carefree geologist of earlier years and in his place there stood a man overwhelmed by the responsibilities of his new office. His friends noticed with concern the change that came over him and, when Ramsay enquired as to what had happened, Jukes responded from Dungarvan, County Waterford, on 14 June 1851:

> I am sorry I have grown so matter-of-fact. I can hardly tell you why it is, but I feel that it is so. I hardly know whether it is the air of Ireland, or the nature of the work here, or what; but certainly much of the zest of life has departed, and nothing but duty and business remain. Not that we are unhappy, but only in a sort of negative state. It may partly proceed perhaps from the utter want of society. We have not made a single acquaintance anywhere.[20]

Change there certainly had been, but Jukes remained a field-geologist of outstanding ability. This talent he was now to deploy in the interests of Irish geology, and under his control the Geological Survey of Ireland became a thoroughly efficient and effective body. All too often historians have presented a picture of post-Famine Ireland as a land of debilitation, gloom, and misery. They have written of economic stagnation and political repression, of rack-rented tenants and absentee landlords, of the national haemorrhage of large-scale emigration, and of the painful throes of a land seeking to discover its true identity. Such a picture doubtless contains much truth, but we see nothing effete if we look at Irish geology as it existed during the middle years of the nineteenth century. Indeed, the twenty years between 1850 and 1870 may with justice be regarded as a golden age for Irish geology. The output of geological papers published in Irish scientific periodicals then reached a level that was not again achieved until the 1960s.[21] Following its banishment from the Dublin Custom House in 1847, the Geological Society of Dublin found for itself a new home inside the walls of Trinity College and there it flourished, becoming the Royal Geological Society of Ireland in March 1864.[22] Within Trinity College itself Haughton held the chair of geology with considerable distinction from 1851 until 1881, and geology also took its place within the newly founded Queen's Colleges in Belfast, Cork, and Galway, the work of Robert Harkness (1816-1878) in Cork and William King (1809-1886) in Galway being of particular note. Back in Dublin there was on view in the Museum of Irish Industry a collection of Irish geological materials such as no modern Irishman has had a chance of seeing since that day in 1922 when the geological galleries of the National Museum of Ireland were disbanded to make way for the civil servants of the newly independent Irish Free State. And let no historian of narrow nationalistic bent protest that all this concern for matters geological was confined to a small minority drawn solely from the social elite; from 1854 onwards geology was one of the subjects covered by courses of public lectures given throughout Ireland under the auspices of the Department of Science and Art. Attendance at many of these courses was free of charge and audiences were often large. Harkness had an average attendance of 110 at his classes in Bailieborough, County Cavan, in August 1860; Wyville Thomson (1830-1882) had an average of 219 in his classes at Waterford in September 1863; Jukes himself attracted audiences averaging 251 to his lectures in Dublin in November 1862; and in Belfast, during the winter of 1862-63, a course given by Ralph Tate (1840-1901) aroused such local enthusiasm for the field-sciences that there was founded the Belfast Naturalists' Field Club which still thrives as the most senior such club in Ireland.[23] At the centre of all this Irish interest in geology there stood the Geological Survey of Ireland, then, as now, by far the largest Irish institution devoted to the earth sciences. Its officers dominated the Irish geological scene as led by Jukes — inspired by Jukes — they between 1850 and 1869 mapped in detail the geology of more than half of Ireland.

Conflict over scales

When Jukes assumed responsibility for the Irish Survey his staff consisted of George V. Du Noyer, Walter L. Willson, and Andrew Wyley as the field-geologists, Joseph G. Medlicott as the office superintendent, and James Flanagan as the Fossil Collector. After inspecting the field-work in progress during the spring of 1851, Jukes was left with some doubts about the competence of his men, and from Waterford he wrote to Ramsay on 31 May:

> Our fellows here seem all good fellows, as good as on the English survey, but sometimes they do astound me not a little and then that leaves an uncomfortable feeling afterwards.[24]

The occasional geological lapses of his field-staff, however, were not for the moment the chief of Jukes's problems. For three years after his arrival in Dublin he was plagued by one difficulty above all others: what scale of map should be employed by the Survey for its field-mapping and upon what scale of map should the Survey publish its findings? It was yet again the old base-map problem — a problem which now makes its final appearance in our story.

One of Jukes's first tasks in Dublin was to oversee the publication of two further geological maps based upon the Ordnance Survey's county index maps. These two were firstly the long-delayed map of County Wexford, and secondly the map of County Dublin, both of which seem to have been published in May 1851. The Wexford sheet is in the same general style as the three earlier sheets of counties Carlow, Kildare, and Wicklow, but the Dublin sheet differs from all the others in that its scale is one inch to one and a half miles (1:95,040) instead of half an inch to a mile (1:126,720), and it retains the somewhat ineffective hachuring featured upon the original Ordnance Survey index map of the county. On the Dublin sheet there is also some evidence of poor editing — perhaps the result of Jukes's inexperience — because manuscript corrections have had to be introduced into the map's key. The two new maps are nevertheless accurate and pleasing pieces of geological cartography, but the public continued to take little interest in the series; by 15 May 1855 there had been sold only 26 copies of the Wexford sheet and a mere 21 copies of the Dublin sheet. Another fact is of interest. By 15 May 1855 the total sales of all five of the county maps amounted to only 155 copies[25], yet by the spring of 1855 the ten-year old Survey had cost around £15,000. Thus each of the county maps which had been sold to the public for only a few shillings, had really cost some £100 apiece to produce.

Following the appearance of the Wexford and Dublin sheets, the next county sheet scheduled for publication was that for County Waterford, and on 10 March 1852 Jukes exhibited a draft manuscript version of this map to the Geological Society of Dublin.[26] But this map was destined never to be published. Its fate was sealed by a dispute which developed between Jukes and De La Beche — a dispute which had its origins early in 1851. Both men were dissatisfied with the small-scale county indexes which had been pressed into service as base-maps for the Survey's publications, but at that basic point Jukes and his chief parted company. For Jukes the detailed six-inch (1:10,560) mapping being conducted by the Irish branch was the chief glory of De La Beche's entire Survey, and Jukes desired nothing less than a nationwide Irish coverage of *published* six-inch sheets. For his part De La Beche quite rightly regarded a national coverage of published six-inch sheets as entirely out of the question. It would have involved the production of some two thousand sheets, each 90 x 60 centimetres in size, and in many places (in the Irish Midlands, for instance) the geology is so uniform that a large number of such six-inch sheets would have carried scarcely so much as a single geological line. From this point De La Beche proceeded to argue that since the six-inch sheets were never going to be published, then six-inch field-mapping itself was a waste of time. Was it not ridiculous to spend months tracing out the minutiae of field phenomena only to lose so much of the carefully plotted detail during the course of reduction to some smaller scale? There was some logic in De La Beche's case. In England and Wales the Survey was both mapping and publishing upon the same scale — upon the one-inch map (1:63,360) — so why, De La Beche mused, should the Irish Survey not both map and publish upon the county index sheets? Such a policy, it seemed, would certainly accelerate the Irish Survey's rate of progress, and De La Beche must have been uncomfortably aware that back in 1843 he had estimated that a new geological map of Ireland could be completed within a period of only ten years. Now the end of that ten-year period — 1855 — loomed uncomfortably close and maps had been published for only five of Ireland's thirty-two counties.

It appears that in February 1851 Jukes did actually receive from De La Beche specific instruction to abandon all six-inch field-mapping and to work henceforth upon what Jukes himself once described as 'those footy little county maps'. His confrontation with De La Beche formed the subject of a letter that Jukes wrote to Ramsay on 3 March 1851.[27] Therein Jukes explained that upon receiving De La Beche's directive his first reaction had been to consider resigning the Irish Local Directorship but, after more thoughtful consideration, Jukes continued:

> I perceive that would be a foolish and also a cowardly way of acting. The interests of the survey here being committed to my charge, I am bound to stick by it and fight for it to the last. Sir H., therefore, gets a letter from me this morning, telling him the six-inch map was one of my greatest inducements to come here; that I was disgusted and disappointed when I found they were not published geologically, and that I am resolved not to rest till they are so published . . . . Taking a strictly 'service' view of the matter, I can show the absurdity of paying £1500 per annum out of the public purse for merely a slightly-amended edition of Griffiths' [*sic*] map. The real question is the amount of *detail* and *accuracy* of work.

That reference to Griffith's map was both apt and timely. Between 1852 and 1855 Griffith was to publish new and greatly improved versions of his quarter-inch map incorporating the results of Patrick Ganly's perceptive field-studies during the previous decade (see p. 75). Clearly it would be acutely embarrassing for Jukes and his men were the Irish Survey to issue sheets depicting little more geology than that already represented upon Griffith's map — a map which, in theory at least, had cost the Exchequer scarcely so much as a single penny.

In the event, Jukes and De La Beche achieved an admirable compromise. It was agreed that the Irish Survey would continue its six-inch field-mapping, but Jukes had to concede that, except in a few mining areas, the six-inch geological sheets would not be published. Instead, publication was to be on a new Irish series of one-inch maps (1:63,360). The Ordnance Survey had been considering the compilation of such a one-inch coverage of Ireland ever since Colby's arrival in Ireland in the 1820s, but it was not until 26 April 1851 that the Treasury finally authorised the Ordnance Survey to begin the construction of a one-inch coverage by reduction from the six-inch sheets.[28] From the 1850s down to the 1880s it remained normal practice for the Geological Survey of Ireland to field-map upon the six-inch scale and then to publish upon the one-inch sheets.

In a presidential address to the Geological Society of Dublin, delivered on 14 February 1855, Jukes explained exactly why he regarded six-inch field-mapping as essential for any well-run national geological survey.[29] The following are the salient and thoroughly cogent features of his case:

1. 'A national geological survey should be conducted with the most minute accuracy possible to be attained', thus obviating the necessity for repeated re-surveying. Further, we must ensure that 'even in wide and barren districts composed exclusively of one particular kind of rock, such as granite or Old Red Sandstone, every square foot of ground be examined, and every exposure of rock carefully marked upon the map'.

2. The field-geologist can locate himself with precision upon a six-inch sheet and the map is itself large enough to accommodate all those observations which in one-inch field-mapping have to be recorded in accompanying note-books.

3. When necessary, detailed geological cross-sections may with ease be drawn directly from the six-inch field-sheets.

4. The accuracy of any smaller-scale maps can easily be tested by reference to the six-inch field-sheets and the explanation for any suspected errors is speedily discerned.

5. A national geological survey should be 'carried on with reference not so much to what may be thought our immediate requirements, as to the wants and requirements of the future'. Six-inch field-sheets can be filed away and kept in perpetuity as 'a great public document of reference'. They will, for example, be available to serve as the basis for the production of such smaller scale geological maps as may be needed, or as the foundation for reports on the mineral potential of properties prepared at the request of individual land-owners.

History has shown Jukes to have been correct. For well over a century those six-inch field-sheets that were his pride and joy have occupied a central position in Irish geology. They and their associated set of fair copies may now be somewhat dated (some of them also fail to achieve that high standard of detail and precision which Jukes enjoined of his staff) but they do remain to this day an invaluable national geological archive. It is to these maps that Survey officers still have recourse in dealing with the innumerable queries about Irish geology which annually flow into the Survey's Dublin headquarters; it is to these maps that any researcher in Irish geology turns before going off to examine the actual rocks themselves. Ireland has good reason to be grateful to Jukes for his insistence upon the maintenance of the six-inch field-survey.

Jukes's mapping programme

Jukes was eager to press ahead with the field-mapping as speedily as possible and soon after his arrival in Dublin he lighted upon the collection of soil-samples as one area where time might be saved. As early as 10 March 1851 he wrote to De La Beche saying: 'Had we not better try to get rid of the soil affair?'. On 18 March De La Beche replied asking Jukes to estimate the amount of time that his men were spending upon soil-sample collection. Jukes responded three days later; he reported that the collection of the samples was occupying each of his men for an average of eight days in every quarter. Collection nevertheless continued, and during 1851 Sir Robert Kane acknowledged the receipt of 398 samples taken from counties Cork, Kilkenny, Tipperary, and Waterford. Between 9 March 1852 and 1 July 1852 a further 142 samples were returned, but then the flow of samples suddenly ceased, although it was not until January 1853 that Jukes and Kane formally agreed that the Survey's involvement in soil collecting should be terminated.[30] It is ironic that many of those soil samples upon which the Survey had expended so much labour — soil samples which had occasioned so much friction between the Survey and Kane — they were never even examined. Today, a century and a quarter later, the boxes containing numerous of the samples evidently lie unopened in the geological store of the National Museum of Ireland.

During Jukes's first few years in Ireland the Survey worked steadily westwards through the south of Ireland from County Kilkenny, through counties Waterford and Cork, into County Kerry. Jukes joined his staff in the field for much of each year and, although he was never entirely happy in Ireland, he did find satisfaction in at least some of his field-work. From time to time the jovial Jukes of earlier years shines through in his correspondence as, for example, in the following extract taken from a letter that Jukes wrote to Ramsay from the little town of Rathdrum, County Wicklow, on 15 June 1853. The reason for Jukes then being in County Wicklow will emerge shortly:

> I am going to get Sullivan[31] down here next week, and we mean to work the very entrails out of all the traps hereabouts, and not leave 'em a single atom of mystery to pride themselves on any longer. Worst of it is, when one has made out the chemistry and mineralogy, one can't see the geology in this confounded rolly-polly wood covered country.[32]

Jukes must here have the sympathy of any geologist who has shared with him the experience of trying to locate exposures in the heavily wooded and deeply incised valley of the Avonmore River in the Vale of Clara upstream of Rathdrum. The region offers some of Ireland's most picturesque landscapes, but for the geologist it can be hard going.

One of the most exciting events of Jukes's early years in Ireland was Flanagan's 1851 discovery of the rich and beautifully preserved fossil-beds in the upper Old Red Sandstone at Kiltorcan in County Kilkenny. Jukes and Edward Forbes visited the site immediately and Forbes subsequently named the magnificent fern fossils *Cyclopteris Hibernicus* (now *Archaeopteris hibernicus (Forbes)*), while, despite Jukes's good natured protests, a large freshwater bivalve shell became *Anodon Jukesii* (now *Archanodon jukesii (Forbes)*).[33] To counterbalance such a dramatic discovery, Jukes had his problems. In 1852 bad weather during the summer virtually halted all the Survey's field activities for two months; in 1854 we find Jukes complaining of the delays in the mapping programme being caused by his men encountering extensive spreads of peat and impenetrable thickets of furze; in 1856 there was another bad summer; and always the staff were wasting long hours tramping between their lodgings and their mapping ground because there was never sufficient money available for the hire of cars or horses.

Despite all these difficulties, the Survey was pressed ahead at a goodly pace, and as early as the summer of 1854 Jukes and his men had reached the mountainous country of West Cork and Kerry. It was there, strangely, that the future course of the new Geological Survey of Scotland was finally decided. De La Beche had resolved to extend the Geological Survey's activities into Scotland and he had deputed Ramsay, the only Scot upon his staff, to go north to break ground in Scotia. De La Beche expected Ramsay to use one-inch maps in the field in Scotland, as was the practice in England, but the Ordnance Survey had just published some six-inch sheets of the area around Edinburgh and, inspired by Jukes, Ramsay was determined to try his hand at a six-inch geological survey of the Edinburgh region. De La Beche disapproved, but Ramsay resolved that as a preliminary to his Scottish enterprise he would visit Ireland to learn exactly how the Irish Survey used its six-inch sheets. As Ramsay explained to De La Beche in a placatory letter:

> It would be a great mistake on my part to omit seeing what they do, and how they do it, in Ireland. It does not follow that the same rules should be applied in Scotland; but whether or no, I want to see how they keep, cut, use, and abuse their maps, what their portfolios are like, how they handle them in the field, and twenty other things that may save us much time and trouble in Scotland, and which only eyesight can instruct upon.[34]

Ramsay arrived in Dublin on 22 August 1854 and he left immediately for West Cork where De La Beche was making what was to prove his final Irish tour of inspection. For once the weather was glorious and the southwest revealed itself in all its rare but magical splendour, with azure skies, blue seas, and richly hued peninsulas fingering out towards the wastes of the Atlantic. In a letter to his wife on 25 August, Ramsay described the events of one of those perfect mornings:

> Yesterday after breakfast, Jukes, Willson, Kinahan, and I drove in a car [from Bantry] to Glengariff; Mrs. Jukes rode out with us for a mile or two on Dolly, and two dogs were also of the party, Carlo the setter, and Tommy the Scotch terrier. We dropped the other men *en route*, and Jukes and I drove on to Glengariff. It is not a town, but a tourists' inn in a lovely valley. It puts one in mind of Loch Lomond, only the water is salt and the hills not so high. There we found Sir Henry. . . .[35]

In Glengarriff the three geologists perhaps took lunch at the hotel kept by Thomas Eccles (lunch was to be had there for one shilling and sixpence) and it was certainly while they were in the village that Jukes, Ramsay and an ailing De La Beche debated the future mapping policy of the Geological Survey of Scotland. There, as they gazed out over Glengarriff Harbour towards Garinish Island and the waters of Bantry Bay, Ramsay finally secured De La Beche's permission to map the area around Dunbar upon the six-inch scale.

The Kinahan mentioned in Ramsay's letter to his wife was George Henry Kinahan (1829-1908)[36] who had joined the Survey as a Temporary Assistant Geologist only three days before Ramsay's arrival at Bantry. The son of a Dublin barrister and the holder of a Trinity College Diploma in Engineering, Kinahan had turned to geology after a year as a civil engineer and he was now in West Cork to be trained by Du Noyer and Wyley amidst the Cork Beds and the Old Red Sandstone. He eventually earned Jukes's approval as 'a hardworking and indefatigable explorer of wild & difficult districts'[37], and he was destined to become the Irish Survey's most turbulent ever character. That Mrs Jukes was present in the area serves as a reminder that Survey officers often went to their field-stations accompanied by their entire families. This was because of the difficulty experienced in securing accommodation within their field territories. As Jukes remarked to Murchison in a letter of 21 December 1855:

> The cabins of the peasantry are of course quite unavailable consisting as they commonly do of one room with neither beds nor furniture.

On the other hand, inns and hotels were few and far between, and as a result married Survey staff commonly rented a house for the field-season and moved in with their families, their furniture, and their pets. Hence the presence at Bantry of Carlo and Tommy; they were evidently members of Jukes's domestic menagerie.

A further insight into the field conditions experienced by Jukes's men is afforded by a few surviving sketches of them in their field attire, the best of these being the sketch of Flanagan taken by Du Noyer at Ferriters Cove, County Kerry, on 9 September 1856 (Fig. 5.2). The relatively conventional, 'country tweeds' style of dress being worn by the subjects of these sketches suggests that before the development of modern protective clothing — of thermal underwear and wind-proof anoraks — field-work under Irish conditions must often have been a rigorous experience. Officers of the Survey were nevertheless expected to work long hours in the field. From the printed rules for the Irish Survey, promulgated on 12 July 1854, we learn that for the six-day working week the field-staff were required to start work not later than 9AM, while in the evening they were to cease operations at a time sufficient to allow them to regain their field-base by 6PM. In winter, however, the geologists had to persist at their labours only until failing light made further field-work impossible. For those working in the Survey's Dublin headquarters life was rather easier. There the day started at 10AM and ended at 5PM except during the four winter months when dusk put an end to all activities at 4PM.

The six-inch field-sheets preserved in the Geological Survey of Ireland afford a fascinating insight into the manner in which Jukes's men performed their arduous duties. The sheets are the standard Ordnance Survey six-inch topographical sheets cut into quarters for ease of handling in the field, each of the field-sheets having a size of 45 x 30 centimetres. Unfortunately by no means all of the field-sheets are signed by their authors so it is often difficult to know whose field annotations are represented upon any particular sheet. The authorship problem is compounded because, in accordance with Jukes's notion that the sheets should constitute a national geological archive, a variety of later geologists have added their own observations to the original field-sheets. Sadly, the geologists making these later additions have not always treated the field-sheets with that reverence which is surely their due. Some members of the Survey involved in revision mapping during the period after 1870, for instance, have desecrated the field-sheets with appallingly crude pencil scrawls, and the presence of untidy additions in ball-pen shows that even more modern geologists are by no means above reproach. So far as the original field-mapping is concerned it is fair to say that the early work in southeastern Ireland was of a very poor standard, but as the surveyors moved through County Waterford and on into counties Cork and Kerry, their work became steadily more detailed and precise. Initially they did little more than follow a few stream sections and plot such outcrops as they saw by walking along some of the principal roads. In consequence an early field-sheet may carry as few as ten recorded outcrops upon a sheet representing an area of 16 square kilometres. Jukes's ideal nevertheless remained a six-inch map upon which virtually every outcrop of solid rock was recorded and described, and as his men moved through Munster this was increasingly the standard to which they themselves aspired.

A number of Jukes's men — Frederick James Foot (1830-1867),[38] for instance, Willson and Wyley — became very proficient field-surveyors, but one of his staff excelled above all the others — Du Noyer. Many of Du Noyer's sheets are beautiful examples of the field-geologists's art and it is noteworthy that Du Noyer was one member of the staff who regularly did sign his field-sheets. He was clearly proud of what he had achieved and that pride was fully justified. Most of his sheets are replete with a wealth of geological detail and are a joy to the artistic eye. In April 1850, when working near Kilmore Quay in County Wexford (Sheet 51/4), he felt the need of a scale larger than the six-inch for his mapping of the Pre-Cambrian rocks at Crossfarnoge Point, so, employing the method of squares, he enlarged his base-map to a scale of two feet to one mile (1 : 2,640). Upon his field-sheets for County Waterford he has introduced a very effective system of hill-shading to emphasise the character of the mountainous terrain, and even smaller landforms, such as the moraines in the corries of the Comeragh Mountains, have been picked out beautifully through the employment of the same technique (Fig. 6.2). Du Noyer was a most accomplished artist — there are many hundreds of examples of his artistic skill in the archives of the Geological Survey of Ireland and elsewhere — and the margins and reverses of many of his field-sheets are embellished with most attractive sketches and watercolours of scenes encountered during his mapping. Most of the illustrations depict sites of geological importance, but some of them are devoid of any geological significance. Sheet 24/3 of County Kildare, for instance, carries a sketch of a busy village fair-day, while

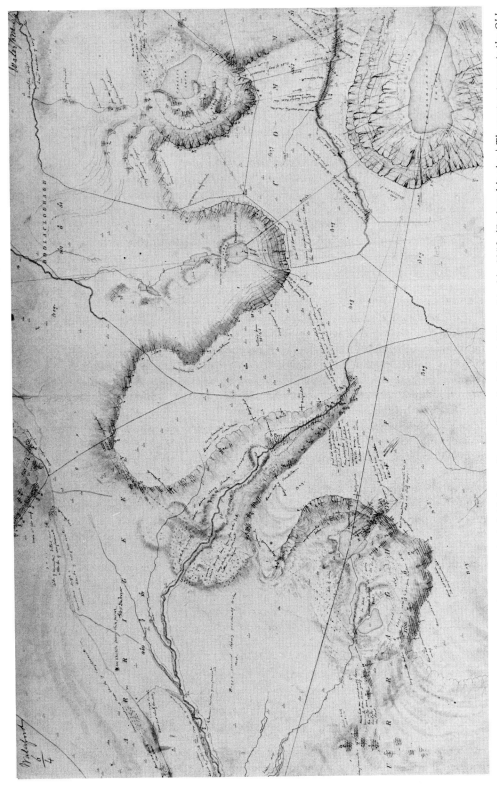

Fig. 6.2. One of George Victor Du Noyer's six-inch field-sheets: County Waterford, Sheet 6/4, mapped in 1850. (Reduced in size.) The map represents the Old Red Sandstone county of the Comeragh and Monavullagh Mountains, and Du Noyer has introduced hill-shading to emphasise the deep corries and to represent the corrie-moraines.

the reverse of Sheet 36/3 of County Kerry bears a drawing of Du Noyer's field-companion, a dog of indeterminate breed, accompanied by the caption 'Mr Buff, very hot after chasing a hare'. It may be that Mr Buff met a sad end because, upon Sheet 66/3 of County Kerry, Du Noyer has inscribed in the waters of Lough Leane near Ross Bay another sketch of his dog accompanied by the laconic phrase 'Alass [sic] poor Buff'. Details such as these give to the sheets a fascinating human dimension and it would be a very insensitive modern geologist who could study the sheets without developing a warm feeling of affinity for Jukes, Du Noyer and their colleagues. Occasionally we find them engaged in a dialogue through the sheets. On Sheet 17/4 of County Wexford the original surveyor has recorded a dyke of finely grained crystalline syenite on the coast near Cahore Point. Alongside the indicated outcrop are three later comments: 'Could not find, J.B.J'; 'Neither could I, G.V.D. 14 Oct. 1862'; 'Dyke quite plain, to be seen a little north of Gilligan's Cove, G.H.K. April 1877'. One other addition to the earlier field-sheets deserves a mention. On the reverse of Sheet 63/3 of County Kerry Arthur Beavor Wynne (1835-1906)[39] has drawn a sketch-map of an area covered by four full six-inch Ordnance sheets and lying southeastwards of Castlemaine Harbour. Upon this sketch-map he has plotted some geological boundaries derived from Griffith's map of 1855. It seems that Wynne, who reported for duty with the Survey on 1 April 1855 in replacement of Wyley, may have been using Griffith's map as a field-guide during his own Survey noviciate.

Administrative changes

On Saturday 11 October 1851 the Great Exhibition at the Crystal Palace finally closed its doors to the public, the exhibition leaving behind it a strong British awareness of the national importance of what were termed 'the fine arts and practical science'. As a direct result of this new consciousness there was established in March 1853 the Department of Science and Art, to which there was assigned the task of developing a nationwide programme for the teaching of science and the arts in an effort to enlist them in the cause of industrial improvement and expansion. Among the bodies taken under the wing of the new Department were both the Geological Survey of the United Kingdom and the Museum of Irish Industry, the latter, in 1854, becoming the Museum of Irish Industry and Government School of Science Applied to Mining and the Arts, following a decision to develop the museum's educational function.[40] Jukes was appointed as the school's first professor of geology (he of course remained the Survey's Local Director) and he soon proved himself to be a very effective lecturer. At least four of those who attended his classes later became officers of the Survey: Richard Joseph Cruise (1842-1895)[41], Alexander McHenry (1843-1919)[42], Joseph Nolan (1841-1902)[43], and Wynne. For the Survey itself the establishment of the Department of Science and Art brought in its train a development of far greater importance. In the wake of the Great Exhibition the climate in official circles was entirely favourable to the Geological Survey. It seemed to offer an obvious area in which science could be harnessed to the interest of industry, and the Treasury therefore sanctioned the augmentation of the Survey's staff on both sides of the Irish Sea. As we have seen (p. 160), when Jukes arrived in Dublin in December 1850, the Survey's Irish staff consisted of the Local Director himself, one Geologist (Du Noyer who had been promoted to this rank in 1849), two Assistant Geologists (Willson and Wyley), one General Assistant (J. G. Medlicott) and one Fossil Collector (Flanagan). Of these Medlicott departed in September 1851 to join Oldham in Calcutta, Wyley resigned in January 1855 to become Government Geologist in Cape Colony, and Willson left in November 1856 to join his four former colleagues who were now in India, Willson explaining in his letter of resignation that it was impossible for him to maintain a family upon his Irish salary in a land where, he claimed, the cost of living had more than doubled in a mere five years.

As replacements and additions to his staff Jukes appointed ten men (six Assistant Geologists, two General Assistants, and two Fossil Collectors) between October 1851 and July

1856 and he must have been well satisfied with the calibre of the men now being attracted to the Survey. Both the General Assistants — John Studdert Kennedy (died 1856) and Henry Benedict Medlicott (1829-1905)[44] — and four of the Assistant Geologists — Foot, Kinahan, Samuel Medlicott (c. 1831-1889), and Joseph O'Kelly (1832-1883)[45] — had been students in Trinity College Dublin, and of these six, five held the College's Diploma in Engineering, the course for which involved attendance at classes in geology. The strangest of Jukes's appointments was that of John Kelly — the very same John Kelly from Borrisokane, County Tipperary, who way back in 1814 had become assistant to Richard Griffith at the Dublin Society (see p. 41). Kelly had passed his entire life in Griffith's service, but in 1853, when ill-health forced his retirement, he found himself without a pension. Jukes evidently took pity on Kelly and in July 1856 he appointed him as an Assistant Geologist expressly to take charge of the Survey's office, and Kelly thus performed for the Survey precisely that function which he had for so long performed in Griffith's Valuation Office.

Not all these appointees remained with the Survey for any length of time. H.B. Medlicott who joined the staff in place of his brother J.G. Medlicott in October 1851, was in 1853 transferred to the British Survey at his own request, and resigned in 1854 to join the Geological Survey of India where, in 1876, he succeeded Oldham as the Superintendent. H.B. Medlicott's replacement, Kennedy, was appointed in November 1853, but he resigned the following year, also bound for the Survey of India. Samuel Medlicott, a third brother of the same family, joined the Survey in September 1854 but resigned in June 1856 to become an Ensign in the 87th (South Cork) Regiment of Militia. (When he left the militia after a few years of service, Samuel Medlicott did something unique in the annals of the Irish Survey — he became a Church of England clergyman.) Finally, Pierce Hoskins, a former schoolmaster from Kinsale who was appointed as a Fossil Collector in October 1854, resigned in the following January, explaining to Jukes that 'my present situation does not answer my expectations'. In consequence of all these appointments and resignations Jukes in January 1857 had a staff of eight men, but of those eight only Du Noyer and Flanagan had to their credit more than two and a half years of service with the Survey. Colby was surely vindicated. Back in the 1840s he had warned that a geological survey manned by civilians would have an unacceptably high rate of staff turnover (see p. 110). There must certainly have been times when Jukes saw himself as administering not so much a national geological survey as a nursery for future officers of the Geological Survey of India.

Through his regular and prolonged visits to Ireland De La Beche had obtained an intimate knowledge of the Irish staff, but his tour of inspection during the summer of 1854 proved to be his last. His health had been failing for some time and he died on 13 April 1855 following a stroke. His successor was appointed on 5 May 1855. He was 'the King of Siluria' — Sir Roderick Impey Murchison. He was no stranger to Ireland. As an Ensign in the 36th Foot he had sailed to the Peninsula from Cork in 1808, and towards the close of the Napoleonic War he had been stationed at Armagh. Later, having turned to geology, he had attended the meetings of the British Association for the Advancement of Science in Dublin (1835), Cork (1843), and Belfast (1852), and during the summer of 1851 he had searched for a base to his Silurian System in the wild country southward of Killary Harbour.[46] In the summer of 1856 Murchison returned to Ireland, this time as Director of the Geological Survey. He arrived in Dublin on 15 August and spent the next few days in the city examining maps and sections at Griffith's home and visiting locations such as the Museum of Irish Industry which, he noted in his diary, was really a display of the 'industry of the world to instruct the benighted sons of Erin'.[47] Outside Dublin his first port of call was Kilkenny to inspect ground being mapped by Willson, and from there Murchison proceeded to Limerick where he stayed at Cruise's Hotel. Thackeray had described that establishment as 'one of the best inns in Ireland'[48] but Murchison observed that, although so celebrated, the hotel really had 'nothing to boast of, save a very civil landlord'. Perhaps the Director was in an ugly mood; the weather had turned

Fig. 6.3. Sir Roderick Impey Murchison, Bart, K.C.B., Director and then Director General of the Geological Survey of Great Britain and Ireland from 5 May 1855 until his death on 22 October 1871. The illustration is based upon a portrait by Henry William Pickersgill (1782-1875). Murchison delighted in receiving titles and honours and he was especially proud of the medal here suspended from his lapel by its blue-bordered crimson ribbon. It is the Military General Service Medal with the three bars for Roleia, Vimiera, and Corunna awarded to him for service in the Peninsula, but issue of the medal was not authorised until 1847 and it therefore bears the effigy of Queen Victoria.

nasty and his efforts to buy a set of waterproof clothing in Limerick had met with no success. From Limerick he went west to Tarbert and thence by steamer to Tralee, there as yet being no rail link into northern County Kerry. In Tralee on 21 August he met up with Jukes, Du Noyer, the seventy-one year old Griffith, and John William Salter (1820-1869) who two years earlier had succeeded Forbes as the Survey's Palaeontologist. The next six days they all spent together examining the geology of the Dingle Peninsula which, as we will see later (p. 182), was presenting the Survey with some perplexing problems. The old sportsman in Murchison was doubtless pleased to note the presence of 'immense large trout, white and brown' in a tarn high up on the slopes of Mt Eagle at the western extremity of the peninsula, but the geologist in him must have been even more delighted by an admission now finally wrung from Griffith. Ever since 1840 Griffith's quarter-inch map had represented the rocks throughout the western half of the Dingle Peninsula as all being of Silurian age. Repeatedly since his visit to the region following the Cork meeting of the British Association in 1843, Murchison had urged Griffith to temper his claims as to the extent of Silurian strata both within the Dingle Peninsula and throughout southwestern Munster generally, but now, in August 1856, Murchison at last secured Griffith's concession that, setting aside the undoubted Silurian strata around Dunquin and Ferriters Cove, it was better to regard the remaining rocks of the western half of the peninsula as being of Old Red Sandstone age rather than as being Silurian.

Griffith's great map had already just entered upon its final state, however, and Griffith's change of mind was never to be given cartographic expression. Perhaps it scarcely mattered because modern opinion holds that so far as the debatable rocks of Dingle are concerned, they are in all probability partly upper Silurian and partly Lower Devonian. Thus the views of Griffith and those of Murchison both enshrined a partial truth (see p. 183).

Murchison left Dingle on 28 August and in bad weather he travelled southwards to the Gap of Dunloe (now known to be a glacial diffluence col) and thence to Muckross. One wonders whether Murchison knew that the churchyard at Muckross contains the mortal remains of Rudolf Erich Raspe (1737-1794), that strange but able German geologist who wrote *Baron Munchausen's narrative of his marvellous travels* and who has the dubious distinction of having been expelled from the Royal Society. At Muckross the weather continued to be bad on 29 and 30 August but for the last two days of Murchison's stay the sun shone and he saw the Lakes of Killarney at their glorious best. Indeed, he was moved to describe Muckross as 'an enchanting place' and it was perhaps with reluctance that on 2 September he joined the train bound for Dublin. The next day he spent with Jukes inspecting ground in County Wicklow and on 4 September he left for Holyhead in the paddle steamer *Cambria* with 'wind southerly; swell heavy; lots of sick people'.

Murchison had not enjoyed his visit. Despite his enchantment with Killarney, his three weeks in Ireland seem to have left him only with memories of inadequate hotels, unappetising food, uninteresting rocks, and day after day of drenching rain. It must also be noted that he shared with De La Beche a deep prejudice against all those manifestations of the Roman Catholic faith which are so widespread in Ireland. Following his return to England Murchison wrote to Ramsay:

> I really must declare that the geology of Ireland is the dullest ('tell it not in Gath') which I am acquainted with in Europe. If St Patrick excluded venomous animals he ought to have worked a miracle in giving to the holy isle some one good thing under ground. But no! everything has had a curse passed upon it. There are as good Cambrian rocks as need be, they are all like the Longmynd, and won't give good slates. Then there are as good Carboniferous Limestone and Millstone-grits as any in Scotia, but it is pitiable to see the miserable small packets of broken culm at intervals of scores of miles, which are dignified by the name of Coal-measures. Then as to mines it is *nil*, except what used to be called the curse of the miner (pyrites). Jukes is a fine energetic fellow, and I made the acquaintance of all his men (inspecting their work), who are really good hardworking youths, who can stand a life no Englishman would tolerate.[49]

When Ramsay met Murchison after his Irish tour the Director exclaimed: 'Catch me going to Ireland again!'.[50] He almost kept his word. He did return for the Dublin meeting of the British Association in 1857, but otherwise, although he remained Director (his title was changed to Director General in 1867) of the Survey until his death in 1871, he seems never again to have set foot upon Irish soil. Murchison's dereliction of his Irish responsibilities threw upon Jukes an increasingly heavy burden of responsibility, and understandably he resented not receiving from his chief that support and assistance which should have been his due.

The one-inch sheets

During Jukes's first year as Local Director — 1851 — the Survey mapped the impressive total of 3,017 square kilometres of fresh territory, and in 1852 a further 2,573 square kilometres were mapped. But in 1853 and 1854 the totals for new territory mapped fell to only 1,741 and 1,923 square kilometres respectively (Fig. 6.4). To some extent this decline was a result of the loss of experienced staff, of inclement weather, of the mountainous character of the terrain now being encountered in West Cork and Kerry, and of the difficulty of finding

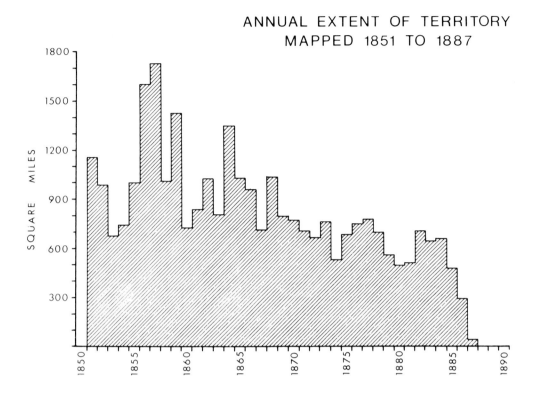

Fig. 6.4. Based upon the manuscript annual reports of the Irish Survey.

field accommodation near to the areas under examination. But there was another reason for the diminished rate of progress — a reason arising from the decision henceforth to publish the Survey's work upon the one-inch scale. Perfectionist that he was, Jukes had decided that much of the Survey's early mapping in Leinster was insufficiently detailed to stand publication on a scale larger than that of the county index sheets. He therefore determined to have large areas of southeastern Ireland re-examined. Jukes and Wyley spent the greater part of the summers of 1853 and 1854 re-mapping old ground in counties Wexford and Wicklow[51], but Jukes's meticulousness left De La Beche feeling exasperated. The southeast was territory that James was supposed to have finished in 1846 and that Oldham had revised in 1847 and 1848, but now here was Jukes insisting upon further revision in the 1850s. In 1855, when the Survey achieved its tenth anniversary, there can have seemed little cause for celebration. According to De La Beche's original estimate that year should have seen the near-completion of the entire geological map of Ireland, but in reality the Survey's series of county sheets had come to an end with only five of the thirty-two counties covered, nothing at all had been published since 1851, and the Ordnance Survey was proceeding only slowly with the production of the new one-inch base-maps. Jukes's own patience must have been sorely taxed. By the beginning of 1855 his men were advancing northwards along a line extending from Berehaven, through Kenmare and Millstreet, to Mallow, and he had in his hands completed six-inch field-sheets for the greater part of Leinster and Munster, yet, in the absence of the one-inch map, he lacked any means of laying the Survey's work before the public. He must have been embarrassed in 1855 when there was call for a geological map of Ireland to be displayed at the year's Exposition Universelle in Paris because there was no official map as yet available. It was the splendid new version of Griffith's quarter-inch map that had to be despatched to France (see p. 75).

Although the production of a one-inch coverage of Ireland had been authorised in April 1851, it was November 1853 before the Ordnance Survey sent the first of the new sheets over to 51 St Stephen's Green to have the geological lines added. Originally the Ordnance Survey intended to cover Ireland in 59 one-inch sheets, each sheet having a size of 90 x 60 centimetres, these being the same dimensions as those of the Ordnance Survey's six-inch sheets. This was far too large a size for convenient use in the field (the Geological Survey, of course, normally cut its six-inch sheets into quarters for use in both the field and the office) and in consequence the Ordnance Survey soon changed its plans by deciding to issue the new one-inch map in sheets a quarter the size of those at first intended. Until October 1858, however, the smaller sheets were regarded as quarter sheets of the larger sheets envisaged earlier, being numbered, for example, 40 S.E. or 47 N.W. Jukes believed, probably wrongly, that he himself had been instrumental in bringing about this change in the Ordnance Survey's policy. He once recounted that late in August 1852 De La Beche had passed through Dublin on his way to the Belfast meeting of the British Association and that while in Dublin De La Beche had enquired of him as to the size he thought the new one-inch sheets should be. Jukes, familiar with the quartered six-inch sheets used by the Survey, responded that their size was the largest that could be handled with comfort in the field and he therefore recommended this to his chief as the most appropriate format for the new sheets.[52]

The first of the new sheets reached Jukes in November 1853; they were the four quarter sheets comprising Sheet 36 and covering a large portion of County Wicklow. After plotting the geological boundaries upon the sheets, Jukes had geological colours added and in this form Sheet 36 was exhibited to the Geological Society of Dublin on 14 December 1853.[53] Jukes despatched the sheet back to the Ordnance Survey in February 1854, and in the following August it returned to his hands, now with all the geological details engraved. This proof map, with colours added, he exhibited to Section C of the British Association meeting at Liverpool that September,[54] and publication should have followed soon thereafter. Instead there ensued almost three years of delay. In his presidential address to the Geological Society of Dublin in February 1855 Jukes blamed the delay firstly upon a shortage of engravers in the Ordnance Survey and secondly upon the military upheaval occasioned by the outbreak of the Crimean War in March 1854,[55] but that the delay was prolonged until 1856 was chiefly the result of a policy change within the Survey itself.

Originally it had been intended to show no drift deposits upon the one-inch maps, a decision which was doubtless influenced by the earlier want of success in trying to depict both the drift and the solid geology upon one and the same map in the county index series of sheets. It should also be noted that in England and Wales it had long been De La Beche's policy to disregard most of the drift mantle and to publish sheets that were essentially representations of the solid geology alone. Perhaps he had never thought very deeply about the problem of the drifts because he had begun his own mapping in southwestern England well beyond the limits of Pleistocene glaciation. But now the issue of drift mapping was raised urgently in the Irish context. On the morning of 2 February 1856 a visitor came to 51 St Stephen's Green to see Jukes. He was John Doyle, the manager of the map department in the shop of Messrs Hodges and Smith which stood just around the corner from 'The Green' at 104 Grafton Street. Doyle wanted to inspect some of the new one-inch sheets, and upon seeing them he immediately expressed disquiet that the new maps did not include a comprehensive representation of the drift overburden. That afternoon he wrote to Jukes formally upon the subject observing that the absence of the drifts was bound to affect sales of the maps adversely because purchasers of Irish estates were invariably interested in the character of the drift mantle present upon their domains. 'In fact', Doyle continued, *'that* is the chief if not the only geological feature of interest to most buyers of the maps'. Jukes was impressed by this commercial argument and he immediately forwarded Doyle's letter to Murchison together with a covering letter of his own indicating his concurrence with Doyle's observations. Jukes further suggested that the drift should be represented by some system of stippling similar to that employed upon the earlier county maps. Murchison was also convinced. He replied on 6 February 1856 agreeing that the drift should be added but leaving to Jukes the decision as to exactly how the task should be executed.

Meanwhile the production of the 'driftless' one-inch sheets had been continuing. During February 1856 Jukes received from the Ordnance Survey the sets of sheets 36, 41, 47, and 53 covering the whole of southeastern Ireland, the sheets all carrying their engraved geological lines and merely awaiting the addition of their geological colours. The decision to represent the drifts now meant that all these sheets had to be returned to the Ordnance Survey for modification. The chief drift deposits that needed to be added were, of course, the extensive spreads of till which obscure the solid geology over so much of Ireland. This till Jukes had decided to depict by means of a stipple without any associated colour-wash and, in returning the sets of sheets 36, 41, 47, and 53 to the Ordnance Survey, he asked them to prepare special stipple plates so that the symbol could be added to the maps already printed. This method was a failure; uneven shrinkage of the paper made it impossible to secure a satisfactory registration of the new plate upon the existing maps. It was therefore decided to scrap all the sheets already printed (a few of them seem to have got as far as having their colours added) and to print fresh copies of the sheets using the original copper plates but with the drift limits and stippling now engraved thereupon. On 24 November 1856 fifty copies of each of the new versions of sheets 36 and 41 were ready for colouring and for this they were despatched to

Fig. 6.5. *Some of the staff of the Survey about 1860. Standing, left to right: Wynne, Denis Mooney (General Assistant and Messenger from 1857 until 1888), O'Kelly, Jukes, and Baily. Sitting: Foot and Du Noyer. Absent are Galvan, Kelly, and Kinahan. Reproduced from an illustration in Robert Lloyd Praeger,* Some Irish naturalists: a biographical note-book, *by kind permission of the Dundalgan Press.*

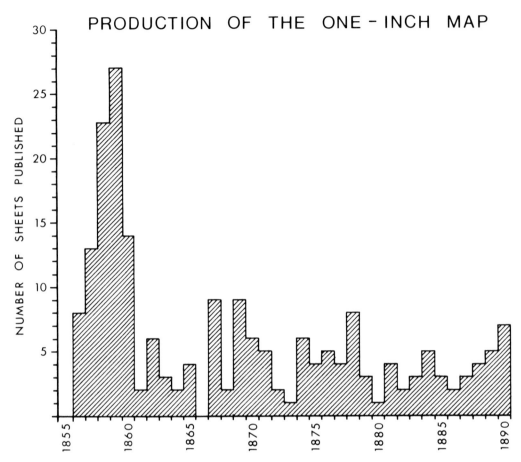

Fig. 6.6. Based upon the manuscript annual reports for the Irish Survey. The dates of publication featured upon the one-inch sheets themselves are not in every case reliable.

London where all the Geological Survey's maps were then coloured by Charles Bone (1808-1875) and his staff so as to ensure a uniformity of the tints employed upon all the Survey's British and Irish maps. By the end of December 1856 the first of these sheets were back in Dublin and in the hands of Messrs Hodges and Smith ready for sale, and by 8 April 1857 Jukes was in a position to present twenty-one quarter sheets to both the Royal Irish Academy and the Geological Society of Dublin. Those copies of the maps destined for sale were priced at from one shilling to two shillings and six pence depending upon the extent of the territory represented. Students of the maps should note that in the bottom left-hand corner of the four quarter sheets comprising Sheet 36 is the erroneous statement that the sheets were all published in June 1855, while in the similar corner of the four quarter sheets of Sheet 41 is the equally erroneous statement that the sheets were published in February 1856. These were evidently the dates upon which it had been hoped to publish the maps before the various delays intervened. Those false dates were an unfortunate blemish, but a trivial one. What really mattered was that almost twelve years after its inception, the Survey had at last begun to publish the series of geological maps which were to constitute the basic national coverage available for sale to the public (Fig. 6.6).

The base-map for all these early one-inch geological sheets is the Ordnance Survey's one-inch 'outline edition' produced between 1856 and 1862[56], an edition upon which topography is represented almost solely by a scattering of trigonometrical points. Between 1855 and 1895 the Ordnance Survey did compile a hachured 'hill edition' of the one-inch sheets, and perhaps Jukes would have been happier with these sheets as his base-map, despite the problems that could have arisen in mountainous areas through geological detail becoming obscured by heavy hachuring. Certainly in 1873 Jukes's successor decided to employ the hill edition as the base-map for all the Survey's subsequent one-inch sheets, but in the 1850s such an option was not available to Jukes because he had to publish geological sheets from Leinster whereas, very inconveniently from his point of view, the Ordnance Survey had decided to make topographical sheets from Ulster the subjects for its earliest essays in one-inch hachuring.

On 16 April 1858 Jukes wrote to the Ordnance Survey observing that their regard of the published one-inch maps as quarter sheets was inconvenient, and he therefore suggested that henceforth the quarter sheets should each be given full sheet status and be numbered consecutively. The implementation of this proposal in the October of that year meant that eventually Ireland would be covered not by 59 large-sized one-inch sheets, but by 205 sheets extending from Sheet 1 representing Malin Head, County Donegal, to Sheet 205 depicting Toe Head, County Cork. Following this renumbering, the eight quarter sheets comprising the original sheets 36 and 41 — the Survey's first published one-inch sheets — became sheets 120, 121, 129, 130, 138, 139, 148, and 149.

Some of the geological work which now began to be published in one-inch form was work of considerable antiquity. When Sheet 137 was being compiled in 1858, for instance, many of the lines were those which had been traced out by De La Beche and Warington Wilkinson Smyth in the early days of the Survey ten years before[57], and throughout sheets 136, 137, 146, and 147 use was made of material derived from Griffith's Leinster Coal District survey of 1809 to 1814.[58] Similarly, Griffith placed his unpublished maps of the Kanturk and Limerick coalfields at the Survey's disposal so that they could transfer to their own maps the outcrops of coal-seams no longer visible.[59] Thus field-work carried out by Griffith and Kelly in the second decade of the century, and by Ganly in the 1830s and 1840s, came to be used by the Survey in the 1850s and 1860s.

The earliest of the new one-inch geological maps appeared more or less contemporaneously with Griffith's quarter-inch geological map in its final form, and the Survey's new sheets clearly invite comparison with Griffith's work. Griffith's was a superb map, but the much larger scale of the new sheets allowed the Survey to introduce for the first time a multitude of refinements in detail. Upon Griffith's map many of the geological boundaries had been drawn by interpolation following the running of a series of cross-country traverses, and the result was geological boundaries displaying a sweeping simplicity. The Survey, however, had been engaged upon detailed exposure mapping and this allowed the boundaries upon the new maps to be drawn with all their true sinuous complexity. Similarly, Griffith had made no attempt to represent drift deposits, whereas the Survey was depicting spreads of ground moraine, river alluvium, and the larger peat-bogs. The one-inch sheets show a multitude of faults which Griffith had not indicated, and another striking innovation was the representation of mineral lodes by gold lines which impart to certain sheets (sheets 139 and 199 for instance) a character reminiscent of that more familiarly associated with Mediaeval illuminated manuscripts. Stratigraphical innovations are also present. Some of the one-inch sheets represent such subdivisions of the Lower Palaeozoic rocks as the Llandeilo (Sheet 111, 1859), the Bala (Sheet 119, 1858), the Llandovery (Sheet 143, 1860), and the Wenlock and the Ludlow (Sheet 160, 1859) and, while the Survey adopted Griffith's tripartite division of the Carboniferous Limestone into an Upper Limestone, a Calp Limestone, and a Lower Limestone, the Survey's intra-Carboniferous limestone boundaries were often very different from those appearing upon Griffith's map. Whether these new intra-Carboniferous Limestone boundaries were in any sense an improvement upon Griffith's equivalent lines is a point to be raised shortly (p. 182), but of one thing there can be no doubt: with the publication of the first of the one-inch sheets Jukes had launched an important series of useful and attractive maps which set a new standard of

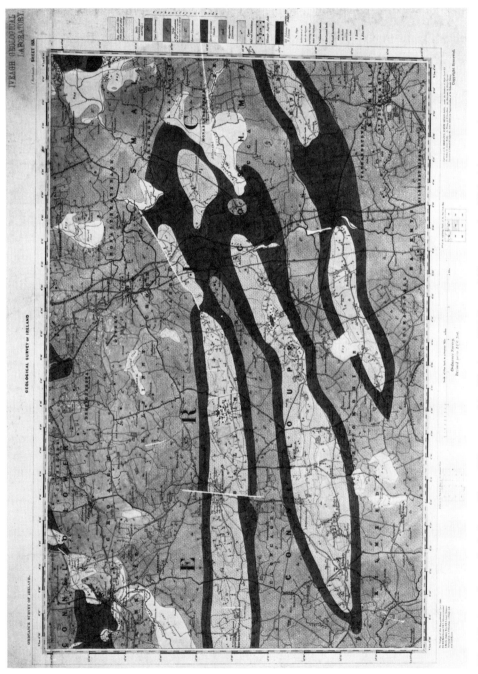

Fig. 6.7. An early one-inch sheet: Sheet 153, surveyed by Kinahan, O'Kelly, and Wynne, and published in 1860. (Reduced in size.) The sheet depicts the Ballingarry Hills of County Limerick formed where a group of inliers of Old Red Sandstone rise through the Carboniferous Limestone. In the extreme northeast a portion of the Limerick Volcanic basin is mapped as 'Felstone passing into Dolerite'. A few bogs and alluvial flats are represented, and most of the sheet is stippled to represent a spread of calcareous till.

accuracy and precision in Irish geological cartography. For thirty years the production of the one-inch sheets remained one of the Survey's prime tasks, and with the publication of the last of the series of 205 sheets in November 1890, the Geological Survey of Ireland found itself reduced to little more than a care and maintenance basis. The Survey's one-inch sheets nevertheless lived on, and for more than a century they have remained fundamental to all studies in Irish geology, although for many years now they have been available for sale to the public only in the uncoloured state; the hand-colouring of the sheets has long since ceased to be economic and is today an all but forgotten art. Only now, in the 1980s, are the one-inch maps scheduled to be superseded. They are to be replaced by a new series of sheets at a scale of 1:50,000, but as yet (1983) the first of these new sheets has still to appear because there has been much delay in the compilation of the necessary base-map. The Survey's cartographic problems over the last fifteen years have been strangely like the cartographic problems that faced James, Oldham and Jukes during the first ten years of the Survey's existence. Only time will tell whether the promised 1:50,000 sheets will achieve in the context of the modern age that same standard of excellence as was attained by the one-inch sheets of more than a century ago.

The memoirs

The final pre-publication editing of the one-inch geological maps was Jukes's responsibility, but from 1855 onwards he was burdened with another and closely related task. It was on 1 December 1855 that Murchison wrote to all the Survey's staff explaining that henceforth he wished every British and Irish one-inch geological sheet to be accompanied by an explanatory memoir. For the benefit of the staff Murchison outlined their new responsibilities as follows:

> You will be pleased to note in the first instance, the physical features of the tract you have surveyed which best exhibit the relations of the rocks. You will then record the names and positions of the localities, whether natural escarpments, river banks, quarries, shafts or adits, which have afforded the clearest evidences; and you will also give, as far as possible, the names of the organic remains and minerals, and point out the lithological structure of the strata, together with their dip and directions.

Each memoir was to be written by the geologist who had surveyed the ground in question, 'his name', as Murchison expressed it, 'being thus publicly associated with our great national undertaking'. It was nevertheless upon Jukes that there devolved the onerous task of editing and co-ordinating the memoirs. The first of the Irish memoirs was published on 28 July 1858. It was written by Jukes, Du Noyer, and Wynne and it describes the area of southern County Tipperary and northwestern County Waterford depicted upon one-inch Sheet 45 S.E. which was shortly to be re-titled Sheet 166. Interestingly one of the deposits discussed in the 27-page memoir is the pipe-clays at Ballymacadam which had been discovered by Griffith in 1821 (see p. 44). This memoir, together with four companions published during 1858, were all described as *Data and descriptions*, while the equivalent works published in 1859 and later were simply styled *Explanations*, but it is as 'the memoirs' that these publications have become familiar to many generations of Irish geologists. In January 1859 Jukes estimated that a geologist could map a one-inch sheet and write its accompanying memoir within a period of twelve months, and by 29 June 1861 he was able to report as published the very creditable total of twenty-six memoirs. There was some delay in compiling memoirs for certain of the one-inch sheets of southeastern Ireland where the mapping had largely been carried out by Oldham, Willson, and Wyley who were no longer members of the Survey's staff, but eventually every sheet in the Irish one-inch series had its accompanying memoir. Many of those memoirs published during Jukes's regime are embellished with attractive illustrations from the hand of Du Noyer or from that of his scarcely less artistic colleague Wynne.

The memoir scheme was never popular with the Survey staff. They saw themselves not as desk-bound, pen-pushing authors of academic monographs, but as rock-hammering field-surveyors ever anxious to get to grips with a fresh piece of ground. Jukes certainly had difficulty in extracting memoir contributions from some of his staff, but his efforts were well worth while if for no other reason than that the necessity of writing the memoirs made the geologists think more carefully about the phenomena that they were investigating and recording upon their maps. In short, the unwelcome literary task of preparing memoirs probably had the effect of making Jukes's men more reflective field-geologists. The classic illustration of this fact is provided by Jukes himself. In 1861 he had to write a section on 'Form of the ground' for the memoir to accompany sheets 176 and 177 — sheets which include the valley of the River Blackwater around Cappoquin in County Waterford.[60] At Cappoquin the Blackwater behaves in a very odd manner. It abandons an easy and obvious eastward strike course over synclinal Carboniferous Limestone to the sea at Dungarvan Harbour, turns southwards and traverses four ridges of hills developed upon Old Red Sandstone anticlines to reach the sea at Youghal Bay. When he completed the memoir Jukes was at a loss to account for the strange behaviour of the river; he noted that it is 'one of the most curious features in the physical geography of the south of Ireland'. But he continued to muse over the problem and late in 1861 or early in 1862 an ingenious explanation dawned upon him as he prepared the memoir to Sheet 194, a sheet covering a region in which the River Bandon displays much the same sort of anomaly as does the Blackwater at Cappoquin. On 28 April 1862 he wrote to Murchison asking his permission to present to the Geological Society of London a paper dealing with the subject and that paper — read first to the Geological Society of Dublin on 14 May 1862 and then to the London society on 18 June[61] — is probably the most influential paper ever prepared by a geologist in Ireland.[62]

In his paper Jukes accepted the then unpopular fluvial doctrine — the notion that topography has been shaped by the long-continued action of rain and rivers — and he went on to argue that, under the influence of the fluvial processes, drainage systems adapt themselves to the underlying geological structures. Thus, he reasoned, in County Waterford the Blackwater below Cappoquin is a surviving relic of an ancient north to south flowing consequent stream that originated upon the newly emerged and southward-dipping surface of southern Ireland, while the Blackwater above Cappoquin is a much younger subsequent stream which has grown by headward extension along the geological strike. This hypothesis Jukes expanded to explain drainage anomalies throughout southern Ireland from the course of the Slaney at Bunclody to that of the Shannon at Killaloe, and his ideas were soon attracting widespread attention. The paper stimulated a revival of fluvialism throughout the British Isles and the lessons that Jukes had adduced from Ireland soon made their impact upon the founders of the great American school of geomorphology which was just beginning to develop. In 1877 one of the greatest of the new generation of American fluvialists — Grove Karl Gilbert — named a peak in the Henry Mountains of Utah 'Jukes Butte' in honour of the man who had reflected upon the course of the Blackwater while writing a memoir for the Geological Survey of Ireland[63]; in 1911 the most influential of all the American geomorphologists — William Morris Davis — paid tribute to Jukes by including Cappoquin upon the itinerary of his pilgrimage to the important geomorphic sites of Europe.[64] There is to be made one final point about Jukes and the fluvial doctrine. Jukes claimed that he had come to embrace fluvialism because his detailed Irish mapping upon the six-inch scale had convinced him that there was no truth in the view, then so widely entertained, that valleys were essentially fissures opened up in the earth's crust as a result of tectonic movements.[65] His six-inch mapping had satisfied him that valleys exist because there are rivers and not rivers because there are valleys; his involvement in the writing of a sheet memoir had led to his preparation of a seminal paper of international significance.

The directive requiring them to prepare memoirs may have served to stimulate the minds of the geologists, but the general public took little interest in the new publications. The normal print-order for the early memoirs was 250 copies but by 29 June 1861 there had been sold in Ireland not more than twenty-three copies of any one of the twenty-six Irish memoirs published by that date. Substantial sales had been expected for the two memoirs covering the Castlecomer coalfield (sheets 128 and 137) and in these cases the print-order had therefore been doubled to five hundred copies, but only sixteen copies of Sheet 128 had been sold by 29 June 1861 and only twenty-three copies of Sheet 137. In the cases of two memoirs (sheets 131 & 132 and 136) only one copy of each had been sold. Even today, more than a hundred years later, substantial unsold stocks of many of the memoirs still lie in the stockrooms of the Irish Ordnance Survey.

The Survey's 1860 field programme was planned upon a modest scale; Jukes, with Murchison's permission, decided to use the year partly for the consolidation of the memoir publication programme and partly for the preparation of the earliest of a new series of longitudinal sections. The first four of these sections (they were styled 'New Series' to distinguish them from the earlier sections published by Oldham in 1848) were published in April 1860 at the six-inch scale, and eventually there were thirty-seven sheets of such sections, the last of them (Sheet 35) appearing in December 1894. One of the early sheets (Sheet 6 of October 1860) displays the Slievenamuck Fault with a throw of 1,200 metres and this Haughton in February 1861 hailed as 'the grandest proved fault in the world'.[66] But throughout this period it was the six-inch mapping and the one-inch publication that remained the Survey's principal concern. During the 1860s the Survey mapped an average of 2,418 square kilometres of new territory every year, the annual average for a single member of the staff being 208 square kilometres of new ground (Fig. 6.4). Throughout the 1860s Jukes's men advanced northwards on a broad front extending across the whole of Ireland from the shores of the Irish Sea to those of the Atlantic, although in 1867 Jukes decided to develop a new mapping enclave in the country around Belfast and well to the north of his main advancing 'mapping front'. It was while engaged upon this new work in Ulster that Du Noyer contracted scarlet fever and he died in Antrim town on 3 January 1869 in the midst of the very country where he had worked with Portlock and the Ordnance Survey thirty years before.

By July 1869 the Survey had published 117 of the 205 one-inch sheets needed to cover the whole of Ireland (Fig. 6.8). Great credit is due to Jukes for maintaining such a steady output of high quality work, but it is interesting to note that he was no longer thinking of the Survey as being involved in an operation of finite duration. When he appeared before a Select Committee on 31 May 1864 and was invited to comment upon the time necessary for the Survey to complete its appointed task, he declined to offer any estimate:

> ... because nobody knows the kind of rocks that may be met with; the rocks in the north-western part of Ireland are of a kind that has never been thoroughly examined anywhere in the world, and nobody knows how long they will take us. Moreover as we go on with the Geological Survey we introduce improvements in the nature of our operations; I must say also that it appears to me that the Geological Survey is indefinitely extensible. The scientific examination of the earth is just as inexhaustible as that of the heavens....[67]

Jukes returned more forcefully to this theme of the Survey as an enterprise in pure rather than in applied science when he published the lecture he had given at the Museum of Irish Industry in December 1866. He then emphasised that while the Survey had immense utilitarian value — its cost, he suggested, would be repaid many times over in economic benefit — it should be recognised that its chief achievement was upon the more lofty intellectual plane through a deepening of man's understanding of the history and structure of his globe:

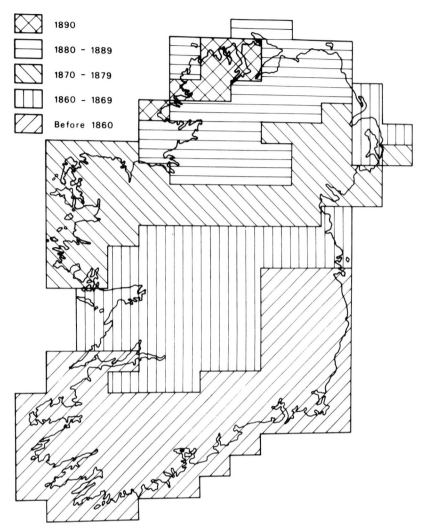

Fig. 6.8. The dates of publication for the 205 sheets constituting the one-inch geological map of Ireland. Based upon the manuscript annual reports for the Irish Survey, the dates of publication featured upon the individual sheets themselves being not in every case reliable. It should be noted that many sheets underwent subsequent revision and it is now difficult to find copies of those sheets in their original form.

Men of science have of late years pandered too much to the utilitarian quackery of the age, and it is time that some one should stand up to protest against it. Government and the House of Commons should be told that science must be supported and encouraged for her own purely abstract purposes, independently of all utilitarian applications.[68]

He developed his argument by drawing an analogy between the Geological Survey and the government-financed Royal Observatory at Greenwich. Was the Royal Observatory not allowed to probe the secrets of the universe far beyond the level necessary for the solution of practical problems in navigation? Then why should the same freedom of action not be granted to the Geological Survey so that it might probe the secrets of earth history to levels far beyond those necessary to permit the answering of mundane enquiries from the coal-miner and the civil engineer? The argument was cogent enough, but it remains an argument to which governments in these islands have usually turned a deaf ear.

Geological problems

Any field-geologist who sets about the compilation of a geological map encounters five types of problem. Firstly, there is the methodological problem: how should he work his ground? Should he make cross-country traverses, should he try to follow geological boundaries, or should he outcrop map? Secondly, there are the environmental problems which hinder his task as an observer. The extent of the drift overburden, the presence of forest, the degree of weathering of an exposure, the growth of lichens upon an outcrop, the angle of the sun, the dampness of the rock surface, even the wetness or dryness of the observer himself — all these factors affect what the geologist sees in the field. Thirdly, there are the problems that arise through the fact that the geologist takes with him into the field certain preconceptions. He believes himself to be studying an interglacial wave-cut platform; he has been taught that granite is a former igneous melt; maps of global tectonics suggest that some ancient suture passes through his ground; and he searches for resedimented conglomerates because he has been told that his rocks constitute a proximal turbidite sequence. Fourthly, there are the problems arising from the need to interpret the observed phenomena. How many phases of deformation are represented in the strata? In what sort of environment were the rocks deposited? Is the entire sequence perhaps inverted? Finally, there are problems of classification. Is this rock a rhyolite or a trachyte? Is this bed of limestone exposed on the floor of a dry valley the same as that present upon a hill-top 5 kilometres away? Is this sill Ordovician or Eocene in age?

Jukes and his men encountered all these problems in varying degrees. The methodological problem they solved early. It was their objective to outcrop map, in theory recording upon their field-sheets every single outcrop of solid rock throughout the length and breadth of this island, although in certain areas even Jukes was prepared to accept a rather lower standard. The following passage is from a memorandum that he wrote in November 1868 and to which Murchison gave his approval:

> As there are wild districts in the north western parts of Ireland in which habitable quarters are very wide apart, while all the rocks in the intervening spaces are essentially of the same character, it seems unnecessary to examine such districts with the minuteness used for the country in general. If lines of section across the strike such as are exposed by river banks, bare ridges, lines of cliffs &c have their geological structure sufficiently examined and described, and if those detached sections be now and there connected by tracing some remarkable bed or band of rocks along the strike from one section to the other, a sufficiently true idea of the geological structure of such a country may be gained.

To the environmental problems there was no such simple answer. As we have seen — and as we will see again — the environmental problems faced by the geologist in Ireland are a constantly recurring theme in Jukes's writings. Moving on to preconceptions, these Jukes's men certainly carried with them into the field — preconceptions acquired from Haughton's lectures in Trinity College, from Jukes's prelections in the Museum of Irish Industry, or from their weeks of training in the hands of their Survey colleagues. Many of these preconceptions seem just as valid today as they did a century ago — the Old Red Sandstone *is* older than the Carboniferous Limestone — but we can now see that other of their preconceptions were false and served merely to cloud the geologists' field-vision. In common with the great majority of their contemporaries, for example, they believed in a recent submergence of the British Isles and to this event they attributed most of those Irish drifts which in reality are of glacial and glaciofluvial origin. In 1862 Du Noyer adduced field evidence in support of the view that southern Ireland had recently been submerged to a depth of 730 metres and that Carrauntoohil and other of the Kerry peaks had formed islands in a sea over which there floated icebergs bearing erratics from the Galway mountains.[69] In 1863 Kinahan, clearly influenced by Darwin's work on coral reefs, divided the Irish eskers into 'fringe eskers', 'barrier eskers', and 'shoal eskers', all of them, he supposed, formed in the waters of the recent transgressive sea[70], and in 1866 he claimed to have found at various levels in counties Clare and Galway a large number of wave-cut platforms trimmed by the retreating waters of the transgression.[71] The fourth class of problems — those of interpretation — were perhaps less exacting for Jukes's men than for the modern geologist because geology was then a young science concerned with broad issues rather than with the minutiae which so often command the attention of the science's modern devotees. In their mapping the Survey men did nevertheless regularly have to address themselves to problems of interpretation. What is the throw on a fault near the Geneva Colliery in the Castlecomer coalfield (Sheet 137)? What is the structure of the Chair of Kildare (Sheet 119)? What is the origin of the curious 'dykes full of breccia' on White Ball Head in County Cork (Sheet 198)? And, of course, why does the River Blackwater behave in so anomalous a fashion at Cappoquin in County Waterford (Sheet 177)?

It was nevertheless the final category of problems — the problems of classification — which most frequently taxed the Survey's staff. The one-inch sheets which they were compiling were essentially stratigraphical documents designed to reveal the age and type of rocks existing at any particular locality, and it was classificatory problems connected with the age and type of rocks which constantly demanded answer. Sometimes the problem was so simple that it scarcely needed to be expressed: 'Is this rock granite or is it schist?' On other occasions the issue might be rather more complex: 'Are these slates of Wenlock or Llandovery age?' In yet other cases the classificatory problem could only be answered in a very arbitrary fashion: 'Where in this drift-encumbered landscape do I draw a boundary between the sandstone exposed to the west and the limestone exposed to the east?' Local classificatory problems such as these were legion, but in Jukes's day three classificatory problems of far wider implication engaged the attention of the Survey: the subdivision of the Irish limestone; the stratigraphy of the Dingle Peninsula; and the age of the slates in southern County Cork. At each of these problems we must glance in turn.

In 1837 Griffith had proposed a tripartite division of the Irish limestone into a Lower Limestone, a Calp Limestone, and an Upper Limestone (see p. 66) and these three categories he had gradually introduced into his quarter-inch geological map. The young Survey adopted Griffith's schema, and in the Survey's early notation the Lower, Calp, and Upper limestones became respectively the d^2, d^3, and d^4 limestones. Here, however, we have a splendid example of the Survey taking the field equipped with a preconception of doubtful validity. Certainly Jukes's men soon found themselves in trouble because in the field it proved difficult to categorise the limestones according to Griffith's model. On many of the one-inch sheets there appears the comment that it had not proved possible to delimit one or other of the three limestones within the area in question, and on other sheets — on Sheet 164, for instance — the colour representing the d^2 limestone is simply allowed to merge into the colour representing the d^4 limestone.[72] Jukes had early recognised that there was neither lithological nor palaeontological justification for a nationwide tripartite division of the limestone[73], and in September 1867 he announced that the Survey would henceforth abandon its attempt to subdivide the limestone.[74] The three categories of limestone did nevertheless continue to feature upon many of the Survey's later one-inch sheets, and even today (1983) the three categories are still depicted in all their deceptive verisimiltude upon the Survey's popular 1:750,000 map of Ireland.

Jukes's second major classificatory problem lay in the area covered by one-inch sheets 160, 161, 171 and 172 — in the remote Dingle Peninsula of County Kerry. The peninsula was by no means unknown to geologists. Charles William Hamilton had described some of its strata in 1838[75]; Thomas Weaver had very crudely depicted the peninsula's rocks upon his 1838 geological map of southwestern Ireland[76]; Murchison and John Phillips had visited the area in 1843; and Ganly had spent a good deal of time exploring the region on Griffith's behalf between 1838 and 1846. In the 1850s it was the Survey's turn to grapple with the Dingle rocks. De La Beche, Forbes, and Jukes were there in 1852, and Griffith, Jukes, Murchison, and Salter were there in 1856, but most of the mapping was executed by Du Noyer and by the still somewhat inexperienced Wynne. The fruit of their labours was published upon the four one-inch maps in 1859.

Within the Dingle Peninsula the local sequence of the rocks, proceeding from the oldest to the youngest, is as follows:

1. At the far northwestern end of the peninsula, in the country around Dunquin, there are steeply dipping mudstones, siltstones, and sandstones all containing fossils indicating for the rocks an upper Silurian age.

2. Immediately to the east, and conformable upon the Silurian strata, is a thick sequence of mudstones, siltstones, sandstones and conglomerates, all looking very much like the Old Red Sandstone found elsewhere in southern Ireland. Despite much searching, however, these rocks have refused to yield any fossils that could be used to confirm their age.

3. Still farther to the east, and resting with marked unconformity upon the indeterminate beds lying to the west, are the grits, sandstones and slates of the undoubted Old Red Sandstone.

Nowhere in the south of Ireland has a major unconformity been discovered within the true Old Red Sandstone sequence and, in view of this, there arose the question of the age that should be ascribed to the rocks which in Dingle come between the fossiliferous Silurian strata and the confirmed Old Red Sandstone. Were they best regarded as belonging to the upper Silurian or to the Old Red Sandstone? Hamilton had made no suggestion as to the age of the rocks; Weaver's map bore little relation to reality; Griffith, despite Murchison's misgivings, had mapped the indeterminate beds as Silurian. But Jukes decided upon a cautious approach. The Survey simply categorised the beds in question as 'the Dingle Beds' of uncertain age, and as the Dingle Beds — the Dingle Group in modern terminology — they have ever since appeared upon geological maps of the region. Jukes's caution was well advised because it now seems probable that the rocks of the Dingle Group are partly upper Silurian (upper Ludlow and Downtonian) and partly Devonian (Gedinnian).

Jukes's third major classificatory problem lay in the far south of Ireland in the area covered by one-inch sheets 191 to 205. Throughout most of southern Ireland the Old Red Sandstone is succeeded first by the thin beds of the Lower Carboniferous Shales and then by the thick sequence of the Carboniferous Limestone, but in southern County Cork there is a marked change in the character of the Carboniferous sediments. There the Old Red Sandstone is overlain not by the Lower Carboniferous Shales and the limestone, but rather by a thick series of arenaceous and argillaceous rocks known to the Survey as 'the Carboniferous Slates' and to more recent geologists as 'the Cork Beds'. A question thus had to be answered: what is the chronological relationship between the Carboniferous Slates in the south and the shales and limestones farther to the north? Jukes decided that the southern slates and the Carboniferous rocks to the north were really all of the same age and that they owed their different characters merely to the fact that the two sequences of rocks had formed in different sedimentary environments.[77] Here he drew upon his personal experience in Australasian waters while in H.M.S. *Fly*. He remembered that off the coast of New Guinea he had seen places where the waters inshore of a barrier reef were clouded with mud, while only a short distance away, seaward of the reef, the waters were crystal clear. Something similar, he suggested, must have happened in County Cork during the Carboniferous, with turbid waters existing in what is now southern Cork while slightly farther north there were clear waters populated with forests of crinoids.[78] Jukes was right — a facies change is indeed involved — but his wilder extension of the concept brought him into conflict with the geological establishment of the day. In papers read to the Royal Geological Society of Ireland and the Geological Society of London in 1865 and 1866 he argued that the slaty and supposedly Devonian rocks of north Devon were in reality of the same age as the Carboniferous Slates of County Cork, and the English rocks were thus really of Lower Carboniferous age rather than being Devonian.[79] He claimed that any real understanding of the true character of the supposed Devonian rocks of Devon must be based upon a study of the rocks of southern Ireland and that such a study would show that:

> . . . the true Devonian rocks are everywhere synchronous with the Carboniferous Limestone, their palaeontological differences depending on *habitat* and *province*, and not upon *time*.[80]

This was tantamount to a denial of the existence of an independent Devonian System, but the controversy that Jukes stimulated was largely concerned with rocks beyond Ireland's shores and it need here concern us no further.

Personal problems

The maps that came from the Geological Survey of Ireland were made by men — men who lazed in the heather atop the Kerry Reeks on sweltering July days; men who flirted with bare-footed colleens wending their way homeward from the two-day August sheep fair at Loughrea; men who took cups of milk warm from brown-eyed cows in the windswept cottages of Connemara peasants; men who shivered in their Mayo lodgings on November nights as they tried to write up their field-notes in the inadequate light shed by two candles and a turf fire. We have their maps, but we will do well to remember the trials and tribulations that they faced in order to give us those maps. Apart from the Fossil Collectors and the General Assistants, a total of thirty-two geologists worked upon the compilation of the one-inch geological map of Ireland between 1845 and the completion of the task in 1890. Presumably most of them, initially at least, felt an enthusiasm for geology, and they must have seen a post with the Survey as one of the very few ways in which they might earn a living from their chosen science. Such was their devotion to geology that some of them refused to lay aside their hammers even during their periods of annual leave; they simply went off to geologise in fresh climes — Foot to Norway in 1866[81], for instance, Richard Glascott Symes (1840-1906) to the Auvergne in 1870.[82] Most of them were evidently reasonably content in their chosen profession because those thirty-two geologists who worked upon the one inch-map each served with the Survey for an average of thirteen years. The profession nevertheless did have its drawbacks and the rigorous life demanded of his Irish staff impressed itself indelibly upon Murchison's mind during his visit of 1856 (see p. 170). The accommodation available at field-stations was often primitive and the semi-nomadic migration from field-station to field-station was severely disruptive of family life. As early as 22 September 1852 Jukes wrote to Ramsay observing of his wife:

> I am almost beginning to fear she will not be able to stand our wandering life much longer.[83]

In 1867, when Du Noyer tried to get a grant to cover the cost of removing his furniture from one field-station to another, he was refused, and Murchison reminded him that every surveyor had to accept the fact that he 'is an ubiquitous observer, who is not to be stationed permanently in any one place'. Again, as Murchison had himself noted, Irish geology is not everywhere wildly exciting. Far too often the solid rocks lie obscured beneath thick spreads of drift, and before the close of the 1850s it must have become clear that there was little prospect of any really startling geological discoveries; Jukes and his men were merely re-investigating ground that had already been well explored by Griffith, by Ganly, and by many another brother of

Sheets of Many Colours

the hammer. It need evoke no surpirse to find that Survey geologists sought occasional relief from their geological work through the study of other aspects of their environment. In 1861 Foot kept a meteorological journal in County Clare and he studied the botany of the Burren[84]; Kinahan turned to archaeology[85]; and Du Noyer busied himself with his sketch-pad and his paints.

And then there was the problem of the Irish weather. Days in both winter and summer can be idyllic, but any geologist who has tried to annotate his field-sheets as the rain drips off his clothes, his hair, his nose, and runs down his pencil so that his map looks increasingly like a piece of disintegrating blotting-paper — such a one must surely sympathise with Jukes's oft-repeated complaints about Ireland's meteorology. This is Jukes writing to his sister from Castlebar, County Mayo, on 29 April 1867:

> I am inspecting work in a more dreary miserable country than I think you could form any idea of without seeing it — miles of brown bog with low gravel mounds at intervals, occupied by wretched fields, separated by low stone walls, with large pools that are only approachable at intervals for boggy marshes, but are called lakes. This morning there is a howling wind and pelting rain, and I have to do thirty miles of that country on an outside car, with only a dirty ill-kept public-house, that calls itself a hotel, at the end of it. Such is my country life in Ireland . . . . I shall be heartily glad when I can escape with a pension.[86]

One of Jukes's staff — O'Kelly — became a victim of chronic asthma and bronchitis as a result of the conditions he encountered while mapping in the uplands of County Tipperary during the winter of 1858-59. In October 1863 he was granted six months' leave so that he could spend the winter in the warmer climate of Malaga, and in 1866 he had to abandon his field career altogether. He was brought to Dublin to take charge of the Survey's office after Kelly's retirement and he eventually died in 1883 at the early age of fifty as a direct result of his respiratory problems.

Two other difficulties faced by the Survey's staff remain to be mentioned. Firstly, there was the problem of salaries. At all levels within the Survey remuneration was on a very modest scale. In the year 1864-65, for instance, Jukes's salary (including the income from his chair) was £575, a Senior Geologist could earn up to £350 and an Assistant Geologist was paid up to twelve shillings per day. For comparison it may be noted that Griffith's pension after his retirement from the Valuation Office and the Board of Works was £1500 per annum. From Oldham onwards virtually all those resigning from the Survey had cited their lowly salary as one of the reasons for their departure. There is a poignant letter addressed to Jukes on 29 January 1863 by Charles Galvan (died 1870) who had replaced Hoskins as a Survey Fossil Collector in 1855. Galvan explains that he cannot support himself, his wife and a child upon his income of six shillings per day, that he would like to open a shop to eke out his Survey wages, but that his frequent changes of field-station seem to preclude his entering retail trade. In particular it was the cost of accommodation which proved so exhausting upon the pocket. Here is Jukes writing to Murchison on 3 February 1863:

> It is often impossible to get any lodgings at all in small towns or large villages, or anywhere else over considerable districts notwithstanding the numerous houses, the inhabitants of which would be supposed from experience in all other countries to be quite ready to let them. The prices asked for very indifferent accommodation are often exorbitant to a ridiculous extent.

Secondly, the 1860s saw a great deal of Fenian activity in Ireland and many a government officer must have wondered about his personal security. On 16 October 1865 Jukes wrote to an officer of the Royal Irish Constabulary about one of the Fossil Collectors — the same Charles Galvan — who was then stationed at Castlecomer:

> He writes me that he is looked upon by the people as a spy on the Fenians and seems in dread of ill treatment by the peasantry. May I ask you to be so good as to tell me whether in your opinion there is really any danger in his traversing the country alone?

Again here is Jukes writing to his sister, this time on 14 March 1867:

> As to what is to take place next Sunday, no one knows; but the military authorities firmly believe there is much work in store for them. There is a strong rumour that we are all to get up murdered next Monday morning, nevertheless Augusta, C., and Miss B. are making preparations to dance at St. Patrick's ball on Monday night.[87]

Sometimes it was the geologists themselves who fell under suspicion of being involved in subversive activities. In October 1866, near Thurles in County Tipperary, W.H. Baily (see p. 187) and the two Fossil Collectors were espied by a conscientious officer of the constabulary who noted that the three strangers looked like anglers or shooting gentlemen yet they carried neither rods nor guns. Clearly his duty demanded further investigation and the geologists were therefore visited at their lodgings and Baily became annoyed at the curt manner in which he was quizzed as to the reason for his presence in the area. The geologists were advised that henceforth they should report to the local constabulary barracks whenever they moved into a new district.

Jukes, as we have seen, found his life in Ireland to be distasteful, but throughout the greater part of the 1850s he nevertheless seems to have enjoyed an amicable relationship with his staff. When Wyley left for the Cape in February 1855 Jukes wrote to him saying:

> In taking leave of you I would wish to thank you for the zeal and industry you have displayed while under my direction and to assure you that I have the highest opinion of your ability & attainments as a man of science & of your honour & integrity, frankness and straightforwardness as a gentleman & a friend.

Before Willson left for India early in 1856 we find him thanking Jukes for:

> . . . the confidential and liberal manner you have always treated me, as my superior officer, and also for the many acts of consideration and personal friendship I have received from time to time at your hands.

In 1856 Jukes revealed his kindly nature by finding a Survey post for Kelly after the latter's enforced and unpensioned retirement from Griffith's Valuation Office, while in 1858 and 1859 Jukes was at great pains to order the affairs of Flanagan when the Fossil Collector was taken mortally ill at Ballyhale, County Kilkenny, where he was staying while collecting in the nearby Kiltorcan quarry. But for some reason, at around the time of Flanagan's death in April 1859, Jukes's attitude towards his staff underwent a change. Initially he merely became increasingly critical of their professional ability. On 2 September 1858, for instance, Jukes complained to Murchison about the quality of Foot's mapping around Killarney and Tralee noting that it was 'very defective and inferior' as compared with Kinahan's mapping, and on 23 May 1860 he refused to sanction a salary increment for Wynne because of the allegedly poor quality of his mapping around Templemore. During the 1860s, however, Jukes's attacks upon his staff became increasingly personal and he was guilty of a series of intemperate outbursts which must have done a great deal to taint the atmosphere of the Survey. The earliest of these outbursts seems to have occurred in June 1860 when, as a result of some trivial misunderstanding, Jukes proceeded to accuse Wynne of gross insubordination. In the margin of the correspondence dealing with the affair Wynne himself many years later wrote the comment:

> Such attacks of irritation were not characteristic – but premonitory of the subsequent sad calamity which befell one of the best of men.

On 9 May 1861 Jukes wrote to Ramsay claiming that Du Noyer 'was at times insane', that he 'had been brought up to me out of the country mad', and that Du Noyer's mother had also gone mad.[88] In April 1862 he accused Du Noyer of removing maps from the Survey's office without permission and he severely reprimanded him for straying beyond the limits of his appointed field-area in order to examine the nearby slate quarries at Killaloe in County Clare. In September 1865 he complained to Murchison of Kelly's eccentricities over six years[89], and on 17 March 1867 he wrote to his chief protesting that since Wyley and H. B. Medlicott had left the Survey in the 1850s, he had never had anybody whom he could trust to perform reliable independent work.[90] Repeatedly in his correspondence with his Survey colleagues in Britain, Jukes refers to the members of his own staff in disgracefully derogatory terms. But then even Murchison did not escape Jukes's acrimony. On 4 February 1864 Jukes wrote to Trenham Reeks at the Survey's London headquarters:

> I have no particular anxiety to come in contact with him during the apparently short time that I am likely to remain under him. I shall not come to London this spring.[91]

Where Murchison was concerned Jukes did perhaps have grounds for harbouring a grievance because of the Director's failure to lend him support through regular Irish visits of inspection, but Jukes's disdainful attitude towards his own staff is quite another matter. The present author confesses to holding Jukes the geologist in the highest regard, but he also has to confess that with the exception of the one case to be discussed below, he knows of nothing to justify Jukes's repeated attacks upon the behaviour, characters, and professional competence of his colleagues. Here, as in all such cases, the attempted defamation is likely to tell us far more about the mental state of the assailant than about the character of his victims. What, then, was wrong with Jukes? Was he suffering from some psychiatric illness? Did he entertain deep anti-Irish prejudices? In May 1861 he certainly felt that there was resentment directed at him 'as an Englishman presuming to rule an Irishman'.[92] Was he jealous of the fact that Du Noyer was the author of a geological map of Ireland first published in 1852[93] or of the fact that Kinahan was becoming widely known in geological circles through his frequent contributions to *The Geological Magazine*? Did he perhaps bear a grudge against some of his staff because they had children — Du Noyer had five and Baily had seven — whereas he and his wife were childless and had been forced to sublimate their parental instincts into a love for animals? It may be that the simplest explanation of Jukes's problem is that he desired nothing more than the care-free life of the single-minded field-geologist, but instead he found himself to be increasingly embroiled with frustrating bureaucratic duties that progressively clouded his judgement and transformed his character. He himself once complained of 'this perpetual strife and atmosphere of contention in which one is obliged to live here' which 'from the perpetual strain upon one's nerves and the gradual sap of one's heart and spirits, ultimately breaks one down'.[94]

In one case alone does Jukes seem to have been fully justified in the strictures that he passed upon a member of his staff — in the case of William Hellier Baily (1819-1888).[95] When Forbes resigned from the Geological Survey in 1854 he was replaced as Palaeontologist by Salter who followed Forbes in being responsible for all the Irish Survey's palaeontological studies. But Salter soon discovered that occasional visits to Dublin were entirely inadequate to allow him to cope with the large numbers of fossils now being collected by Jukes and his men, and Salter was therefore released from his Irish responsibilities as from the close of 1856. Jukes now urged that the Irish Survey should be allowed its own Palaeontologist, and in a letter to Murchison on 28 October 1856 he observed that at 51 St Stephen's Green the backlog of fossils awaiting examination filled no less than 30 large boxes and 176 drawers. As a result, Willson having resigned to go to India, the Bristol-born Baily was sent over from England to replace him in July 1857, Baily being accorded the title 'Senior Geologist (Acting Palaeontologist)'. Baily was soon at work, examining thousands of fossils each year, but he very rapidly became a disgruntled man. He wanted the title 'Palaeontologist in Ireland' coupled with an appropriate salary but, despite the strong support of Jukes (and of Jukes's successor), London repeatedly refused to sanction Baily's promotion. When he died in August 1888 in Rathmines, Dublin, after a lingering illness, he was still merely a Senior Geologist (Acting Palaeontologist). Jukes viewed London's refusal to advance Baily as a slight upon the entire Irish Survey and he adverted to the matter in the most vigorous of terms in the memoir to sheets 187, 195 & 196 published in 1864. But while Jukes strenuously supported Baily's claims to promotion, it is quite clear that Baily repeatedly proved himself to be a very difficult colleague. Jukes might perhaps have handled him with greater tact, but in their repeated altercations it was almost invariably Baily who was in the wrong. Many times Baily had to be reprimanded for failing to perform his duties adequately, and some of the memoirs had to be published without their fossil lists because of Baily's recalcitrance. One typical fracas between the two men took place in August 1861. Baily, who was always short of money, invited Jukes to perjure himself by certifying that some drawings of fossils that Baily was preparing for Jukes's Museum lectures were finished when in reality they were still incomplete. Baily explained that he needed the fee involved so that he might attend that year's British Association meeting in Manchester where he hoped to read a paper. Jukes very properly refused to sign the certificate in question and at this Baily, already smouldering because of his lack of promotion, blazed up in a fit of fury. On 27 August Jukes wrote to Murchison complaining that Baily was refusing to perform his duties adequately and protesting that Baily had been 'grossly insolent, and finally set me at complete defiance'. He had commanded that Jukes should never again address him save in the course of official business and Jukes, not unreasonably, now informed Murchison that he could no longer tolerate Baily's behaviour. In consequence Murchison suspended Baily from his duties as from 2 September for 'gross insolence and insubordination' and he was threatened with dismissal, but the culprit eventually apologised and he was reinstated late in October.

In Britain during the 1860s there was some concern at the slow pace with which the Geological Survey was proceeding in all parts of the British Isles, and it was decided to expedite the work by augmenting the Survey's staff. On 24 November 1866 Murchison was informed that as from 1 April 1867 the authorised establishment of the Irish branch of the Survey would be as follows: Jukes, whose title was now to be Director instead of Local Director, Murchison himself having been elevated to the status of Director General; one District Surveyor, this being a new rank; four Senior Geologists; and up to twelve Assistant Geologists. Within the civil service, as elsewhere, there had now dawned what Gladstone had termed 'the age of examinations' and all candidates for these new junior posts in the Survey had to secure a nomination, pass a geological test, and then pass the appropriate examination set by the Civil Service Commissioners. Symes in 1863 had been the first candidate appointed after jumping the examination hurdle — he failed at his first attempt finding the examination 'more severe than I had anticipated' — and candidates sitting the Commissioners' examination in 1867 had to display their proficiency in handwriting and orthography, in arithmetic (including vulgar and decimal fractions), in English composition, in drawing, in Euclid (Book I), and in the use of common mathematical instruments.

In April 1867 Jukes had a staff of five geologists — Du Noyer, Kinahan, O'Kelly, Baily, and Symes — and most men in Jukes's position must surely have been delighted at the proposed three-fold increase in their establishment. But not so Jukes. The Survey expansion of 1867 shows him again in an unfortunate light because he used the occasion to utter a deluge of further disparaging remarks about his staff. On 5 February 1867 Ramsay wrote to Archibald Geikie:

> Poor Jukes is in a sort of semi-despair about all this business, and considering that he will be adding ten more Irishmen to his already Irish lot, I don't wonder at it. His chief man [Du Noyer?] lately informed him that he had given up taking and recording dips,

as he found it to be useless! Jukes simply longs for the day when he will be able to retire, from age, and wear out the fag-end of his days, unworried by Irishmen and Boiler-men, and I considerably sympathise with him.[96]

On 15 June following Ramsay wrote again to Geikie:

> Jukes is exceedingly fidgety. He has not a man in Ireland that he can trust to training others. Also, they are all so unruly, that without rules (printed) every one will be in rebellion.[97]

Ramsay can hardly be accused of having misrepresented Jukes's sentiments because that same summer Jukes himself wrote to Geikie as follows:

> I think I am getting a good batch of new men, worse than some of the old ones I could not have especially my two Seniors Du Noyer & Kinahan who can neither spell nor write English, & whose excessive bumptiousness & conceit are in the nth power of their ignorance. I have been very near reporting Kinahan to the Dr. General & asking his dismissal from the service this spring.[98]

The same sort of message went to London; on 28 April 1868 Jukes wrote to Reeks:

> But at present I feel decidedly against more ignorant assistants. Had poor Foot lived I might have got through better but at present my two seniors Du Noyer & Kinahan are almost at open war with me & all correspondence is stilted official stuff.[99]

It must be re-emphasised that the Survey's correspondence files, the field-sheets, the memoirs, and the wider geological literature contain little to justify such calumny; what we are obtaining here is not a perceptive insight into the professional bankruptcy of the Irish Survey, but rather an insight into the imaginings of Jukes's afflicted mind and, as we will discover shortly, that mind was now in a very real sense a sick mind.

Despite some misgivings, Jukes decided that Du Noyer, his longest-serving member of staff, should become the first Irish District Surveyor, although he complained to Murchison that while Du Noyer was an accomplished draughtsman and an accurate observer, he could not be trusted with any independent work because of 'his want of all powers of generalisation and logical reasoning'.[100] Du Noyer's tenure of the post was tragically brief because of his death from scarlet fever in January 1869 (see p. 179), and Jukes, with some reluctance, agreed that Kinahan should be Du Noyer's successor as District Surveyor as from 1 March 1869. Eight new members of staff were appointed between 1867 and 1869: R.J. Cruise, Frederick William Egan (1836-1901)[101], Hugh Leonard (1841-1909)[102], William Benjamin Leonard (c. 1850-1876), Joseph Nolan, William Acheson Traill (1844-1933)[103], James Liley Warren (c. 1844-1872), and Sidney Berdoe Neal Wilkinson (1849-c.1927). Of these eight, only three had university qualifications — Egan, Traill, and Warren were all graduates of Trinity College and were either diplomates or licentiates of the College in engineering — and it may be that university graduates were reluctant to enter for competitive examinations of the type now being demanded of candidates for Survey appointments. Certainly the posts themselves can hardly have looked very attractive. The pay was only seven shillings per day, the posts were only as *Temporary* Assistant Geologists (the government clearly still looked forward to the day when, the mapping completed, the Survey could be disbanded), and there was no entitlement to superannuation. The eight men appointed attained varying degrees of proficiency as field-geologists — Egan was perhaps the best, W. B. Leonard was probably the worst — but none of them ever achieved much eminence in the world of geology at large. During the 1860s the Geological Survey of Ireland was certainly far less successful in recruiting men of outstanding talent than was the Geological Survey of Scotland where such figures of international repute as James Croll, James Geikie, John Horne, and Benjamin Neeve Peach all joined the ranks between 1861 and 1867. Sadly, one of the chief characteristics of Jukes's recruits of the 1860s was a propensity towards premature mortality. Warren died in 1872, seemingly of tuberculosis. W.B. Leonard was drowned at Belmullet in 1876. Hugh Leonard injured himself in a cliff fall while on Survey duty at Ballycastle, County Mayo in 1875, and he had to retire in 1881 as unfit for further duty, although his death did not occur until February 1909. Finally, in the summer of 1899 Egan was thrown from a car in which he was travelling somewhere in Leinster with Sir Archibald Geikie, the then Director General, and his Survey colleague McHenry, and after a severe illness Egan died in January 1901 as a direct result of his injuries. If Jukes's men really were accident prone, then it now has to be revealed that this was a trait that their chief shared with them in all too full and tragic a measure.

The death of Jukes

On 1 July 1867 a new geological map of Ireland was published by Edward Stanford in London and by Messrs Hodges and Smith in Dublin. Its scale was almost one inch to eight miles (1:506,880), and Jukes was its author.[104] The southern half of the map he had based upon the work of the Survey — presumably he had Murchison's permission to use Survey material in this manner — while the northern half is derived from Griffith's map of 1855. The new map was both an attractive piece of geological cartography and a handy conspectus of Irish geology, but behind the map there lurked a sad fact: its author was a dying man. From 1864 onwards Jukes was the victim of a steady decline. Some — wrongly it seems — attributed his deterioration to an excessive indulgence in tobacco[105] (his wife claimed that he never smoked at all during the last fifteen years of his life), while others saw him as an overzealous martyr to a particularly demanding science.[106] Among his intimates, however, there was little doubt as to the true cause of his problems: they were the result of an accident that had occurred on 27 July 1864. Jukes had then been staying at Kenmare, County Kerry, on one of his regular tours of inspection, and when leaving his inn that morning (was he perhaps staying at the Lansdowne Arms?) he slipped on the staircase and plunged headlong onto the flagstone floor beneath. Initially he seemed to have suffered little ill-effect, and after a few minutes delay he continued with the day's programme as planned. But complications soon followed. During the following month he began to suffer from severe headaches and he was forced to take sick leave as from the beginning of September 1864. A continental excursion was resolved upon as a suitable therapy, but while geologising at Coblenz at the end of September he suffered a seizure and became partially paralysed. Murchison arranged for him to be granted extended leave until the end of January 1865, but when Jukes returned to his Irish duties in February 1865 he was a shattered man experiencing long fits of deep depression, regular if temporary bouts of amnesia, and periodic blackouts. His judgement was seriously impaired and he experienced difficulty in reaching administrative decisions. This sad state of Jukes's health must be borne in mind in considering his increasing want of tact in his dealings with his staff. The drowning of Foot in a skating accident on Lough Key in January 1867, and Du Noyer's death two years later, were events that plunged Jukes into periods of the deepest gloom, and his situation was exacerbated by the problems he had with Baily and by an increasingly acrimonious correspondence with Kinahan in 1867 and 1868 about both the nature of Kinahan's duties and his mapping of the geology of Connemara. Kinahan claimed — and with some justice — that the instructions he had received were 'a mass of contradictory confusion', but this particular altercation is chiefly of interest as a presage of those occasions to come when Kinahan's increasingly petulant outbursts would blast the peace of the Survey and thoroughly poison its atmosphere.

In the spring of 1868 Jukes collapsed while travelling with Kinahan in a car somewhere in the wild country around Clifden, County Galway, and he was taken back to his inn unconscious.[107] Mrs Jukes was summoned from South Wales and Jukes's physician was called from Dublin, but by the following morning Jukes had recovered sufficiently to permit his continuing with his programme of inspection. On 5 March 1869 he suffered a similar collapse at Holywood, County Down, but on this occasion it was two days before he was fully restored to consciousness.[108] Again he struggled to regain his former self and on St Patrick's Day 1869 he wrote to Archibald Geikie:

> I believe I was living too low, and am now taking a glass of claret at lunch & two or three at dinner. They say that for the last two or three days I seem better than I have for the last two years, so I may hope perhaps eventually to get something like what I used to be.[109]

Pleasant though the claret 'cure' may have been, Jukes was really now beyond any assistance that could be rendered by either medical science or the vintners of Bordeaux. He had just finished the editing of one-inch sheets 82 (Clogher, County Louth) and 95 (Lough Corrib, County Galway) when renewed illness struck him down in April 1869 and he had to be granted six months' leave of absence. This time there was no question of his leaving Ireland to seek recovery through continental travel. As he reclined in the warmth of a Dublin spring did he perhaps muse over the path that fate had decreed for his life? Did he perhaps rue the fact that back in 1850 he had allowed the devious De La Beche to inveigle him into an acceptance of the Irish Local Directorship? For Jukes, Ireland had proved to be a land of sadness and tribulation; now it was to be the land of his death. Never again would he stroll from his home in Upper Leeson Street, over the canal bridge and to the Survey's office in The Green; never again would an Irish rock splinter beneath his hammer; never again would he announce to his friends, in that favourite phrase of his, that he had 'spifflicated' some problem in Irish geology. He died on 29 July 1869 aged only fifty-seven years. In an act symbolic of their twenty years of discomfort in Ireland — in an act regretted by his Dublin friends — his wife took his body back to his native Warwickshire for burial. He lies in Selly Oak, Birmingham, in the still surprisingly secluded graveyard of St Mary's parish church where grey squirrels scurry about among tottering and deeply weathered memorials cut from the local sandstones. Sadly, his grave cannot now be located, and it is to Ireland that we must turn in search of his chief memorial. That memorial is not the work of some monumental mason but is, rather, the geological work inspired by Jukes himself — the 117 one-inch sheets published under his direction, the 55 sheet memoirs which he edited, and the 1,100 or so manuscript six-inch field-sheets compiled during his regime. Jukes repeatedly argued that those six-inch field-sheets would prove — must prove — to be of inestimable value as a national geological inventory, and their incessant use by generation after generation of later geologists has amply vindicated Jukes's faith. Only last year (1982) the field-sheets covering the Irish Republic were removed, county by county, from the Milners' Patent Fire-Resisting safes which since May 1891 have been their home, and taken thence into a photographic studio, there to be photographed so as to ensure an even greater degree of permanency for the achievements of Jukes and his colleagues. Jukes must have felt a deep satisfaction could he have known that more than a century after his death the field-sheets prepared under his direction would still be esteemed as the fundamental archive enshrining our basic knowledge of Ireland's geology. About Jukes's death there is to be mentioned one final point — a point which his sister preferred to suppress when writing his biography, a point which his obituarists diplomatically omitted to mention, but a point which now should surely be known. His death certificate reveals that he died not in his Upper Leeson Street home, but at Hampstead House in the Dublin suburb of Glasnevin. Hampstead House was a private lunatic asylum; Jukes had been admitted upon the authority of his wife and two physicians on 8 May 1869.

REFERENCES AND NOTES FOR CHAPTER SIX

1. Some sources for the life of Jukes: ADB; DNB; DSB: Cara Amelia Browne, *Letters and extracts from the addresses and occasional writings of J. Beete Jukes*, London, 1871, pp. xx + 596; R.A. Bayliss, The travels of Joseph Beete Jukes, F.R.S., *Notes Rec. R. Soc. Lond.* 32 (2), 1978, 201-212; *Q. Jl. Geol. Soc. Lond.* 26, 1870, xxxii-xxxiv; *Geol. Mag.* 6, 1869, 430-432.
2. J.B. Jukes, Address to the Geological Section, *Rep. Br. Ass. Advmt. Sci.*, Cambridge, 1862, part 2, 54.
3. John Willis Clark and Thomas McKenny Hughes, *The life and letters of the Reverend Adam Sedgwick*, Cambridge, 1890, vol. 2, 490.
4. Joseph Beete Jukes, *Excursions in and about Newfoundland, during the years 1839 and 1840*, London, 1842, in two volumes, pp. x + 322 and iv + 354; *General report of the geological survey of Newfoundland*, London, 1843, pp. 160.
5. Thomas Webster (1773-1844) was appointed to the post.
6. Joseph Beete Jukes, *Narrative of the surveying voyage of H.M.S. Fly, commanded by Captain F.P. Blackwood, R.N.*, London, 1847, in two volumes, pp. xiv + 423 and viii + 362; *A sketch of the physical structure of Australia, so far as it is at present known*, London, 1850, pp. 95.
7. IGS 1/4.
8. RP and Browne, *op. cit.*, 1871, 272, (ref. 1).
9. Archibald Geikie, *Memoir of Sir Andrew Crombie Ramsay*, London, 1895, 105.
10. Archibald Geikie, *Life of Sir Roderick I. Murchison*, London, 1875, vol. 2, 328.
11. His salary as Local Director was raised to £400 in 1854 but it was almost immediately reduced to £300 when Jukes was appointed to the chair of geology in the Museum of Irish Industry. The salary of the latter post was £100 and not until April 1857 was Jukes again allowed the total combined salary of £500.
12. DLBP, Jukes to De La Beche 20 September 1850.
13. Browne, *op. cit.*, 1871, 461, (ref. 1).
14. DLBP.
15. RP. Jukes to Ramsay 2 December 1850.
16. MP, Jukes to Murchison 23 November 1850.
17. Trinity College Board Minutes, Trinity College Dublin. The candidates were Haughton, Jukes, and John Scouler (1804-1871). Scouler evidently received no votes.
18. RP and Browne, *op. cit.*, 1871, 464, (ref. 1).
19. RP.
20. RP and in part in Browne, *op. cit.*, 1871, 462, (ref. 1).
21. G.L. Herries Davies, The earth sciences in Irish serial publications 1787-1977, *J. Earth Sci. R. Dubl. Soc.* 1 (1), 1978, 1-23.
22. G.L. Herries Davies, The Geological Society of Dublin and the Royal Geological Society of Ireland 1831-1890, *Hermathena* no. 100, 1965, 66-76.
23. A. Albert Campbell, *Belfast Naturalists' Field Club: its origin and progress*, Belfast, 1938, pp. 44.
24. RP.
25. *Geological Survey (Ireland) return relating to skeleton maps of the Geological Survey of Great Britain and Ireland*, H.C. 1854-55 (278), XLVII, 2.
26. J.B. Jukes, Sketch of the geology of the county of Waterford, *J. Geol. Soc. Dubl.* 5 (2), 1852, 147-159.
27. RP and Browne, *op. cit.*, 1871, 461-462, (ref. 1).
28. John Harwood Andrews, *A paper landscape: the Ordnance Survey in nineteenth-century Ireland*, Oxford, 1975, 229-239.
29. *J. Geol. Soc. Dubl.* 6, 1853-55, 272-283.
30. In this chapter all documents not given a specific location will be found in the archives of the Geological Survey of Ireland. Kinahan offered a different account of the termination of soil collection but there seems to be no evidence to support his version. See George Henry Kinahan, *Superficial and agricultural geology. – Ireland: no. 2. – soils*, Dublin, 1908, xxvii.
31. William Kirby Sullivan (c. 1825-1890), Chemist to the Museum of Irish Industry.
32. Browne, *op. cit.*, 1871, 485, (ref. 1).
33. E. Forbes, On the fossils of the Yellow Sandstone of the south of Ireland, *Rep. Br. Ass. Advmt. Sci.*, Belfast, 1852, part 2, 43.
34. Geikie, *op. cit.*, 1895, 216-217, (ref. 9). Robert Bryden Wilson, *A history of the Geological Survey in Scotland*, Edinburgh, 1977, pp. 24.
35. Geikie, *op. cit.*, 1895, 216-217, (ref. 9).
36. DNB; SIN; *Geol. Mag.* N.S. dec. 5, 6, 1909, 142-143; *Ir. Nat.* 18, 1909, 29-31.
37. IGS 1/360, Jukes to Murchison 17 March 1867.
38. SIN; *Geol. Mag.* 4, 1867, 95-96.
39. SIN; *Q. Jl. Geol. Soc. Lond.* 64, 1908, lxii; *Geol. Mag.* N.S. dec. 5, 5, 1908, 143.
40. The annual reports of the Geological Survey were henceforth published within the annual report of the Department of Science and Art.
41. *Irish Times* 29 October 1895.
42. SIN; *Q. Jl. Geol. Soc. Lond.* 76, 1920, lx-lxi; *Geol. Mag.* N.S. dec. 6, 6, 1919, 336; *Ir. Nat.* 28, 1919, 102-104.
43. *Geol. Mag.* N.S. dec. 4, 9, 1902, 288.
44. DNB; *Q. Jl. Geol. Soc. Lond.* 62, 1906, lx-lxi; *Geol. Mag.* N.S. dec. 5, 2, 1905, 240; *Nature, Lond.* 71, 1904-5, 612.
45. *Geol. Mag.* N.S. dec. 2, 10, 1883, 288.

46. J.B. Archer and G.L. Herries Davies, Geological field-sheets from County Galway by Patrick Ganly (1809?-1899), *J. Earth Sci. R. Dubl. Soc.* 4 (2), 1982, 167-179.
47. MP, *Germany, Scotland, Ireland, &c 1853-56*.
48. M.A. Titmarsh [William Makepeace Thackeray], *The Irish sketch-book. 1842*, chapter XIV.
49. Geikie, *op. cit.*, 1875, vol. 2, 261-262, (ref. 10).
50. *Ibid.*, vol. 2, 262f.
51. J.B. Jukes and A. Wyley, On the structure of the north-eastern part of the county of Wicklow, *J. Geol. Soc. Dubl.* 6, 1853-55, 28-44, 282.
52. Joseph Beete Jukes, *Her Majesty's Geological Survey of the United Kingdom, and its connection with the Museum of Irish Industry in Dublin, and that of Practical Geology in London: an address*, Dublin, 1867, 10f.
53. Jukes and Wyley, *op. cit.*, 1853-55, (ref. 51).
54. J.B. Jukes, The one-inch map of the northern part of the County Wicklow, geologically coloured, *Rep. Br. Ass. Advmt. Sci.*, Liverpool, 1854, part 2, 87.
55. *J. Geol. Soc. Dubl.*, 6, 1853-55, 272-283.
56. Andrews, *op. cit.*, 1975, 229-239, (ref. 28).
57. *Explanation of Sheet 137 ... of the maps of the Geological Survey of Ireland*, Dublin, 1859, 16, 28, 29, etc.
58. *Ibid.*, 26, 28, 29, 32 etc.; *Explanations to accompany Sheet 128 ... of the maps of the Geological Survey of Ireland*, Dublin, 1859, 24f.
59. *Explanations to accompany Sheet 175 of the maps ... of the Geological Survey of Ireland*, Dublin, 1861, 25f; *Explanations to accompany sheets 163, 174, and part of 175 of the maps of the Geological Survey of Ireland*, Dublin, 1861, 16f. This use of material was reciprocal; on his 1855 map Griffith acknowledges his use of material derived from the Geological Survey's published maps.
60. *Explanations to accompany sheets 176 and 177 of the maps of the Geological Survey of Ireland*, Dublin, 1861, 7-9.
61. *J. Geol. Soc. Dubl.* 10 (1), 1862-63, 51, 52, 72-74; On the mode of formation of some of the river-valleys in the south of Ireland, *Q. Jl. Geol. Soc. Lond.* 18, 1862, 378-403.
62. Herries Davies, *op. cit.*, 1978, 16, (ref. 21).
63. Grove Karl Gilbert, *Report on the geology of the Henry Mountains*, Washington, 1877, pp. x + 160.
64. W.M. Davis, A geographical pilgrimage from Ireland to Italy, *Ann. Ass. Am. Geogr.* 2, 1912, 75-76.
65. Jukes's letter in *The Reader*, 10 December 1864.
66. *J. Geol. Soc. Dubl.* 9 (1), 1860-61, 215.
67. *Report from the Select Committee on scientific institutions (Dublin)*, H.C. 1864 (495), XIII, 143, See also a letter in the GSI archives, Jukes to Murchison, 7 December 1865.
68. Jukes, *op. cit.*, 1867, 21, (ref. 52).
69. G.V. Du Noyer, On the evidence of glacial action over the south of Ireland during the Drift Period, *The Geologist*, 5, 1862, 241-254.
70. G.H. Kinahan, The eskers of the Central Plain of Ireland, *J. Geol. Soc. Dubl.* 10 (2), 1863-64, 109-112.
71. G.H. Kinahan, Ancient sea margins in the counties Clare and Galway, *Geol. Mag.* 3, 1866, 337-343.
72. See also *Explanations to accompany sheets 163, 174 and part of 175*, Dublin, 1861, 7f.
73. Joseph Beete Jukes, *The student's manual of geology*, Edinburgh, 1862, 513-514; On the subdivisions of the Carboniferous formation in central Ireland, *J.R. Geol. Soc. Irel.* 2 (1), 1867-68, 1-12.
74. *Explanations to accompany sheets 96, 97, 106, and 107*, Dublin, 1867, 4.
75. C.W. Hamilton, An outline of the geology of part of the County of Kerry, *J. Geol. Soc. Dubl.* 1 (4), 1838, 276-285.
76. T. Weaver, On the geological relations of the south of Ireland, *Trans Geol. Soc. Lond.*, second series, 5 (1), 1838, 1-68.
77. *Explanation of sheets 187, 195, and 196, of the maps ... of the Geological Survey of Ireland*, Dublin, 1864, 32-37.
78. *Ibid.*, 35f-36f.
79. J.B. Jukes, Notes for a comparison between the rocks of the southwest of Ireland, and those of north Devon, and of Rhenish Prussia, *J.R. Geol. Soc. Irel.* 1(2), 1865-66, 103-138, 138-143; On the Carboniferous Slate (or Devonian rocks) and the Old Red Sandstone of south Ireland and north Devon, *Q. Jl. Geol. Soc. Lond.* 22, 1866, 320-371; *Additional notes on the grouping of the rocks of north Devon and west Somerset; with a map and section*, Dublin, 1867, pp. xxii + 15.
80. Jukes, *op. cit.*, 1867, xviii, (ref. 79).
81. F.J. Foot, Notes on a tour in Norway in the summer of 1866, *J.R. Dubl. Soc.* 5, 1870, 96-114.
82. R.G. Symes, On the "Geology and extinct volcanoes of Clermont, Auvergne", *J.R. Geol. Soc. Irel.* 3 (1), 1870-71, 16-23.
83. Browne, *op. cit.*, 1871, 478, (ref. 1).
84. F.J. Foot, Meteorological journal kept at Ennistimon and Ballyvaughan, County of Clare, during the year 1861, *J.R. Dubl. Soc.* 3, 1860-62, 327-341; On the botanical peculiarities of the Burren District, County of Clare, *Proc. R. Ir. Acad.* 8, 1861-64, 136; On the distribution of plants in Burren, County of Clare, *Trans. R. Ir. Acad.* 24, 1871, 143-160.
85. G.H. Kinahan, Notes on some of the ancient villages in the Aran Isles, County of Galway, *Proc. R. Ir. Acad.* 10, 1866-69, 25-30; Notes

on a crannoge in Lough Naneevin, *ibid.*, 31-33.
86. Browne, *op. cit.*, 1871, 557-558, (ref. 1).
87. *Ibid.*, 556.
88. IGS 1/360.
89. IGS 1/16, Jukes to Murchison 25 September 1865.
90. IGS 1/360.
91. IGS 1/8.
92. IGS 1/360, Jukes to Ramsay 9 May 1861.
93. *Geological copy. Fraser's travelling map of Ireland, shewing all the towns, lakes, rivers, roads and railways with the distances marked between all the towns, railway stations, and other important places.* Published by James McGlashan, 50 Upper Sackville Street, Dublin. Copies have been seen dated 1852, 1854, and 1859. The scale is 1:633,600.
94. Browne, *op. cit.*, 1871, 458, (ref. 1).
95. SIN; *Q. Jl. Geol. Soc. Lond.*, 45, 1889, proceedings 39-41; *Geol. Mag.* N.S. dec. 3, 5, 1888, 431-432, 575-576.
96. Geikie, *op. cit.*, 1895, 295, (ref. 9).
97. *Ibid.*, 295.
98. *Geikie Papers*, Gen 525.
99. IGS 1/360.
100. IGS 1/360, Jukes to Murchison 17 March 1867.
101. *Ir. Nat.* 10, 1901, 47.
102. *Geol. Mag.* N.S. dec 5, 6, 1909, 191.
103. *Belfast Telegraph* 6 July 1933.
104. Joseph Beete Jukes, *Geological map of Ireland*, London and Dublin, 1867.
105. *Irish Builder*, 11, no. 236, 15 October 1869, 232.
106. *Geol. Mag.* 6, 1869, 430-432; 8, 1871, 565-569.
107. IGS 1/360, Jukes to Reeks 28 May 1868.
108. *Geikie Papers*, Gen. 525, Jukes to Geikie 17 March 1869.
109. *Ibid.*

7

THE TASK COMPLETED

1869-1890

The County Kerry quarter six-inch Sheet 21/2 depicts the Carboniferous country located upon the Listowel-Killorglin Lowland northwestwards of Tralee. The quarter sheet was mapped for the Geological Survey about 1857 by Frederick James Foot. As he worked the ground around the village of Kilflynn, lying at the foot of Stack's Mountains, did Foot perhaps meet a seven-year-old boy, the second son of the new owner of nearby Crotta House? The boy was tall for his years, fair skinned, golden-haired, and possessed of bright blue eyes with which he gazed in a penetrating fashion which even his closest companions found strangely disconcerting. We are told that he was a serious minded and inquisitive child, and what more natural than that he should have paused to enquire of this stranger why he was chipping fragments from the local rocks and peering at them through a hand-lens? Could they have known it, fate held in store for these two individuals a common destiny — death by drowning. Foot met his end in the skating accident upon Lough Key, County Roscommon, on 17 January 1867; the boy, who had grown up to become Field-Marshal Earl Kitchener of Khartoum, was drowned when H.M.S. *Hampshire* was mined off Marwick Head in the Orkney Islands on 5 June 1916.

If that encounter between Foot and Kitchener really did take place in a Kerry lane about 1857, then it was merely the first of several encounters that Kitchener was to have with officers of the Geological Survey of Ireland. One of those later encounters took place in a location far removed from Kitchener's native Kerry — in Shepheard's Hotel, Cairo, in November 1883. The London-based Palestine Exploration Fund was then financing an expedition to map the topography and geology of the Wadi Araba lying between the Dead Sea and the Gulf of Aqaba. As leader of the party the Exploration Fund had selected Edward Hull, the man who in 1869 had succeeded Jukes as Director of the Geological Survey of Ireland, and Major Kitchener had agreed to spend his leave with the expedition deploying his considerable skills as a topographical surveyor. The selection of Hull as leader of the expedition seems odd; he had no experience of desert travel and his previous knowledge of the region was derived largely from his assiduous studies of the Old Testament. The expedition nevertheless completed satisfactorily its appointed task, and according to Hull the relations among members of the party were entirely harmonious. In particular he claims to have found Kitchener 'a most agreeable companion'. But we now know that Kitchener did not reciprocate Hull's sentiments. It seems that the soldier took a strong personal dislike to the geologist and he was contemptuous both of Hull's ignorance of the Arab and of Hull's persistent fear that the expedition was on the point of being annihilated in an Arab ambush. Perhaps the heroic Kitchener — the future victor of Omdurman and Paardeberg — could have no respect for an 'old womanish' geologist who collected chips of porphyry but who, presumably, was unable to deal with a jammed Martini-Henry, who enthused over ammonites but who in a tight corner was unlikely to be able to defend himself with his fists as Kitchener had done to save his own life at Safad in 1875. Yet even a geologist has to admit that there does indeed seem to have been something second rate about Hull. His autobiography — admittedly written very late in life — suggests him to have been a sycophant of shallow intellect. He displays a naivity of approach to a variety of issues and, occasionally, a blatant ignorance which others might have preferred to conceal. He admits, for instance, to having lived in Dublin for many years before discovering — and then through the agency of a visitor — that Trinity College — Hull's own *alma mater* — possessed a world-famed manuscript known as the Book of Kells! Yet, whatever his weaknesses, this was the man who directed the affairs of the Geological Survey of Ireland from 1869 until 1890 and who, despite many difficulties, brought the six-inch mapping of Ireland to its successful conclusion.

The new Director

Edward Hull was born in Antrim town on 21 May 1829, the son of a Church of Ireland curate.[1] Educated at schools in Edgesworthstown, County Longford, and Lucan, County Dublin, he in 1846 entered Trinity College Dublin to study for the Diploma in Engineering. There, through Thomas Oldham's lectures, he became fascinated by geology, although he tells us that initially his encounter with the science was disturbing because until then his conception of the history of the globe had been based exclusively upon his studies in *Genesis*. He took his engineering diploma in 1849 and his B.A. in the following year, but the great period of Irish railway construction was almost over and his efforts to secure an engineering appointment in Ireland came to nothing. Somewhat at a loose end, he in the autumn of 1849 visited 51 St Stephen's Green to take his leave of Oldham, and the professor suggested that Hull should approach Sir Henry De La Beche to inquire about the possibility of a post with the Geological Survey of England and Wales. Armed with Oldham's letter of introduction, and supported by Sir Roderick Murchison whose sister was married to Hull's cousin, Hull tackled De La Beche and as a result Hull on 1 April 1850 became a Temporary Assistant Geologist with the Survey of England and Wales. He was trained for six months under Jukes in North Wales, and thereafter he was sent to map in England where the geology of the South Lanca-

Fig. 7.1. Edward Hull in his seventies, from an illustration in his autobiography.

shire coalfield became his principal preoccupation.

Initially Hull seems to have achieved an acceptable standard in his work for the Survey, but after a while his mapping became increasingly slipshod. Following an inspection of some of Hull's ground, Ramsay, the Local Director for England and Wales, noted that Hull's work was getting 'worse and worse', and De La Beche instructed Ramsay to secure Hull's resignation as an alternative to dismissal. The kindly Ramsay nevertheless interceded upon Hull's behalf, but there was little improvement in the quality of Hull's work. In 1858 he was passed over for promotion to the rank of Geologist[2] despite his family connection with Murchison who by then had succeeded De La Beche as the Survey's Director. Perhaps this setback persuaded Hull to mend his ways; in 1859 he certainly obtained the promotion that had been denied to him in the previous year and during the Survey expansion of 1867 he received further advancement when he was appointed to the Geological Survey of Scotland with the rank of District Surveyor. In Scotland he was given responsibility for the mapping of the Lanarkshire coalfield and he took up residence in Glasgow, but his stay there was brief. In April 1869 the deterioration of Jukes's health necessitated his being granted six months' leave of absence from the Geological Survey of Ireland and it was clear that somebody had to be appointed to run the Irish Survey in Jukes's absence. Murchison selected Hull for the task. Hull was appointed as Jukes's locum-tenens on 28 April 1869 and he sailed over to Ireland on 6 May, just two days before Jukes was committed to Hampstead House. When Jukes died on 29 July 1869 Hull was clearly well placed to secure the succession, and in October 1869 he was indeed named as Jukes's successor both as Director of the Geological Survey of Ireland and as professor of geology in the institution which had been the Museum of Irish Industry but which was now become the Royal College of Science for Ireland.[3]

In Ireland folklore is often preferred to real history and Hull's appointment to the Irish Directorship has become the subject of a folklore which has persisted within the Irish Survey until very recently. According to the folklore Hull was an English intruder upon the Irish scene; he was an incompetent geologist; he was transferred from Scotland because Archibald Geikie, the Director of the Scottish Geological Survey, had found him to be an incubus; he secured the Dublin post solely as a result of nepotism arising from the nuptial link between his own family and that of Murchison; and his appointment to Ireland frustrated the legitimate ambitions of George Henry Kinahan, the senior member of the Irish staff who, had he been given the opportunity, would have made a far better Irish Director than did Hull. The truth of the matter is that Hull was not an English intruder – he was, as we have seen, Irish both by birth and by education. Hull may have had his early problems in the Survey – and he certainly was not entitled to stand in the front rank of Survey geologists alongside the Geikie brothers, Jukes, or Ramsay – but he was nevertheless now an enthusiastic geologist with nineteen years of Survey experience behind him, and for the two most recent of those years he had been second in command of the Scottish Survey. And what can be said of Kinahan's supposed claims to the post? Already, by 1869, it was becoming clear that Kinahan was mentally a very unstable individual. He was largely incapable of balanced geological judgement and he was entirely incapable of establishing a normal rapport with his Survey colleagues. As we will discover shortly (pp. 216-222), his persistent tantrums were to become a sordid and notorious feature of the Irish Survey. To those who might argue that all would have been different had Hull never appeared in Dublin and had Kinahan been appointed to the Directorship, one can only respond that when Kinahan did find himself in charge of the Irish Survey during Hull's absence in the Middle East in the winter of 1883-84, his actions merely confirmed the fears of those who regarded him as entirely unsuited to hold an executive office. In short, Murchison must be adjudged to have taken a wise decision in passing over Kinahan's claims to the Directorship, and since he had to find his new Director outside Ireland, then there was within the Survey no candidate better qualified than was Hull by both birth and experience.

Certainly there was no financial inducement drawing Hull to Dublin; his salary as District

Surveyor in Scotland was £440 per annum whereas the salary of the Irish Director was only £400 per annum. In view of the loss of salary occasioned by his promotion, the authorities magnanimously agreed to allow Hull to retain his former salary in his new post.[4] During his tenure of that post the traditions established under Jukes were well maintained and there came from the Irish Survey a steady stream of new one-inch sheets and sheet memoirs. Hull himself was a prolific writer with many geological papers and books to his credit, his books including widely used reference works on the coalfields of the British Isles and on building and ornamental stones, together with two volumes arising from his Middle Eastern travels in Kitchener's company.[5] Attention to literary detail, however, was not his forte and from time to time he received stern rebukes from London because of his carelessness in editing some of the Irish Survey's publications. The same cavalier attitude towards minutiae is evident in some of his geological work. He enjoyed making sweeping generalisations; he relished the invention of new stratigraphical models of far-ranging application. Sometimes a breathtaking naivity led him to adopt simplistic solutions to problems that really possessed great complexity. But for one innovation Hull must receive the highest credit: he early recognised the importance of the petrological microscope for the study of rocks in thin section. On 16 May 1871 he wrote to Murchison asking him to approve the following items for expenditure:

Jordan's Lapidary Machine for slitting & polishing specimens	£10. 10. 0
Microscope suitable for examining slit specimens of minerals	£15. 0. 0
Achromatic condenser	£ 1. 0. 0
Polarizing apparatus	£ 1. 12. 6
	£28. 2. 6

Soon notes by Hull upon the microscopic character of the local rocks began to appear in the Irish sheet memoirs (the first such notes were in the memoir to one-inch Sheet 48 published in May 1872) while in periodicals such as *The Geological Magazine* and the *Journal of the Royal Geological Society of Ireland* we find papers by Hull dealing with the petrology of the Lambay Island porphyry, the County Limerick volcanics, and the Galway and Wicklow granites.

There are two other facets of Hull's character which we can be confident that he would have wished to see mentioned. Firstly, he was a staunch Conservative and Unionist entirely out of sympathy with any Irish aspirations towards Home Rule. Secondly, he was an equally staunch supporter of the Church of Ireland, and during his years in Dublin he took an active part in the affairs of The Island and Coast Society of Ireland, a Church of Ireland body founded in 1833 'for the education of children and promotion of Scriptural truth in remote parts of the coast and adjacent islands'.[6]

The players in the final act

When Hull assumed responsibility for the Irish Survey its headquarters were still located at 51 St Stephen's Green. It was a pleasant building in a fashionable part of Dublin, but from the outset friction had resulted from the placing of the Survey under the same roof as the institution which had now become the Royal College of Science for Ireland. But quite apart from the difficulties inherent in cohabitation, the Survey faced another serious problem within 51 St Stephen's Green — its accommodation was far too cramped for its needs. Within the building the Survey had to house a rapidly expanding number of six-inch field-sheets, substantial stocks of published one-inch sheets, longitudinal sections, and memoirs, and large numbers of rocks and fossils awaiting attention. In March 1864 shelf space had to be found for General Portlock's substantial geological library which his widow presented to the Survey[7], and following the Survey's expansion in 1867 working space had to be provided for numerous new members of staff upon those occasions when they left the field to pursue their labours in Dublin. The increased educational role of the Royal College of Science made it impossible for the Survey to expand within 51 St Stephen's Green; clearly new premises had to be found for the Survey, but since the Survey's fossil and rock collections were to remain within the College's museum, it followed that any new premises had to be as close to The Green as was possible. Jukes had begun the search for a new home in August 1865 but it was left for Hull to complete the task. Initially, in October 1869, he wanted to transfer the Survey to a house in Upper Merrion Street, the street in which the Duke of Wellington is said to have been born, but in the event the Survey went to 14 Hume Street, a street of terraced Georgian houses just around the corner from St Stephen's Green. Hume Street may have lacked association with historical

Fig. 7.2. Number 14 Hume Street, Dublin, the headquarters of the Geological Survey of Ireland from March 1870 until the present day. The street was laid out in the 1760s and the construction of its houses was evidently completed before 1784, in which year Sir Richard Griffith was born across the street in number 8. The three windows at the first-floor level are those of the room which Hull and all succeeding Directors of the Survey have used as their office. From a photograph by Terence Dunne.

figures of the magnitude of the Iron Duke but, strangely, number 14 stands just across the street from the house in which Sir Richard Griffith was born in 1784. The move to Hume Street took place in the spring of 1870; 31 March 1870 was Hull's first full day in his new office and on 8 June he wrote to Trenham Reeks in London:

> We are now completely settled in our new offices which are the envy & admiration of all beholders.

There was certainly nothing temporary about the move; 14 Hume Street has remained the headquarters of the Geological Survey of Ireland down to the present day although a move to new premises is scheduled for 1983.

Murchison never set foot inside the new Irish office; he suffered a stroke on 30 November 1870 and he died on 22 October 1871. His successor as Director General was Andrew Crombie Ramsay who for so many years had been Jukes's friend and confidant. Sadly, Ramsay was past his best when he received the appointment in March 1872, but he nevertheless took the Irish responsibilities of the Director General far more seriously than had Murchison. Ramsay visited Ireland regularly to make tours of inspection and during many of these tours it was Irish Pleistocene landforms which particularly engaged his attention. Ever since spending his honeymoon in Switzerland in 1852 he had been a devoted student of all aspects of the glacial legacy, and in 1859 he had pioneered the view that glaciers have played a major erosive role in shaping many of the world's landscapes. The glaciated Wicklow Glens, the moutonnée landscapes of Connemara, the moraines of the Midlands, and the wide scattering of drumlins, eskers, and kames all served to make Ireland a land well suited to Ramsay's geological taste. His first visit as Director General occupied almost four weeks in September and October 1872 and it took him along the Survey's mapping-front from Carlingford and the Mourne Mountains, through the drumlin swarm of northern Ireland, to Boyle and Sligo.[8] Writing to Archibald Geikie about this tour Ramsay observed:

> I have seen all the staff but two, and a very nice set of fellows they are.[9]

In view of this opinion, based upon first-hand experience, one wonders what Ramsay now made of all that denigration of his 'Irish lot' which had featured so regularly in the private correspondence of the ailing Jukes. But then it does have to be admitted that Ramsay seems to have employed somewhat unusual criteria in evaluating his Irish staff. After meeting Traill, 'a fine young fellow', at Warrenpoint in 1872 Ramsay wrote in a letter:

> He looks something like what I did when I joined the Survey, only he is much handsomer, sings a great deal better, but cannot jump so high.[10]

Sir Andrew Ramsay — he was knighted in 1881 — resigned from the Survey on 31 December 1881 and he was immediately succeeded by the Survey's third successive Scottish Director General — by Archibald Geikie, who had been Hull's Director in the Geological Survey of Scotland from 1867 until Hull's departure for Dublin. Geikie, who received his own knighthood in 1891, was one of the most internationally famed geologists of his generation and he too took his Irish responsibilities very seriously. He claimed to have visited Ireland at least once every year during his term as Director General from 1882 until 1901,[11] and in his autobiography he confessed 'that in no part of my duties did I find more pleasure than in those which took me to Erin'. It was during a visit in 1899 that he was involved in the serious carriage accident in which Egan received injuries which eventually proved fatal (see p. 190). Geikie revived a Survey tradition which had been dormant since the days of De La Beche: he became personally involved in the investigation of Irish geological problems. The memoir to Sheet 17, for instance, contains a discussion of the rocks exposed on the shore of Lough Swilly near Manorcunningham written by Geikie and dated 23 March 1885, while the memoir to sheets 31 (in part) and 32 of 1891 contains a chapter by Geikie devoted to the Archaean rocks

Fig. 7.3. Sir Andrew Crombie Ramsay, Director General of the Geological Survey from 15 March 1872 to 31 December 1881. In the autumn of 1878 'an acute nervous affection in his left eye' rendered necessary its removal. Reproduced from an illustration in The Geological Magazine *N.S. dec. 2, 9, 1882.*

Fig. 7.4. Sir Archibald Geikie, K.C.B., O.M., Director General of the Geological Survey from 1 January 1882 to 1 March 1901. He was a man of short stature. Reproduced from an illustration in his autobiography.

of the Lough Derg inlier near Pettigo in County Donegal.

Hull thus received a full measure of support from the two Directors General under whom he served after 1871, but where his own subordinates were concerned the situation was less satisfactory. When he took over the Irish Survey in the summer of 1869, Hull found himself possessed of a somewhat inexperienced staff. Of his twelve geologists only four had more than two years of service to their credit; all the remainder were men recruited as a result of the 1867 expansion. Of the four, one was Baily, who was disgruntled as a result of his still being merely a Senior Geologist (Acting Palaeontologist), one was O'Kelly who since 1866 had been largely restricted to office work as a result of his bronchial problems, and one was Kinahan whose temperamental instability was seriously diminishing the reliability of his field-work. The sole geologist of experience who could be relied upon was Symes. The men appointed as a result of the expansion of 1867 were of course rapidly learning their trade but, as we noted earlier, none of them was ever to achieve much distinction within the earth sciences and in any case Warren died (seemingly of tuberculosis) in November 1872, W.B. Leonard died during the summer of 1876, Traill resigned in August 1880 to join the staff of the new Giant's Causeway, Portrush and Bush Valley Railway and Tramway, and Hugh Leonard retired in 1881 as unfit for further duty following injuries received in a fall while on duty at Ballycastle, County Mayo, in 1875. As additions and replacements four appointments as Temporary Assistant Geologist were made to the Survey during the 1870s. Firstly, in July 1870, there was Edward Townley Hardman (1845-1887), a former Drogheda Grammar School boy and a graduate of the newly-founded Royal College of Science in Dublin.[12] Secondly, in May 1871, there was William E. L'Estrange Duffin, the son of Maghera Rectory in County Down, a graduate of Trinity College Dublin and one of the College's licentiates in engineering, but a man who remained with the Survey only until February 1874 when he resigned to become County Surveyor in the Western Division of County Limerick. Thirdly, in June 1874, there was James Robinson Kilroe[13] (1848-1927), in later years the author of a much consulted soil geology of Ireland.[14] Fourthly, in February 1875, there was William Fancourt Mitchell (born 1845) who proved to be one of the Survey's more unfortunate appointments; in a confidential report of 1889 Geikie described him as 'a most unsatisfactory member of the staff'.[15] Finally, and most interestingly, there was the appointment of Alexander McHenry.[16] Born in County Antrim in 1847, the son of a policeman, McHenry attended school in Dublin under the noted educationalist and scholar Patrick Weston Joyce (1827-1914). After leaving school he joined Jukes's geological class at the Museum of Irish Industry and his enthusiasm and patent ability soon brought him to Jukes's notice. Upon Jukes's recommendation he in January 1861 was appointed to the Survey as a Fossil Collector in replacement of the deceased Flanagan. In that role his admirable qualities early became apparent to all his colleagues; he was given unofficial promotion to the part-time post of Acting Assistant Palaeontologist, and in December 1872, following Warren's death, Baily, Kinahan, O'Kelly, and Symes presented to Hull a memorial urging that McHenry should be given the now vacant post. Hull concurred, but the Civil Service Commissioners insisted that McHenry must pass the appropriate qualifying examination. He attempted the examination in June 1874 and failed, but the four memorialists again espoused his cause and the eventual outcome was a waiving of the examination requirement in McHenry's case. He was appointed as a Temporary Assistant Geologist as from 26 March 1877. In his new rank he fully lived up to his colleagues' expectations and in Geikie's confidential memorandum of 1889 he noted that McHenry 'is, on the whole, the most efficient member of the staff'.[17] Geikie was rarely lavish in his praise for others, but in his autobiography he singled McHenry out for especial mention as:

> ... one of the most helpful and efficient travelling companions I have ever had the good fortune to meet with. He undertook all the arrangements as to trains, cars, hotels, and communications; always had his wits about him, and being personally agreeable and conversable, he made even the longest drives and the poorest inns less irksome.[18]

At the time of his death in Dublin on 19 April 1919 McHenry was almost the last survivor of the Survey team of Jukes's day.

There remain to be mentioned two final appointments, both made in the 1880s and both the result of deaths within the Survey. When O'Kelly died in 1883 Hull brought back to the Survey a veteran of earlier years — Arthur Beavor Wynne. Wynne had been appointed to the Survey in 1855 but he had resigned in 1862 in order to join the Geological Survey of India. There he laboured for twenty years, but in 1883 ill-health forced his return to Ireland and in the same year he found himself reinstated with the Irish Survey to take charge of the Hume Street office, a duty he continued to perform until the completion of the last of the one-inch sheets in 1890. The other appointment arose from Baily's death in 1888. Hull then claimed that the Irish Survey no longer had need of a Palaeontologist because:

> Little now remains to be done in Irish Palaeontology except to add to the collections from time to time from new openings.

He therefore recommended that Baily should be succeeded not by another palaeontologist, but by a petrologist. The proposal was accepted and to this new post in 1888 there was appointed John Shearson Hyland (1866-1898),[19] a former student at University College Liverpool who had earned his Ph.D. at Leipzig under the renowned Ferdinand Zirkel for a dissertation on the rocks of Kilimanjaro.

During the later 1870s Hull had at his command a staff of fourteen geologists — the largest staff ever possessed by the Geological Survey of Ireland until the expansion of the Survey during the 1970s. But although large in number, the geological achievements of the staff were less impressive than might have been expected. To some extent this was a result of the relative inexperience of most of Hull's men during the 1870s, and it must also be re-emphasised that, for whatever reason, many of the men recruited to the Irish Survey never developed any very high intellectual attainments. When in March 1889 Geikie wrote his confidential review of the ten geologists remaining upon Hull's staff he expressed himself as reasonably satisfied with the work of only five (Egan, Hyland, Kilroe, McHenry, and Nolan) and he adjudged three (Cruise, Kinahan, and Mitchell) to be unworthy of retention within the Geological Survey of the United Kingdom once the primary mapping of Ireland was completed.[20] It may be, however, that Hull's men would have been rather better geologists had their general working conditions been more satisfactory. Their promotional prospects, for instance, were poor. Cruise, Egan, Kilroe, and Mitchell were all appointed as Assistant Geologists between 1867 and 1875, but even by 1890 none of them had secured promotion to the rank of Geologist. And then there was the case of Baily. Despite his incessant pleas to Jukes, to Hull, and to London, he died in 1888 still retaining the rank of Senior Geologist (Acting Palaeontologist) to which he had been appointed more than thirty years earlier. Salaries, too, were a source of constant complaint. Even as late as 1888 the salary of an Assistant Geologist was still only 14 shillings per day, and married Survey officers must have found their financial situation particularly precarious. It is perhaps significant that Traill married in 1879 and resigned in 1880. The meagre salaries may also explain the fact that some members of the staff developed the habit of leaving a trail of debts behind them as they moved from one field-station to another. On 27 July 1875 Hull issued a memorandum to his staff deploring such behaviour and threatening dismissal to any officer who disgraced the Survey in so reprehensible a manner. The practice nevertheless seems to have persisted and many are the letters of rebuke that Hull addressed to Cruise because of his failure to satisfy a variety of creditors. One of Baily's letters to Hull offers an interesting insight into this problem of Survey officers running themselves into debt. In the letter, dated 4 August 1875, Baily expressed the belief that Survey salaries and allowances were on so modest a scale that many officers must have had to take refuge in the Bankruptcy Court had it not been for the possession of modest private incomes.

There were also the environmental problems of life with the Survey. Meteorological conditions in Ireland were no better in Hull's day than they had been in Jukes's day, and in some of the annual reports — in 1872, for instance, and in 1883 — bad weather is invoked in explanation of low annual returns of ground mapped. There remained, too, the problem of the Spartan accommodation that Survey officers commonly had to accept when in the field. On 7 April 1876 Kinahan wrote to Hull from Ballina, County Mayo, about W.B. Leonard who was mapping the county's remote northwestern corner from a field-station at Glenamoy:

> It seems to me that Mr Leonard has shown considerable zeal for the public service, by locating himself and family, as I found him, in a miserable hovel with a clay floor, that he might be near his work.

But then it has to be observed that where W.B. Leonard was concerned, Kinahan was perhaps not entirely an impartial witness because Kinahan's sister Katherine was to marry W.B. Leonard's brother Hugh on 24 October in the same year. Sometimes in remote districts the geologists obtained the use of shooting lodges, and during the summer of 1875 they were allowed accommodation in the coastguard station at Ballycastle, County Mayo. It was from there that they mapped most of the wild country between Broad Haven and Killala Bay. In other regions there was just no accommodation of any type available, and in January 1872 the Survey spent £25 upon a tent of size sufficient to sleep four men and expressly intended for use in the surveying of counties Galway and Mayo. Interesting social problems arose when Wilkinson had to use the tent while mapping in the Cuilcagh Hills on the borders of counties Cavan and Fermanagh. On 22 April 1876, Wilkinson — he was the son of Lieutenant Colonel Berdoe Wilkinson of the Royal Engineers — wrote to Hull protesting that it was quite impossible for him to share the tent with his servant! Later, in 1884, the Survey bought a pre-fabricated hut from the Royal Irish Constabulary. The hut was erected first at Malin More at the far western end of the Slieve League Peninsula of County Donegal. There it served as a field-base during the summer of 1884, and there it survived the winter of 1884-1885. In March 1885 the hut was dismantled and removed to Fintown, County Donegal, to be re-erected upon a plot of land behind the post-office. There the hut ended its brief association with the Survey. It was severely damaged by storms during the winter of 1885-1886, and on 18 March 1886 the local Royal Irish Constabulary sergeant reported the structure to be beyond repair. After some hard bargaining the ruin and its contents were sold to the Fintown postmaster for £8 in May 1886.

Alongside these natural, environmental problems there also have to be remembered the unnatural risks to their personal security faced by the Survey's officers. Jukes's administration had co-incided with the Fenian troubles; Hull's administration co-incided with the agrarian outrages of the Land War. The Survey staff were government officers and their safety in rural Ireland was by no means guaranteed. Hull alludes to the problem in his annual report for 1881, and his own autobiography contains a chapter entitled 'The Irish Reign of Terror'. He himself was staying at the Great Southern Hotel in Killarney on the morning of Sunday 7 May 1882 when there arrived the news of the previous evening's murder of Lord Frederick Cavendish and Thomas Henry Burke in Dublin's Phoenix Park, and the following day Hull was horrified when in his first-class compartment aboard the Dublin train a gentleman — he proved to be the fugitive agent for a Kerry estate — produced a revolver with which he proposed to defend himself should there materialise the threatened attempt upon his life.[21] The risk of Hull finding himself caught in an ambush must have seemed just as real in Ireland as it did during those weeks with Kitchener in the Wadi Araba. Perhaps Kinahan was a more valorous soul than was Hull, but even he felt endangered and in the winter of 1880-1881 he tried to persuade the Survey to buy him a Winchester repeating rifle! Among the *Kinahan Papers* in the Royal Irish Academy[22] there certainly are a number of certificates entitling Kinahan to carry firearms within various of the Irish counties. Admittedly Kinahan often indulged in a little duck shooting — in the 1860s his sporting activities brought him into conflict with some County Galway landlords — and it may be that in obtaining some of his firearms certificates he had nothing more sinister in mind than a tasty *caneton rôti*, but how many duck shooters take to the water armed with five- or six-chambered revolvers? Such are the weapons that Kinahan was authorised to carry. When McHenry had the task of conducting Geikie through the southwest of Ireland in the spring of 1884, he regaled the Director General with stories of his own and his colleagues' experiences during the recent troubles. As they stood in a quiet County Cork lane at the close of one of these harrowing tales, Geikie remarked how fortunate it was that the more settled state of the country now rendered it unnecessary for Survey officers to carry arms. At this McHenry smiled as he withdrew a revolver from a holster hidden beneath his coat.[23]

Fresh scenery

When Hull assumed responsibility for the Survey in 1869, one-inch geological maps had been published for the whole of Ireland to the south of a line running from the shores of Galway Bay northward to Lough Mask and thence eastwards to the coast of the Irish Sea in County Louth (Fig. 6.8). In the summer of that year the field-staff was distributed as follows: Hugh Leonard was mapping the granite and schist islands of County Galway; Cruise was at Roundstone and Clifden; Warren was in the schist country at Letterfrack; Kinahan was at Recess where he and his family had been living for some years in an old police barracks; Nolan was on the shores of Lough Mask at Toormakeady; Symes was in the drumlin-drift country at Westport; Wilkinson was on the Carboniferous Limestone at Swinford; W.B. Leonard was in the Longford-Down Massif at Kingscourt; Egan was amidst the Silurian rocks at Banbridge; and Traill was at the foot of the Mourne Mountains at Newcastle, County Down. Each December the officers had to report how much new ground they had mapped during the year, together with the total length of geological boundary or coastline that they had examined. Taking one year at random — 1879 — we find that Egan, Hardman, Nolan, Symes and Wilkinson had each mapped between 77 and 98 square miles (200 and 254 square kilometres) of new ground, while the Survey as a whole had examined 1,038 miles (1,670 kilometres) of coastline or geological boundary running through the newly mapped territory. In studying the annual reports a modern reader must be struck by the frequency with which ill-health interfered with the work of the officers. Warren was on sick leave from October 1871 until his death on 25 November 1872. Hugh Leonard fell while examining a cliff at Ballycastle, County Mayo, in 1875 and received concussion which largely incapacitated him from further work, although he remained with the Survey until 1881 when he retired with a disability pension of £135:1:0 per annum. During the winter of 1875-1876 W.B. Leonard was for more than three months laid low at Glenamoy with 'a partial paralysis of the hand and bronchial attack', and he was drowned while bathing at Belmullet on 23 August 1876. He left a widow and four young children, but so far as the Survey was concerned he was no great loss; his work had been severely criticised and he was under threat of dismissal. Early in 1880 Wilkinson was out of service for four months as a result of his contracting typhoid. In 1882 Kinahan was a victim of rheumatic gout, the result of 'excessive work and exposure'. On 13 April 1883 O'Kelly died in Dublin of bronchitis, and in the same year Egan had to take nine weeks' sick-leave because of sciatica brought on, he claimed, by overwork. On 30 April 1887 Hardman died of typhoid in the Adelaide Hospital in Dublin, and on 6 August 1888 Baily died in Dublin while on sick-leave. Thirty-two geologists worked for the Survey during the period from 1845 to 1890, and of those thirty-two, no less than seven, from Foot in 1867 to Baily in 1888, died while still in Survey harness. Of the seven, only Baily had passed his sixtieth birthday. So much, perhaps, for the fine, healthy, out-door life of the nineteenth-century field-geologist.

Despite these problems, the work continued. Slowly, year by year, the mapping-front was pushed northwards; the new one-inch sheets were published (Fig. 6.6). In those sheets Hull introduced few innovations; he was well content with the traditions so securely established by Jukes. The Survey's now well-tried system of mapping worked as follows.[24] Each geologist was assigned his ground by the Director and thither he was despatched equipped with two sets of the six-inch sheets covering his area and two sets of the one-inch Ordnance Survey sheets of the same district. As we saw earlier (p. 165), the six-inch sheets were each cut into four quarters for ease of handling and one set of these — the 'working sheets' — was employed in the field, while the other set — the 'duplicate sheets' — was retained as a fair copy. Upon the working sheets the surveyor was instructed to record neatly every exposure of rock; in 'obscure districts' he was under orders to explore trenches dug to receive the foundations of buildings and to examine ditches around the margins of fields. In those areas, however, where the surveyor's problem was not so much a scarcity of rock as a plethora of outcrop — areas such as the karstic terrain of the Burren or the glaciated country of Donegal — there they were permitted to base their conclusions upon the running of a number of closely spaced traverses. As much detail as possible was recorded upon the six-inch sheets but sketches and additional details were entered in field-notebooks.[25] All the field-work was recorded upon the working sheets in pencil, but in the evening the surveyor was supposed to re-inscribe his day's work in Indian ink so that his annotations would retain their legibility come rain, come storm. Similarly, in the evenings, or upon wet days when out-of-door activity was impossible, he had to transfer all his work from the working sheets to the duplicate sheets, the latter being intended for preservation in perpetuity as the Survey's basic archive of Irish geology. But there was no question of destroying the often field-worn working sheets once the transfer of information had been made; these sheets, too, were destined for preservation in the Survey's Dublin headquarters.

From time to time at his field-station the surveyor was required to transfer the essentials of his work to the two sets of one-inch sheets with which he was provided, and in this manner he was able to see his detailed six-inch mapping in its broader regional context. One of these sets of one-inch sheets was for the surveyor's own use; the other set was to be kept up-to-date and available for despatch to Dublin when requested so that the Director or Director General could periodically review the progress and significance of the work. When the mapping of any one-inch sheet was complete, it was the surveyor's task to send the sheet to Dublin for the Director's approval and, approval granted, the geological sheet was returned to the surveyor together with another clean copy of the same Ordnance Survey one-inch sheet. To that sheet the surveyor now had to add all the details, symbols, and lines (geological boundaries were drawn in carmine and faults were represented by strong blue lines) which he wished to see upon the eventual published geological sheet. Next, the annotated sheet went to the Ordnance Survey for the geological information to be engraved and shortly thereafter the surveyor received his first proof of the new geological sheet. Upon this proof he was required to mark any desired changes in red and then to tint with sepia those portions of the sheet which were to receive the drift stipple (see p. 173). When these final details had been engraved by the Ordnance Survey, the surveyor received his second proof and to this he had to affix geological colours in the approved manner. This proof then became the 'pattern copy'; it was the prototype from which the published sheet was prepared.

Where the one-inch sheets were concerned, Hull introduced only one innovation worthy of particular mention. Jukes of necessity had employed as a base-map the Ordnance Survey's one-inch outline edition upon which little attempt is made to depict relief (see p. 174), but as they moved into northern Ireland the geologists found themselves in a region for which there were available the Ordnance Survey's one-inch hachured sheets of the relief edition. Hull, doubtless anxious to illustrate the relationship of geology to topography, in 1873 decided to change the Survey's one-inch base-map from the outline edition to the relief edition for all the sheets lying to the north of a line drawn roughly from Dundalk to Clew Bay. The first geological sheet upon a hachured base — Sheet 28 — was published in October 1874.[26] Three geological sheets based upon the outline edition — sheets 21, 29, and 36 — had already been published for districts lying to the north of the Dundalk to Clew Bay line, and in order to preserve a uniformity of style for northern Ireland the original versions of these three sheets were withdrawn and new versions based upon the hachured map were published between 1876 and 1883. In some cases the new hachured geological sheets have a somewhat mirky, overburdened appearance (sheets 26 and 43 for instance), but most of the hachured maps are extremely handsome specimens of geological cartography and especially fine examples are sheets such as numbers 10, 15, 34, and 58 (Plate 4).

By the close of 1879 forty-nine one-inch sheets had been published under Hull's direction, and of these it is sheets 59, 60, 70, and 71 which hold the greatest scientific interest. These four sheets, published between August 1875 and December 1876, depict the Tertiary Slieve Gullion ring complex located in County Armagh at the western end of the Caledonian Newry Granite — a granite body, incidentally, which the Survey believed to have resulted from the extreme metamorphism of the local Silurian strata. The complex is some 10 kilometres in diameter, and its tough igneous rocks form a striking circle of rugged hills centred upon Slieve Gullion itself. The ground was mapped by Egan, Nolan, and Traill, and they performed their task well, but they failed to differentiate adequately between the various types of igneous rock present in the region. As a result they were largely oblivious to that remarkable annular pattern of the geological outcrops which later geologists were to bring to light.[27] Thus, while Egan, Nolan, and Traill were aware that they had lighted upon something unusual,[28] they failed to appreciate that here was a type of structure which was altogether unknown to geological science. Not until ring structures had been discovered in the west of Scotland did William Bourke Wright (1876-1939), himself a former officer of the Geological Survey of Ireland, point out in 1924 that the Irish surveyors had partially mapped an identical structure almost half a century before.[29]

In addition to the one-inch sheets, the Survey under Hull maintained a steady output of the accompanying sheet memoirs. Some of those memoirs had now become substantial monographs. The 1878 memoir describing northwestern County Galway (sheets 93 and 94), for instance, contains 177 pages (it sold at 9 shillings) and the 1891 memoir on northwestern County Donegal (sheets 3, 4, 9, 10, 15, and 16) was of equal size. Hull was himself a regular contributor to the memoirs, and when Traill resigned from the Survey after mapping the Mourne Mountains but before completing the memoir (sheets 60, 61, and 71), it was Hull who personally undertook to remedy the deficiency. Hull also continued the series of six-inch longitudinal or horizontal sections inaugurated by Jukes in 1860 (see p. 179), and publication of the series of thirty-seven sections was finally completed in December 1894. Where these sections were concerned Hull did feel compelled to modify the Survey's field technique. In Britain, under both De La Beche and Murchison, the absence of any six-inch Ordnance Survey sheets had forced the geologists to undertake the laborious task of mapping the topography along the line of their longitudinal sections by means of chains and theodolites. It was largely this type of work which had engaged Hull in North Wales during his six months' apprenticeship under Jukes in 1850. In Ireland, however, James, Oldham and Jukes had never found it necessary to survey their sections in the British manner; possession of the admirable six-inch maps had allowed them to derive their topographical detail directly from the Ordnance Survey sheets. But now, in the mountainous districts of the west of Ireland, Hull concluded that the six-inch sheets carried insufficient altitudinal detail to permit the construction of accurate topographical sections. We therefore find him writing to Murchison on 11 April 1871 asking him to sanction the purchase of two 5-inch theodolites and some steel chains, and presumably in running many of their sections from that time onwards the Irish Survey adopted the British Survey's methods. One other innovation of Hull's regime must be mentioned. Jukes had originally hoped that it would prove possible to publish six-inch geological sheets, but De La Beche had declared such cartographic lavishness to be entirely beyond the realms of financial possibility. Under Hull, however, between 1874 and 1877, there were published ten six-inch geological sheets covering the Connaught and Tyrone coalfields and the iron-ore mining district near Larne in County Antrim.

Early in 1878 Hull felt the moment opportune for the issue of a new geological map of the whole of Ireland. Perhaps he and his publisher were not entirely oblivious to the fact that the British Association was due to meet in Dublin in August 1878 and that in consequence there would be many visiting geologists anxious for guidance in their study of Ireland's rocks. Indeed, that year's visit of the British Association created a bonanza in Irish geology because not only did it summon forth from Hull a new map of Ireland, but it stimulated both Hull and Kinahan to complete substantial tomes devoted to Irish geology.[30] Hull's new map was really a fresh edition of Jukes's map of 1867 (see p. 190). It was published by Edward Stanford on 30 March 1878; its scale is one inch to about eight miles (1:506,880) and it sold for thirty shillings dissected, mounted upon linen, and in a slip-case.[31] Upon the map Hull admitted that in plotting the geology of those areas not yet mapped by the Survey – County Donegal was the principal region in this category – he had sought guidance from Griffith's map of 1855, and this was to be the last occasion upon which the author of a geological map of Ireland was forced to seek his inspiration in Griffith's masterpiece. Hull's map incorporates many refinements of detail not present upon Jukes's map of 1867, but a comparison of the two maps reveals between them two principal differences. Firstly, Jukes had depicted all the Carboniferous beds above the limestone simply as Coal Measures, but in the light of his considerable experience of coalfields throughout the British Isles, Hull now claimed to be able to divide the Irish Upper Carboniferous strata into the Yoredale Shales at the base, then the Millstone Grit, and finally the Coal Measures at the top.[32] The extent of the Coal Measure strata was thus far smaller upon Hull's map than upon that of Jukes and, as a Unionist, Hull perhaps took some satisfaction in reminding the Nationalists that Ireland was not a land richly endowed with coal-seams. Some of those Nationalists, perhaps misled by the extent of the Coal Measure strata upon Jukes's map, had been arguing that Munster contained one of the largest coalfields within the British Empire but that British entrepreneurs had refused to exploit Munster's mineral riches for fear of endangering the prosperity of the English coalfields. By showing the true and very limited extent of Munster's Coal Measure strata, Hull was demonstrating that it was the harsh realities of Irish geology, rather than British self-interest, which explained the non-existence of Irish counterparts to Bradford, Gateshead and Wigan.

The second 'improvement' introduced by Hull in his map of 1878 merits our closer attention; it forms one facet of the wider episode to be discussed in the section that follows.

Old scenery revisited

From its earliest days the Survey from time to time undertook the revision of ground for which maps and memoirs had already been published. In 1853 and 1854, for example, Jukes and Wyley had re-examined much of counties Wexford and Wicklow (see p. 171); in 1861 and 1862 Jukes had supervised a re-investigation of the traps of counties Waterford and Wicklow; and in 1873 Hardman was sent down to the Leinster coalfield to re-map the Upper Carboniferous strata in accordance with Hull's new tripartite classification. During the 1870s Hull resolved to launch the Survey upon another major piece of revision – a piece of revision which might at the best be described as unwise and which might at the worst be castigated as absurdly foolish. It was the most unfortunate episode in the Survey's nineteenth-century history and it was something that a geologist of Jukes's calibre would surely never have permitted to happen.

As we have noted already, Hull was not one to be concerned with the painstaking investigation of detailed field evidence. Rather was he constantly falling victim to the seductive charms of grandiose schemes of speculative synthesis. His spacious approach to the earth-sciences is epitomised by the title of a book that he published in 1882: *Contributions to the physical history of the British Isles. With a dissertation on the origin of western Europe, and of the Atlantic Ocean.* He possessed a mind that was determined to impose order upon the natural world. Above all he was dangerously convinced of his own ability to resolve problems which had for long perplexed even the most distinguished of his predecessors. During the 1870s he was clearly exercised by the unresolved problem of the Dingle Beds lying far to the rear of the Survey's current mapping-front. He saw the problem as needing to be mopped up; the age of those enigmatic rocks must be determined once and for all. Further, he was convinced that since Jukes's death the Survey's mapping of the area around Killary Harbour (sheets 83 and 84) had shed important fresh light upon the age of the Dingle Beds. In the Dingle Peninsula, as we have seen (p. 183), the stratigraphical sequence consists of fossiliferous upper Silurian rocks overlain conformably by the unfossiliferous Dingle Beds which, in their turn, are overlain with marked unconformity by the undoubted Old Red Sandstone. Hull believed that around Killary Harbour the Survey had located rocks — the 'Mweelrea Beds' and the 'Salrock Slate Series' — which, on the basis of lithology, could be correlated with the Dingle Beds but which, unlike the Dingle Beds, contain a modest abundance of fossils. Those fossils were regarded as being of upper Silurian age and that, he reasoned, must be the age of the enigmatic rocks of the southwest. (He was evidently unaware that Griffith had advanced much the same argument back in January 1845.[33]) After due reflection upon the entire problem represented by the succession in the southwest, Hull arrived at the following three conclusions:[34]

1. Because of their conformable relationship to the upper Silurian rocks, and because of the evidence from Killary Harbour, the so-called Dingle Beds must themselves be of upper Silurian age.
2. Jukes and other geologists had pointed out that the rocks of the Iveragh Peninsula lying to the south of Dingle Bay — rocks known as the 'Glengarriff Grits' — were lithologically identical to the Dingle Beds, but because of the absence of any unconformity between the Glengarriff Grits and the Old Red Sandstone reposing upon the Dingle Beds, the Grits had been mapped as Old Red Sandstone rather than as Dingle Beds. This, Hull decided, was an error. On the basis of their lithological similarity, he concluded that the Glengarriff Grits and the Dingle Beds must be grouped together and that both the sets of strata must be assigned to the upper Silurian.
3. Since there is no unconformity to be found in the sandstones and grits to the east of the Iveragh Peninsula, then it follows that an extensive tract of country in Kerry and West Cork — a tract which the Survey had mapped as Old Red Sandstone — must now instead be assigned to the upper Silurian.

Hull first gave cartographic expression to these conclusions in his map of Ireland published in March 1878, but at that time he evidently had very little field experience of the rocks in question. His 'solution' to the problem of the Dingle Beds was merely a piece of 'arm-chair' geology devised in the Director's room in Hume Street and in the Dublin suburbs where the Hull family resided at 5 Raglan Road. But clearly the proposed new classification had to be considered in the light of the field evidence so, in September 1878, after the close of that year's Dublin meeting of the British Association, Hull set off for the southwest accompanied by O'Kelly and McHenry. His choice of companions is interesting. O'Kelly, with his bronchial problems, can hardly have been the most vigorous of field-workers, while McHenry, who had been promoted from Fossil Collector to Assistant Geologist only the previous year, was in no position to restrain his Director should he suspect him of trying to view the field evidence through the distorted lens of a preconceived theory. Guided by the field-sheets prepared in Jukes's day, Hull and his companions visited sites around Dingle, Killarney, Kenmare, Sneem, and Glengarriff, and a mere few days of investigation sufficed to convince Hull that his new interpretation of the local stratigraphy was indeed valid. He tells us that O'Kelly and McHenry were likewise convinced that the truth was at last discovered,[35] but we can only wonder at a mentality which allowed Hull to imagine that in a few brief days he could achieve that understanding which had eluded Ganly, Griffith and Jukes despite their painstaking studies extended over many years. Hull nevertheless boldly announced his new interpretation to the Geological

Society of London on 9 April 1879, and he secured Ramsay's permission to send McHenry back to Munster to map out the detail of the new classification. That revision occupied McHenry for the years 1879, 1880, and 1881, and it was a revision which caused raised eyebrows at Westminster. For reasons of economy the government wanted the one-inch maps of the British Isles to be completed as speedily as possible and questions were asked in the Commons as to why revision was being undertaken in Ireland in advance of the completion of the primary mapping.[36] Those questions were fully justified. McHenry's revision was a waste of time.

At a first glance Hull's re-interpretation of the Munster stratigraphy looks plausible enough, but more mature reflection should be sufficient to convince the geologist that here lies a baited trap. If he swallows the attractive idea that the Dingle Beds are equivalent to the Glengarriff Grits and that both are of upper Silurian age, then there must follow a series of concomitant problems the resolution of which will lead the geologist into deeper and deeper trouble. The perceptive Jukes doubtless saw the nature of the trap and avoided it[37]; the less discerning Hull fell into the trap and soon found himself in serious difficulty. The character of the trap is extremely simple. Throughout Munster the rocks which the Survey had mapped as Old Red Sandstone are succeeded with evident conformability by rocks of Carboniferous age. But if the strata beneath the Carboniferous beds are regarded as upper Silurian rather than as Old Red Sandstone, then how is the observed conformable relationship to be explained? Surely it is absurd to believe that most of Munster could have remained unaffected by diastrophism throughout those aeons during which thousands of metres of Old Red Sandstone were being deposited elsewhere within these islands. In fact it was patently obvious that in some parts of Munster major earthmovements had indeed occurred during the interval between the Silurian and the Carboniferous. Even Hull had to allow that in the Dingle Peninsula undoubted Old Red Sandstone rested upon the Dingle Beds with marked unconformity. As he struggled to extricate himself from this dilemma, Hull had resort to various contrivances which do nothing to enhance his geological status in the eyes of posterity.

On his map of March 1878 Hull groups the Dingle Beds and the Glengarriff Grits together without specifically claiming the rocks as being of upper Silurian age. The problem of the conformable relationship between these strata and the overlying Carboniferous rocks he handled in two ways. Firstly, he introduced narrow outcrops of Old Red Sandstone between the Dingle Beds and the synclinal Carboniferous strata around the Kenmare River and Bantry Bay, thus implying that there had been a continuity of sedimentation from the Silurian into the Carboniferous (Fig. 7.5). Secondly, he introduced a line running through northwestern County Cork delimiting the Dingle Beds to the northwest from the Old Red Sandstone to the southeast, and he thus confined the problem of a continuity of sedimentation from the Silurian into the Carboniferous to that area lying northwestwards of the line. This line in northwestern County Cork, together with the lines separating the Dingle Beds from the Old Red Sandstone around the Kenmare River and Bantry Bay, were clearly added to the map at a late stage because they are all manuscript lines and not a part of the map's engraved format. Indeed, the map bears trace of Hull's slap-dash editorial methods because he has allowed the index number '4', indicating Old Red Sandstone, to survive in two localities within the area which was now coloured as Dingle Beds. Many users of the map must have been puzzled as to what was intended.

Hull derived all the manuscript lines delimiting his Dingle Beds directly from Griffith's quarter-inch map and none of them has any reality in nature. They are all lines which Ganly had 'invented' for Griffith so that the grits and conglomerates of West Cork and Kerry could be represented as Transition Clay Slate (1839-40) and then as Silurian (1840 onwards), and separated from the similar rocks farther to the east which were represented as Old Red Sandstone. In short, Griffith's map had carried these lines because they were necessary to extricate Griffith from much the same stratigraphical difficulty as that in which Hull had now enmeshed himself. That boundary which Hull revived in order to separate a Dingle Bed region in the west from an Old Red Sandstone region in the east has already been mentioned as the worst line

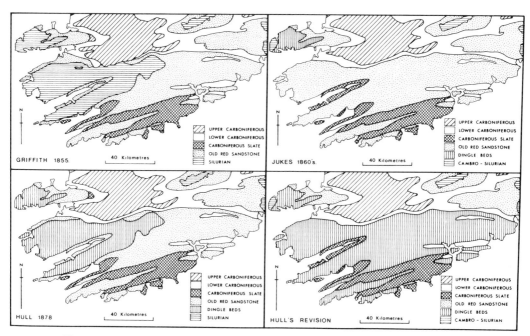

Fig. 7.5. Changing stratigraphical interpretations of the geology of southwestern Ireland.

upon Griffith's map (see p. 70); it was essentially a topographical boundary which Ganly had plotted in some haste during bad weather in January 1839, and it was a line in which even Ganly himself had soon lost all faith.[38] Yet this was the line upon which Hull in 1878 chose to bestow a fresh lease on life.

Perhaps Hull felt uneasy about his delimitation of the Dingle Beds upon the 1878 map because when he visited the southwest in September 1878 he seems to have been approaching his stratigraphical problem in a slightly different manner. If it could be demonstrated that there was an hitherto unrecognised unconformity at the base of the Carboniferous, then he was free to regard all the pre-Carboniferous rocks of the southwest as being of upper Silurian age, and he could remove those dubious-looking outcrops of Old Red Sandstone around the Kenmare River and Bantry Bay which had featured upon both Griffith's quarter-inch map and his own map of 1878. Discovery of this postulated unconformity became the prime objective of Hull, O'Kelly and McHenry as they toured the southwest in the autumn of 1878, and Hull claimed to have found that for which he sought. Initially he admitted the elusive character of the unconformity, but he claimed to have satisfied himself that in the district around Sneem, on the northern side of the Kenmare River, there was indeed an unconformity to be discerned at the base of the Carboniferous.[39] Slightly later he was less guarded about his achievements during the tour of September 1878:

> The general result was that, both along the shores of Kenmare and Glengariff [*sic*] Bays, we found the clearest evidence of a great *hiatus* between the Glengariff beds and those which immediately overlie them in those districts; resulting in the entire absence of the Old Red Sandstone at the base of the Carboniferous beds.[40]

In the light of his supposed discovery, Hull believed that the way was now open for him to eliminate the Old Red Sandstone from the whole of western Munster lying to the south of Dingle Bay. He now read the stratigraphical sequence as consisting simply of the upper Silurian Dingle Beds or Glengariff Grits overlain directly by the lowest beds of the Carboniferous sequence, with the Old Red Sandstone intervening between the upper Silurian rocks and the Carboniferous strata only within the Dingle Peninsula. Henceforth Hull intended that the Survey should represent the rocks of most of western Munster as upper Silurian rather than as Old Red Sandstone, and he saw himself as bringing the Survey around to a view very similar to that which had been enshrined in Griffith's quarter-inch map ever since 1840. Hull thought that Griffith should be one of the first to learn of the Survey's conversion. Towards the close of his southwestern tour in September 1878 Hull wrote a letter from Eccles's Hotel at Glengarriff to Griffith in Dublin announcing the discovery of a major post-Silurian hiatus in the stratigraphy of the southwest, but it was a letter that Griffith was never to receive.[41] The 'father of Irish geology' had died on 22 September shortly before the letter was delivered to his home at 2 Fitzwilliam Place. Perhaps Hull was in any case mistaken in believing that Griffith would have welcomed the news from Glengarriff; during his visit to Dingle with Murchison in 1856 Griffith had come around to the view that the Dingle Beds were perhaps after all better included with the Old Red Sandstone rather than with the upper Silurian strata exposed at Dunquin (see p. 169).[42]

Having satisfied himself of the existence of a basal Carboniferous unconformity in the southwest, Hull saw himself as facing only one further problem in connection with his stratigraphical re-interpretation: how far eastwards in Munster did the upper Silurian strata extend into the ground which had formerly been mapped as underlain by the Old Red Sandstone? What Hull now wanted to find was a major unconformity – a major unconformity overlooked by all previous investigators of Munster's geology – separating the upper Silurian rocks to the west from the overlying Old Red Sandstone to the east. If the unconformity could be located in the vicinity of that boundary which Ganly had drawn for Griffith in 1839, and which Hull had used upon his map of 1878, then so much the better, but in 1879 Hull was already admitting that the upper Silurian rocks extended 'farther east than the meridian of Cork, and may possibly be found entering the sea about Youghal Bay'.[43] The location of the unconformity, and its tracing out upon the ground, were clearly tasks that would occupy a great deal of time – far more time than Hull personally could devote to the project. It was for this reason that he secured Ramsay's permission to send McHenry down to Munster to revise the mapping during 1879, 1880, and 1881, and during 1882 Cruise was also sent south to undertake revision in County Cork.

Presumably McHenry was under orders to find the elusive unconformity and a soul-destroying task it must have proved because McHenry and his colleagues seem to have had little sympathy with the re-interpretation that Hull was requiring them to enforce. In 1880 Kinahan reported that during conversation after their visit to the southwest with Hull in September 1878, O'Kelly had expressed serious doubts as to whether his Director had read the field evidence aright, while McHenry had preferred to draw a diplomatic veil of silence over the entire episode.[44] It was not until 1905 – fourteen years after Hull's departure from Ireland – that McHenry summoned up the courage to confess in print that he had no faith in the re-interpretation which he had been required to impose upon the rocks.[45] The Survey's archives reveal that in June 1879, just two months after Hull presented his re-interpretation to the Geological Society of London, Baily, McHenry, and Richard Clark (1853-1933) – Clark had taken over McHenry's former post as Fossil Collector – were despatched to the southwest to search for fossils that would confirm the upper Silurian age of the Glengarriff Grits. But the trio brought no comfort to Hull: in his report dated 18 June – did Hull notice that it was Waterloo Day? – Baily affirmed that on palaeontological grounds the Glengarriff Grits could not possibly be of Silurian age. They could only be Old Red Sandstone. As ever (see pp. 216-222), Hull's most vigorous critic was the turbulent Kinahan. He attacked his Director's schema through the pages of a variety of geological periodicals and in a forthright paper presented to the British Association at Swansea on 26 August 1880.[46] That particular eruption was perhaps the cause of Hull's precipitate departure from the Swansea meeting to undertake a spontaneous excursion to the less actively volcanic environment represented by the Auvergne.[47] To both Ramsay and Geikie, Kinahan protested that his own early mapping in the southwest was being mutilated to make it conform to Hull's new model, and in a letter to Ramsay on 8 June 1879 he objected to McHenry being:

> ... engaged drawing fanciful boundaries which may make nice looking maps, but are an outrage on geology.

In his regular confrontations with Hull, Kinahan was by no means always in the right, but in opposing Hull's re-interpretation of the Munster stratigraphy Kinahan was entirely justified. Hull had launched the Survey upon a wild-goose chase. Modern geologists would agree with Kinahan that in the far southwest there is no great hiatus between the Glengarriff Grits and the overlying Carboniferous rocks, and even at Sneem, Hull's type locality for the unconformity, Hull's successors have been entirely unable to discern that unconformity which Hull claimed was present.[48] The same is true farther to the east. The major unconformity between the supposed upper Silurian strata of the west and the Old Red Sandstone of the east — the unconformity for which McHenry had to search — was just a chimaera. Nowhere between the Iveragh Peninsula in the west and the Lismore-Youghal region in the east is there any regionally significant stratigraphical discontinuity. The rocks throughout this tract of country either all had to be upper Silurian or else they all had to be Old Red Sandstone, and since Hull had already advocated an upper Silurian age for the rocks in the west, he had no alternative but to reclassify as upper Silurian all those rocks of eastern County Cork which previously had been mapped as Old Red Sandstone. It was there, in eastern County Cork, that the full absurdity of Hull's re-interpretation became painfully apparent; it was there that the trap was fully sprung.

The town of Fermoy (Sheet 176) stands upon the Carboniferous rocks flooring the North Cork syncline. A few kilometres to the southwest of Fermoy anticlinal sandstones and conglomerates rise above the syncline to form the Nagles Mountains; a few kilometres to the northeast of Fermoy seemingly identical sandstones and conglomerates rise above the syncline to form the Kilworth and Knockmealdown mountains. The rocks rising on either side of the syncline, let it again be emphasised, are lithologically identical, but according to Hull's re-interpretation the rocks to the south of Fermoy were Dingle Beds or Glengarriff Grits of upper Silurian age, while the rocks to the north of Fermoy were Old Red Sandstone of Devonian age. The rocks to the south were to be designated as b^6 and tinted with a muddy grey-brown; the rocks to the north were to be designated c^2 and tinted with a rich red-brown. It was ridiculous but Hull had no alternative. He did allow the Dingle Beds to extend as far east as the shores of the St George's Channel on Sheet 188, but farther to the north, in the Fermoy region, he this time could find no *modus vivendi* for himself by invoking another eastward extension of his upper Silurian rocks because to the east the beds forming the Kilworth and Knockmealdown mountains rest upon true Silurian strata with marked unconformity. The Kilworth and Knockmealdown rocks thus *had* to be of Old Red Sandstone age. Hull was asking the geological world to extend its credulity to the acceptance of three suppositions. Firstly, that while the pre-Carboniferous rocks on either side of the North Cork syncline might appear to be identical, this was an illusion; the rocks to the north of the syncline were of Old Red Sandstone age while those to the south were of upper Silurian age. Secondly, that within the width of the North Cork syncline — a distance of a mere 5 kilometres — the entire Old Red Sandstone sequence of southern Ireland — a sequence containing thousands of metres of strata — simply died out. Thirdly, that while it was admittedly difficult to discern in the field, there was really a major discontinuity intervening southward of Fermoy between the supposed upper Silurian rocks and the overlying younger strata on the southern side of the North Cork syncline. McHenry had failed to find any real unconformity to support Hull's conception of the Munster stratigraphy so the Survey had to invent an unconformity just as Ganly had invented an unconformity for Griffith in 1839. Adjudication as between the relative merits of two fictitious unconformities is perhaps absurd, but it can at least be said that Ganly's unconformity followed a topographical boundary. Hull, on the other hand, was inviting his geological contemporaries to indulge in what was virtually an act of blind faith entirely unsupported by field evidence of any type.

All the published one-inch sheets depicting those parts of Munster revised by Cruise and McHenry between 1879 and 1882 now carry an inscription such as 'Revised in 1880 by Alex. McHenry M.R.I.A.', and a reader might be misled into believing that the sheets had been published soon after the date of the revision. The truth is more complex. A new edition for the Killarney region (Sheet 173), incorporating revisions by Cruise and McHenry, was published in 1883, and in the same year revised versions of sheets 171, 174, 175, and 176 were engraved by the Ordnance Survey. Publication of those four sheets was deferred, however, because Geikie, the new Director General, wished to examine the revision in the light of the field evidence. In the spring of 1884 he therefore toured the southwest in company with McHenry himself, and Geikie's verdict was that publication of the revised maps, whether engraved or not, should for the moment be delayed. It would be satisfactory to be able to record that a few brief days in the field had been sufficient to convince an astute Geikie of the speciousness of Hull's re-interpretation, but in reality such was not the case. In 1884 Geikie delayed the publication of the revised one-inch sheets not for scientific reasons, but simply upon the grounds of political expediency. At the time he was under severe governmental pressure to bring the entire geological survey of Britain and Ireland to a speedy conclusion, and he knew that his political masters would object strongly if they saw the Irish Survey diverting some of its energies into the preparation of revised maps in advance of the completion of the outstanding one-inch sheets of northern Ireland. Once the last of those sheets had been published — an event which took place in November 1890 — then the situation was changed and, with Geikie's approval, the geologists in Hume Street — Hull himself had now gone into retirement — began to prepare the revised Munster sheets for publication. There were thirty sheets involved (sheets 171, 172, 174-177, and 182-205) and on 12 November 1890 a first batch of fifteen of the revised sheets was despatched from Hume Street to the Ordnance Survey for engraving. On many of the sheets only small changes had to be introduced, but the Ordnance Survey moved slowly and not until 5 November 1892 did revised versions of the last of the thirty sheets arrive back in Hume Street. Next the geologists prepared coloured pattern copies of the revised sheets, and on 18 November 1892 twenty-two of the pattern copies were sent over to London for Geikie's approval. It was at this juncture that skeletons began to be found in cupboards.

On 29 November 1892 Geikie wrote to Dublin expressing his disquiet with certain features of the geology being depicted upon the pattern copies, and he soon fastened upon the Fermoy sheet — Sheet 176 — as the most dubious sheet in the set. On 28 January 1893 he wrote to Hume Street as follows:

> Would you ask McHenry to draw me a rough section of what he supposes to be the structure of the ground along the eastern margin of Sheet 176.... I specially want to know the evidence for the sudden disappearance of the Old Red Sandstone southwards when it is so copiously developed north of the Fermoy–Lismore syncline.

McHenry replied on 31 January 1893; the truth came out at last. Firstly, he confessed that during his revision he had never seen even so much as a hint of an unconformity intervening between Hull's upper Silurian Dingle Beds and the overlying Carboniferous strata:

> In the very many sections that I examined the two sets of rocks were perfectly conformable. In no instance did I notice what could be looked upon for a moment as otherwise.

Secondly, he admitted that ever since 1880 he had been convinced that the pre-Carboniferous rocks on either side of the North Cork syncline were really identical. But, he explained, Hull had told him that the geology lying northward of the syncline was none of his concern, and

McHenry, the policeman's son and former Fossil Collector, now protested to the Director General that 'my duty was to simply obey my instructions'. So much for the geological map as an exercise in objective observational science.

Faced with such admissions, Geikie had before him only one possible course of action: as in 1884 he decided to withhold the revised sheets, but this time his motive was not political expediency but rather a desire for scientific rationality. The annual report of the Irish Survey for 1893 contains the following note:

> The publication of revised editions of a large number of maps in the South of Ireland has been postponed — further consideration being deemed necessary by the Director General.

The revised version of Sheet 173 had been published in 1883 (see p. 214) and some copies of at least two of the other four sheets engraved that year — sheets 174 and 175 — also seem to have reached the public sometime between 1883 and 1892, although the publication of these sheets was never officially announced. But what was the fate of the other sheets revised by McHenry between 1879 and 1881? None of them was ever published. On 27 October 1894 the geologists in Hume Street received from Geikie a directive informing them that henceforth all the rocks which Hull had reclassified as Dingle Beds were to revert to being coloured as Old Red Sandstone. Jukes's interpretation was to be restored. Early in the present century the problem of the Munster stratigraphy was again brought prominently before the Survey, firstly during an investigation of the drifts of the Cork region in 1903,[49] and secondly during a revision of the geology of the Killarney-Kenmare region (sheets 173 and 184) in 1911-1912.[50] These two surveys highlighted surviving inconsistencies upon the Munster sheets, and as a result the then Director of the Irish Survey, Grenville Arthur James Cole (1859-1924), prepared new versions of all the thirty sheets affected by Hull's 1879-1882 revision. All the sheets revised by Cole were issued on the eve of World War I bearing the inscription 'Re-edited 1913'. Since the Survey had perpetrated the ill-advised Munster revision, it was only just that in the fullness of time the Survey should itself have published a damning critique of Hull's conception of the region's stratigraphy. This happened when W.B. Wright prepared the new memoir on the Killarney and Kenmare region — a memoir which was evidently completed by July 1916 but a memoir which, because of changing political conditions in Ireland, remained unpublished until 1927.

One question about Hull's revision of the Munster stratigraphy remains to be asked. Why did he devise and cause the Survey to carry through so doubtful a geological project? The answer may be that around 1878 he had a strong personal reason for wishing to impress the geological world through the solution of some problem which had long been taxing his predecessors. During the 1870s it must have been clear that Ramsay was unlikely to remain Director General for very long. Now if the next Director General was to come from within the ranks of the Survey, then Hull must have seen himself as one of three contenders for the office, the other two candidates being Archibald Geikie, then the Director of the Geological Survey of Scotland, and Henry William Bristow (1817-1889), the Senior Director for England and Wales. Hull had been longer in the service than Geikie and he was a better known geologist than Bristow. Perhaps he believed that by publishing a few seminal papers he would strengthen his claims upon the Directorship General and go some way towards countering the very considerable reputation that Geikie had already achieved through the employment of his own facile pen. Is it entirely fortuitous that between April 1877 and January 1882 Hull read to the Geological Society of London five rather grandiose papers? In one of those papers he made a bid for geological immortality. He attempted to establish the concept of a Devono-Silurian stratigraphical episode[51] and he thus sought to join that elite band consisting of those few geologists who have been founders of geological systems. The last two of these five papers were read to the society on 11 January 1882, just four weeks and three days after Geikie in Edinburgh had received a letter from the Earl Spencer inviting him to accept the Directorship General. The succession decided — the prize lost — never again during his tenure of the Irish Directorship did Hull publish a paper in *The Quarterly Journal of the Geological Society of London*. If it was personal ambition which actually clouded Hull's geological judgement during the 1870s, then as a scientist he must of course stand condemned. One must nevertheless have a sneaking sympathy for Hull as a man. If he really was anxious to escape from Dublin to London, then one aspect of his desire to do so is easily understood. He was having serious problems of a personal nature in Hume Street. Kinahan was proving intractable and was resolved to make his Director's life as intolerable as possible.

War in the wings

From the very year of its inception, down to 1890, the Geological Survey of Ireland was the scene of a succession of personality clashes. Henry James had taken exception to Trevor Evans James; Oldham had fought with McCoy; and Jukes had crossed swords with Baily and Du Noyer. But all such affairs pall into insignificance as compared with the feud between Hull and Kinahan which shattered the tranquillity of Hume Street over a period of twenty years. It was a conflict that achieved notoriety in scientific circles throughout the British Isles; its ramifications extended into the geological literature and into the council chambers of such bodies as the Royal Irish Academy, the Royal Geological Society of Ireland, and the British Association; and the pouring of oil upon the troubled Irish waters became one of the prin-

Fig. 7.6. Mr and Mrs George Henry Kinahan in the 1860s. They were married in St Thomas's church, Dublin, on 12 November 1855, fifteen months after Kinahan joined the Survey. Mrs Kinahan, the former Harriette Anne Gerrard from County Westmeath, died on 18 May 1892 and she is buried with her husband at Avoca, County Wicklow. From a photograph in the Kinahan Papers *in the library of the Royal Irish Academy, here reproduced by kind permission of the Academy.*

cipal tasks of both Ramsay and Geikie during their respective terms of office as Director General. No history of geological mapping in Ireland can afford to overlook this most unfortunate episode.

Kinahan was a large, tough, shaggy individual, the very epitome of the popular image of the geologist as a virile, hammer-swinging, cross-country-striding sort of man, careless of the elements as he grapples with the secrets of earth-history in nature's remotest fastness. Like Edward Battersby Bailey (1881-1965) of the Scottish Survey, Kinahan would surely have preferred to wade chin-deep across a river rather than detour a few metres upstream to a bridge; like his contemporary, the Irish naturalist Richard Manliffe Barrington (1849-1915), Kinahan was surely capable of seating himself upon a submerged boulder in some mountain torrent in order to partake of his lunch and to test the mettle of a field-companion. Geikie, himself a man of small stature, once described Kinahan as being 'not the kind of antagonist one would care to encounter in a personal scuffle'.[52] From October 1876 until October 1882 Kinahan and his family lived at Avoca Lodge, Avoca, in the mining district of County Wicklow, and there he was known to the locals as 'the big miner'. After his death in Dublin in 1908 his body was taken to Avoca and it was the Wicklow miners who carried him to his last resting place in the graveyard of the Church of Ireland where his headstone is a rugged granite erratic. Of his devotion to geology there can be no doubt; almost forty years of continuous activity in the field gave him a knowledge of Irish geology which perhaps only Ganly and Griffith could have rivalled. Throughout his life there came from his pen a steady stream of papers, pamphlets, and books. The pamphlet which he wrote in association with Maxwell Henry Close on the glaciation of the country northward of Galway Bay contains a fine map of the region's glacial landforms[53]; his *Manual of the geology of Ireland* of 1878 is a minor classic of Irish geological literature; and his *Economic geology of Ireland*, published between 1887 and 1889, remains to this day an invaluable work of reference.[54]

Some of Kinahan's work for the Survey was admirable. On 24 March 1868, for instance, Jukes wrote to Kinahan:

> On examining the six-inch maps of the Galway district which you have sent me I have only to give every praise to the neatness of the execution of the work and the care with which the observations seem to have been made.

On the other hand much of his work — especially his later work — was idiosyncratic, confused, and deficient in scientific logic, and many of his field-sheets are crude and untidy. Some of his mapping the Survey even refused to accept. In June and November 1884 Geikie and Hull inspected Kinahan's revision mapping of sheets 148 and 149 around Gorey in County Wexford and they found the work so unsatisfactory that they ordered the destruction of the type for the memoir that Kinahan had written to accompany the two sheets — a memoir which was already at page-proof stage. On 21 December 1885 Geikie wrote to Hull about Kinahan's work upon Sheet 17 in County Donegal:

> With regard to the Map, I do not think the alterations are yet satisfactory. Mr Kinahan's conception of the structure of the district is so entirely different from those of his colleagues, and I must add from my own, which I formed after examination of his ground with him, that I do not think any further progress can be made with the Map until you take an opportunity of going over the disputed ground with him, and with one or more of the officers who have worked the adjacent tracts.

In less serious vein, but indicative of his slap-dash approach, there is this story of Kinahan which circulated widely within the British Survey. One day, in a work-room in Hume Street, Kinahan encountered two of his colleagues comparing their maps of adjacent ground and trying to relate their respective lines. 'Joinin'-up, eh?', observed Kinahan, 'If I see throuble brewin', I just dhraw a hell of a fault along the whole confounded frontier, and cut off every blessed line'.[55] Some of Kinahan's favourite ideas were sadly outmoded. In 1869 he argued, as had many eighteenth-century geologists, that once a landscape is clothed with grass it experiences accretion rather than denudation.[56] In 1875 he argued that cirques were largely a result of marine erosion during a recent submergence.[57] In 1878 he claimed that 'in Ireland, in general, the rivers are due to the valleys, not the valleys to the rivers',[58] and even as late as 1893 he chose to regard as 'crustal shrinkage fissures' such County Wicklow glacial spillways as The Scalp and The Devil's Glen.[59]

In temperament Kinahan was quite unbalanced. Sometimes he was placid, pliable, and possessed of a child-like innocence. This was the mood in which Geikie discovered him when he went down to Avoca on 30 June 1882, and that evening Geikie found himself returning to Dublin by train with an armful of flowers as a gift from the Kinahans to the Hulls.[60] But for much of the time Kinahan was a victim of black moods associated with fits of anger during which he was clearly both unstable and intolerable. Jukes had trouble with him from about 1866 onwards and few of his Survey colleagues were immune from his vituperation. In 1872 Kinahan and Cruise spent three months together running sections in Connemara where, as regulations allowed, they hired two assistants and shared the cost involved. Later, in his capacity of District Surveyor, Kinahan applied to the Survey for a refund of the money outlaid and when the reimbursement was made he pocketed the entire sum and refused to concede Cruise's entitlement to any of the money. In 1877 Hugh Leonard was sent down to County Wexford to assist Kinahan in his revision of Sheet 158 around Enniscorthy. Kinahan demanded his immediate withdrawal as being 'perfectly useless', yet Leonard was his own brother-in-law. When McHenry was despatched to Cork and Kerry in 1879 to begin the Munster revision, Kinahan protested that his own early mapping was 'being put into the hands of ignorant officers'. In 1883-1884, when, as the second-in-command, Kinahan took charge of the Survey during Hull's absence in the Middle East, he fell out with Wynne who had just retired from the Indian survey and taken charge of the Hume Street office. The cause of the altercation was Wynne's efforts to re-organise the storage of the six-inch field-sheets, and when Hull and Geikie read the resultant correspondence they were appalled. Geikie always tried to be sympathetic towards Kinahan, but on 25 February 1884 even he wrote to Hull observing that Kinahan had written to Wynne:

> ...in a tone that seems purposely meant to be insolent and irritating, and is as far removed as possible from the language which should be used by a superior officer to those under his charge. The perusal of this correspondence has painfully convinced me that in future it will be impossible for me to allow Mr Kinahan to occupy any position of authority over his colleagues.

Henceforth Kinahan addressed Wynne merely as 'the Office Keeper' instead of by his proper title of 'the Resident Geologist'.

It was nevertheless Hull himself who was cast in the role of Kinahan's principal *bête noir*. In April 1869 Hull and Kinahan were both District Surveyors although Hull was two years Kinahan's senior in the rank. Kinahan deeply — and understandably — resented Murchison's sending of Hull from Scotland to take charge of the Irish Survey during Jukes's final illness. This intruder might have been Irish by birth and education, but he knew little of Irish geology, and Kinahan may well have been aware that in his earlier years Hull had been regarded as a Survey officer of very inferior calibre. At his remote field-station in Recess, County Galway, Kinahan fumed at the thought of this interloper seated in that Dublin chair which, he believed, was rightfully his. He, the most experienced of Ireland's practising geologists, had been cruelly slighted; his personal pride was grievously injured. Within hours of learning of Jukes's death on 29 July 1869, Kinahan wrote to Murchison asking for Hull's removal and for the recognition of his own claims to the Irish Directorship. Another similar letter followed on 8 August,[61] but to all such pleas Murchison turned a deaf ear. He was evidently satisfied — and rightly so —

that Kinahan was quite unsuited for the post to which he aspired. But while Kinahan had to admit that there was nothing that he could do to force Murchison's hand, he was fully alive to the fact that Murchison and his successors were powerless to prevent him from making Hull's tenure of the Directorship as unpleasant an experience as was possible. And that was what Kinahan now set about doing. Hull's presence in Ireland was for more than twenty years like a thorn buried deeply in Kinahan's flesh; the pain made him angry and sometimes left him mentally unhinged.

Hull himself sometimes lacked tact and forbearance in his dealings with Kinahan, and occasionally the barbs of his tormentor drove him into fits of ill-advised fury. On 9 January 1882, for instance, he wrote a letter to *The Geological Magazine* criticising some of Kinahan's views and couched in such terms that one can only express surprise that the editor — Henry Woodward — should have accepted the letter for publication. Kinahan must have found wry satisfaction in the fact that two months later Hull had to publish a retraction of the letter expressing 'regret for having allowed myself to pen it'.[62] Hull was thus by no means an entirely blameless element within this turbulent relationship, yet there can be no doubt but that the guilt lay overwhelmingly upon Kinahan's side. In view of the relentless and severe provocation that Hull experienced, the wonder is not that he occasionally lost his equilibrium, but rather that for most of the time he managed to keep the work of the Survey moving forward so smoothly and so effectively. We must sympathise with the request that Hull made to Geikie on 25 August 1882: he had himself just lost the opportunity of escape from Ireland through promotion to the Directorship General, and in despair he wrote to Geikie inviting him to remove Kinahan from Ireland to either the Survey of Scotland or that of England and Wales. But Geikie chose to take no such action.

Trouble between Kinahan and Hull erupted within only weeks of Hull's arrival in Dublin, and on 9 August 1869 Kinahan's truculence earned him a reprimand from Murchison. Thereafter the next twenty years saw a series of continual skirmishes which resulted in Kinahan receiving repeated warnings from the Director General, many of those warnings involving threats of dismissal if he failed to mend his ways. Indeed, on a number of occasions he was informed that he would have been dismissed had the Director General not been aware of the hardship that must befall his wife and family were he to be deprived of both his post and his pension. Any blow by blow discussion of the encounters between Hull and Kinahan would require a full book; all that can be offered here is samples illustrative of the four types of tactic employed by Kinahan during his conduct of the war.

Firstly, he used obstructionist tactics. In September 1869 he delayed the proofs for the memoir to Sheet 95 by insisting that he had to revise the punctuation. In January 1871 he explained that the pressure of his other work was making it impossible for him to complete indexes for certain of the memoirs for which he was responsible. In the summer of 1876 he tried to hold up publication of Sheet 93 by protesting that certain important faults had been omitted. In December 1881 he explained to Hull with mock sincerity that indoor work upon his maps was impossible at Avoca during the winter because the configuration of the Vale of Avoca resulted in the winter sun affording him with insufficient illumination. In September 1886 he protested that he was unable to prepare a certain sheet memoir because he had been sent too few sheets of foolscap, while in January 1888 he complained that he was unable to colour his maps because he had been sent only inferior brushes and he needed sable-haired brushes! And then there was the case of the memoir to sheets 138 and 139 covering southern County Wicklow. These two sheets had been published in 1856 but there was no accompanying memoir, so while he was stationed at Avoca around 1880 Kinahan was under instruction to remedy this lacuna. His revision of a part of the two sheets was inspected by Geikie and Hull in the Avoca–Aughrim–Woodenbridge–Arklow district on 27 and 28 March 1884 and they refused to accept Kinahan's interpretation of the local structure. In a report dated 3 April 1884 Geikie wrote:

> The revision of the maps by Mr Kinahan appeared to me altogether reckless and not warranted by any evidence to be obtained on the ground itself.... If the work were everywhere as worthless as where I have personally tested it, the gravest consequences would follow.

Hull therefore took upon himself the task of completing the missing memoir, making use of some of Kinahan's material with appropriate acknowledgement. Kinahan was furious. He fought a delaying action over the memoir, and during Hull's absence from Dublin in March 1888 he sent to the Director's room for various maps and documents relating to the memoir, modified the proofs of the memoir to his liking, and then sent them off to London to be printed. When Hull returned to Hume Street there took place in the Director's room on 28 March a particularly ugly scene at which Cruise, McHenry, and Wynne were the witnesses, it having been Hull's policy since April 1876 never to interview Kinahan without a third party being present. That particular Wednesday morning, however, Hull may have regarded his three colleagues not so much in the role of passive observers, as in the role of a potential bodyguard should physical violence occur, because we are told that Kinahan struck a defiant pose before storming from the room uttering threats. In fairness to Kinahan it should be noted that he may have felt a rather poignant relationship with the memoir in question. His son Gerrard had turned to geology and some of his observations had found a place within the memoir, but Gerrard himself had been killed by a poisoned arrow on 27 May 1886 while participating in a mineral survey in the Niger basin of West Africa.

Secondly, Kinahan employed the tactic of studied insolence. On 3 March 1871 he began a letter to Hull: 'I am in the receipt of your letter dated in hieroglyphics'. In his official report upon his activities during 1872 he claimed to have spent his time in mapping, in the running of sections, and in trying to defend his character 'from the calumny of Mr Hull'. On 8 November 1881 he wrote to Hull observing that he was an incompetent Director and that he ought to resign. Should he decline to accept this advice, then Kinahan wanted to know how much of his salary Hull was prepared to make over to him for the performance of those duties which were beyond Hull's ability. When he submitted his quarterly financial return for the second quarter of 1887, Edward Best at the Survey's London headquarters noted that Kinahan had 'had the brainless folly to write the remark which you will see'. Ramsay was the shocked witness of one of Kinahan's episodes of monstrous insolence. In August 1877 Ramsay and Hull were touring County Wicklow and they arranged to meet Kinahan at the beautifully situated Woodenbridge Hotel on 17 August, Kinahan having been instructed to come down from Avoca bringing his field-sheets together with any correspondence that might have arrived for either Ramsay or Hull.[63] He appeared bearing Ramsay's mail but he had deliberately left his field-sheets behind and upon Hull enquiring about his own letters, Kinahan retorted angrily: 'Am I to be Mr Hull's postman!' Ramsay noted in his diary that Kinahan was 'outrageously savage in his conduct to Hull', and he continued:

> Altogether he was so insulting to Hull & so sulky with me, that after a ½ hours thought, I told him I had decided not to go further in that direction & he might return to his quarters.

Thirdly, Kinahan employed the tactic of regular disobedience. Repeatedly he refused to obey Hull's orders, raising against them a multitude of what Geikie once termed 'frivolous objections'. He refused to conform to the Survey's rules for the nomenclature of rock types. He refused to sign for maps that he borrowed from Hume Street. He flouted the Survey's rules by publishing papers without first obtaining permission. He objected when he was required to move from one field-station to another, and his removal from Avoca to Letterkenny, County Donegal, in October 1882 was the cause of particular acrimony. In County Donegal in 1884 he wanted to map the Fanad Peninsula but it was a part of the ground which Hull had already assigned to Cruise. Kinahan nevertheless went off with the six-inch sheets of Fanad and he proceeded to map the region. In October 1888 he had to be severely reprimanded for wasting time in the unauthorised mapping of a part of the Inishowen Peninsula which had already been surveyed by Cruise.

Finally — and here the Irish Survey's dirty linen was displayed for all to see — Kinahan sought publicly to subvert Hull's reputation as a geologist. Whenever Hull entered into print on some aspect of Irish geology, there commonly followed a reply from Kinahan expressing his vigorous disagreement with Hull's conclusions. Often, it seems, the burden of Kinahan's geological writings owed just as much to personal malice as to the principles of scientific reasoning. When Hull spoke to the Royal Geological Society of Ireland on 11 January 1871 on the subject of the Ballycastle coalfield, Kinahan replied in the December issue of *The Geological Magazine*.[64] When Hull wrote about the drifts so magnificently displayed in the coastal cliffs southward of Killiney in County Dublin, Kinahan wrote claiming that Hull had misunderstood what he had seen.[65] When Hull proposed his re-interpretation of the stratigraphy of Munster, Kinahan mounted a major counter-attack through the pages of a number of journals including those of the *Transactions of the Manchester Geological Society*.[66] At first sight the involvement of that particular journal in an Irish dispute might seem incongruous, but the wily Kinahan had doubtless discovered that Hull had been an honorary member of the Manchester society since 1874. When Hull turned to microscope petrology, Kinahan followed suit,[67] and while they both protested ignorance of the fact that the other was pregnant of a book, they did both in 1878 publish rival volumes devoted to the geology of Ireland, with Hull's study appearing slightly the earlier and receiving the better reviews. In 1879 Kinahan found himself strongly opposed to the stratigraphical interpretation of the Curlew Mountains and the Fintona district that was being adopted by the Survey under Hull's direction, and he was incensed when Hull denied him access to the Survey's relevant field-sheets. He therefore went off to the Royal Irish Academy, where he was a Member, and he persuaded an innocent Academy council both to establish a committee to examine the stratigraphical issues involved and to provide £50 for the committee's field activities during the summer of 1879. The members of the committee were Kinahan and Baily, and in his new role as a member of an Academy committee Kinahan again tried to get his hands upon the field-sheets in question. It was now Hull's turn to be angry, and the Academy backed down speedily when Hull explained how it had allowed itself to become caught up in what was essentially a personal vendetta.[68] The final episode of this particular fracas took place in the spring of 1880. Kinahan was due to report the committee's findings to the Academy on 12 April and he had the audacity to send Hull a formal invitation to be present at the meeting.[69] Hull excused himself on the ground of having a prior engagement, but it is interesting to note that without Kinahan's invitation Hull would have had no right to attend the occasion. Although a Fellow of the Royal Society since 1867, Hull was never elected to Membership of the Royal Irish Academy. How far Kinahan was able to engineer Hull's exclusion from the Academy it is now impossible to say.

Today Kinahan is one of the best remembered figures of the nineteenth-century Irish Survey, and there is a widespread belief that he played a major role in the Survey's mapping programme. Such a view stands in need of revision. In Jukes's day Kinahan certainly was one of the Survey's stalwarts; his name features upon twenty-seven one-inch sheets issued between his joining of the Survey in August 1854 and Jukes's death in July 1869. But with the advent of Hull, Kinahan's role was changed. From July 1869 down to the autumn of 1890 Kinahan's name appeared upon only eleven new one-inch sheets, and of these no less than seven had been surveyed either in whole or in part during Jukes's tenure of the Directorship.[70] For comparison, Symes's name appears upon twenty-nine new sheets issued during Hull's regime. Throughout the greater part of that regime, Kinahan was stationed not at the Survey's mapping-front but far to its rear. As the Survey's District Surveyor he should have been at the mapping-front supervising the younger men and bringing to bear upon their problems the wisdom derived from his long experience. Instead he worked in isolation. Hull stationed him far to the south of the mapping-front, first at Wexford (1872 to 1876) and then at Avoca (1876 to 1882), and from these stations his assigned duty was merely revision mapping. Not until Kinahan was transferred from Avoca to Letterkenny in October 1882 did he again find himself working alongside his colleagues in the mapping of virgin territory. Until then it seems to have been the policy of Hull (and perhaps of the Director General himself) to keep Kinahan at work as far away as was possible from the ground being investigated by his brother officers. Perhaps there was a fear that his abrasive character might disrupt the progress of the work, and when Hull went to the Middle East in the winter of 1883, leaving Kinahan in charge of the Survey, Geikie was at pains to emphasise that there was no need for Kinahan to transfer himself to Dublin, and in particular it was stressed that it was entirely unnecessary for him to visit his colleagues to inspect their work. For his part, from 1869 onwards, Kinahan was perhaps resolved to do for the Survey nothing which might in any way redound to Hull's credit. As we have seen (p. 220), much of his revision work in Leinster was rejected by both Geikie and Hull as unworthy of publication. The suppression of his revision of one-inch sheets 138, 139, 148, and 149, and of the accompanying memoirs that he had written, meant that during his six years of residence at Avoca he contributed virtually nothing to the Survey's progress. He undoubtedly possessed geological talents, but there is no escaping the conclusion that from 1869 onwards Kinahan was to the Survey more of an incubus than an asset. Geikie appears to have held Hull in low regard and he always seems to have had some sympathy for Kinahan, but on 5 March 1889, when Geikie wrote his confidential report on the future of the Geological Survey of Ireland, even he could offer nothing more than the following curt verdict:

> Mr Kinahan is no longer either bodily or mentally capable of doing useful work for the Survey, and ought to be retired.

The curtain falls

In January 1880 the Irish Survey still had to complete and publish thirty-nine of the 205 one-inch sheets necessary to depict the geology of the whole of Ireland. The rocks remaining to be mapped ranged in variety from the grey Carboniferous limestones of County Sligo to the black basalts of County Antrim, but at the heart of the surviving *terra incognita* there lay that ancient complex of gnarled schists, gneisses, and granites which is County Donegal. No Irish county presents the geologist with a greater challenge. Nowhere do the rocks display a higher degree of intricacy; nowhere do the rocks impose upon the geologist a more stringent test of both his intellectual capacity and his physical stamina. Hull and his men would doubtless have liked to make a slow, cautious, and reflective approach to this their final major problem. Time was essential, but time, alas, was not on their side. When it suits them, government departments can have long memories, and Ramsay, Geikie, and Hull all found that De La Beche's estimate made in 1843 was by no means forgotten in Whitehall: a geological survey of Ireland should occupy little more than ten years. By 1880 the Survey had been at work for thirty-five years and yet there was still a substantial tract of ground awaiting examination. Now the authorities grew impatient; Whitehall wanted the Irish Survey brought to a speedy conclusion. On 6 September 1880 his superiors addressed to Ramsay a letter complaining of what they saw as a want of energy in the prosecution of the Geological Survey's operations throughout the British Isles and protesting that since 1870 the annual amount of territory being mapped had shown a steady decline despite the very considerable augmentation of the staff.[71] On 26 March 1881 Ramsay was informed that the Survey was to undertake no further revision without first obtaining permission from higher authority, and on 11 July 1881 Ramsay was told that his superiors had found quite unacceptable his estimate that the Irish Survey needed a further ten years in which to complete its task. The pressure was on; this was a very different world from that in which James, Oldham and Jukes had spent ten years in producing nothing more than five of 'those footy little county maps'.

In order to expedite affairs Hull introduced three changes into the Irish Survey's programme. Firstly, he abandoned his plans for publishing fifty-four further six-inch sheets covering mining districts in Queen's County, and in counties Antrim, Carlow, Kilkenny, and Tipperary. Secondly, he stopped the preparation of duplicate sets of the six-inch sheets; henceforth the working copies of the six-inch sheets were for some years the only ones available. Thirdly, all the revision work in progress was halted apart from some minor revision carried out by Cruise in Munster during the summer of 1882. McHenry was sent up to County Antrim, and Kinahan — under protest — was removed from Avoca first to Letterkenny and then to Rathmelton on the shores of Lough Swilly. Thus by the autumn of 1882 Hull had his entire force deployed in the north at grips with the rocks of Ulster.

One further modification of the Irish Survey's programme was not of Hull's making; it was the result of a directive from on high. In their search for means of stimulating the geologists to greater efforts, officials within the Science and Art Department had recently suggested to Ramsay that the work of the Survey might be accelerated through the substitution of one-inch field-mapping for the six-inch field-mapping that was now being practised throughout the British Isles. Ramsay doubted the wisdom of such a change, but he was now on the verge of his retirement, a tired man reluctant to do battle with his superiors. As a result we find Hull sending a circular to all his staff on 9 August 1881 explaining that he had received instructions requiring future field-work to be conducted upon one-inch sheets instead of upon the familiar six-inch sheets. Henceforth six-inch field-mapping would be permitted only in areas of unusual geological complexity or in regions possessing important mineral resources. The principle for which Jukes had fought so strenuously in the 1850s was thus lost largely by default in the 1880s. But for the Science and Art Department it was a dubious victory. Hull's men claimed, with some justice, that far from accelerating their work, the use of one-inch field-sheets would retard their progress. In particular, they protested that they would be wasting a great deal of time trying to fix the location of exposures upon the one-inch sheets, a task which was so simple upon the six-inch scale. As a result there arrived in Hume Street a steady stream of requests from the field-staff seeking permission to be allowed to continue with six-inch mapping. The first four such applications arrived in August and September 1881 from Egan in Coleraine, Hardman in Sligo, Kinahan in Avoca, and Symes in Ballymena. All these requests were granted, and by June 1882 more than half of the staff had received Hull's permission to continue with the traditional six-inch field-mapping.

As the surveyors moved into County Donegal their thoughts very naturally turned to the lessons learned by other geologists who had grappled with areas of similar complexity elsewhere. In a paper read to the Geological Society of London on 20 February 1861, Robert Harkness, of Queen's College Cork, had observed that the Donegal rocks display certain similarities to the rocks of the Northwest Highlands of Scotland,[72] and this point was not lost upon Hull. On 7 May 1880 he wrote to Ramsay asking for permission to visit the Northwest Highlands to see whether the lessons of Highland stratigraphy had any relevance to the geology of Donegal. Approval granted, the visit took place later the same month, Hull taking Symes with him, and in Scotland they travelled under the guidance of the then Director of the Scottish Survey, Archibald Geikie. A train of the Highland Railway deposited the visitors at the little station of Garve, near Dingwall, whence they travelled northwestwards over the Moine Schists and the Durness Limestone to reach the Torridonian Sandstone in the vicinity of Ullapool. From there they moved northwards to Inchnadamff and then on to Scourie and Rhiconich to inspect the Lewisian Gneiss. Finally, they turned southeastwards to cross the Moine Schists via Loch Shin in order to regain the Highland Railway at Lairg. They thus made two parallel traverses of the Northwest Highlands some 50 kilometres apart and they also saw a good deal of the fascinating geology exposed along the western seaboard. But it was hardly time well spent; it was a case of the blind leading the blind. The interpretation which Geikie presented to Hull and Symes was the interpretation which he and Murchison had offered in 1861[73] — an interpretation which involved belief in an eastward younging succession from a Fundamental (Lewisian) Gneiss at the base, up through a Cambrian (Torridonian) Sandstone, to a Lower Silurian series containing the Durness Limestone near the base and the Crystalline (Moine) Schists at the top. The concept of a stratigraphical sequence in which sedimentary strata such as the Torridonian Sandstone and the Durness Limestone were overlain by schists had long troubled certain geologists, but not until Charles Lapworth (1842-1920) began his now classical studies in Sutherland in 1882 — studies based, be it noted, upon six-inch mapping — did it become clear to all that the apparent stratigraphical succession in the Northwest Highlands was an illusion. The Law of Superposition was there inapplicable. The Crystalline Schists were really ancient rocks imported into the region from the east along low-angle thrust-faults.[74]

When Hull reported upon his Scottish journey to the Royal Geological Society of Ireland on 20 December 1880 no thoughts of massive overthrusting were present in his mind.[75] Instead, he had returned from Scotland nagged by one question above all others: were there to be found in Ireland ancient rocks equivalent to the Fundamental Gneiss which he had seen so splendidly displayed in the heavily glaciated terrain around Scourie and Rhiconich? The possible presence of such ancient rocks — they were then usually referred to as Laurentian rocks — in various other parts of the British Isles had been a much debated issue over the previous twenty years, but on 20 December 1880 Hull informed his audience that during the following summer he would be going to County Donegal in an effort to resolve the question of whether there was present an Irish counterpart to the Fundamental Gneiss of the Scottish Highlands. Upon his map of 1855 Griffith had already indicated the high antiquity of certain rocks in the west and north of Ireland by designating them as 'Azoic (Primary System)', but the apparent absence of Cambrian strata from the greater part of Ireland rendered difficult the positive identification of any still older Pre-Cambrian rocks. Hull was nevertheless confident when in May 1881 he set off for County Donegal accompanied by Symes and Wilkinson. At the time, Hull subscribed to the view — a view then widely held among Irish geologists — that the Irish granite bodies were largely the result of high-grade metamorphism rather than being a product of the intrusive emplacement of magma. Indeed, at around this period, Hull commonly referred to areas such as Iar-Connaught or the Gweebarra River basin of Donegal as being underlain by gneiss rather than by granite. Influenced by this theory, and guided by Griffith's map (although over twenty-five years old it was still the only detailed geological map of Donegal) Hull, Symes, and Wilkinson spent almost two weeks examining the southeastern margins of the Donegal Granite from Fintown to Glen and the northwestern margins of the Granite from Creeslough to Dunlewy. Hull's conclusion was that what Griffith had represented as granite was really a body of Laurentian gneiss identical to the Fundamental Gneiss of the Scottish Highlands, while the adjacent schists Hull categorised as being of Lower Silurian[76] age and thus equivalent to the Crystalline Schists of Scotland.

Hull announced his 'discovery' of the existence of Laurentian rocks in Donegal in *Nature* on 26 May 1881 and he went on to argue that the 'gneisses' of the Ox Mountains, Belmullet, and Iar-Connaught were also probably all of Laurentian age.[77] So far as Belmullet and Iar-Connaught were concerned, Hull's decision came too late — the one-inch sheets covering those regions were already published — but Laurentian rocks did feature upon five one-inch sheets published after 1881. They appeared first upon Sheet 26 published in 1882, the strata designated as Laurentian being the rocks of the Tyrone Igneous Group, and in 1884 Sheet 55 represented the rocks of a portion of the Ox Mountain inlier as 'Metamorphic Beds of uncertain age (probably Laurentian)'. The other three sheets showing Archaean or Laurentian rocks are numbers 24, 31, and 32 covering the Lough Derg inlier near Ballyshannon. The rocks of the inlier, represented as quartz schist when the three sheets were first published between 1885 and 1888, became Archaean gneiss when new versions of the sheets were issued in 1891-1892. The precise ages of all these rocks remains to this day somewhat uncertain, although there can be no doubt but that the Survey was entirely correct in regarding these rocks as being among the oldest of Ireland's geological foundations. Strangely, when the six sheets covering northwestern Donegal — sheets 3, 4, 9, 10, 15, and 16 — were published in 1889 and 1890, the Donegal Granite was not depicted as a gneiss of Laurentian age. This was because during a tour of Donegal with Geikie in October 1887, the field evidence had caused Hull to abandon his earlier belief that the Donegal Granite was essentially of metamorphic origin. He now accepted that the Granite was an intrusive rock consolidated from an igneous melt, and as such it could only be younger than its enclosing rocks.[78] If, as he still believed, the schists encircling the Donegal Granite were rocks of Lower Silurian age, then there was no way in which the Granite itself could be regarded as being of Laurentian vintage.

Hull's 'discovery' of Laurentian rocks in Donegal may speedily have proved to be nothing more than a chimaera, but the Survey did make two significant and unexpected finds during the surveying of County Donegal. Firstly, while mapping the Fanad Peninsula in 1885 the surveyors came upon a downfaulted outlier of Old Red Sandstone which had featured upon no earlier geological map of the region.[79] Secondly, in 1887 Kilroe found two tiny outliers of Lower Carboniferous sandstone atop the quartzites of Slieve League at an altitude of 600 metres.[80] Geikie was taken to see the outliers of 24 October 1887 and they made upon him a profound impression as indicating the scale of the denudation experienced by Ireland's Carboniferous strata. From the outliers he drew the inference:

> ... that the Irish coal-fields, now so restricted in extent, once spread far and wide over the hills of Donegal, from which they have since been gradually denuded. Truly the woes of Ireland may be traced back to a very early time, when not even the most ardent patriot can lay blame on the invading Saxon.[81]

Geikie can hardly have been aware of the fact, but many years earlier another geologist had spent his Christmas encamped upon that identical spot awaiting the clear visibility necessary to allow him to conduct his triangulation observations. His name was Joseph Ellison Portlock; the year was 1827 (see p. 96). Had Portlock in his hours of enforced idleness perhaps lighted upon those very outliers which now so fascinated Geikie?

The death of O'Kelly in 1883 and of Baily in 1888 did little to hamper the advance of the Survey. More serious was the secondment of the able Hardman to Western Australia 'on special colonial service' in February 1883. He returned to Ireland in October 1885 but only eighteen months later he was struck down by typhoid, his constitution having been undermined by a chest infection recently developed as a result of working in bad weather around Woodenbridge in County Wicklow whither he had been sent to prepare a memoir to sheets 138 and 139, Kinahan's essay on the subject having been declared unacceptable. It was Hardman's widow who eventually received £500 as a gratuity from the Western Australian government in recognition of her husband's part in the discovery of the Kimberley goldfield. As the 1880s progressed the extent of the unmapped ground was steadily diminished. In 1884 Symes took the Survey's tent out to Rathlin Island, where he got himself into trouble for spending £1:15:0 of the Survey's money on the hire of a boat in which to sail around the island. In 1885 Egan, McHenry, and Symes completed the mapping of the country around the Giant's Causeway, where Susanna Drury had deployed her artistic talents so brilliantly a hundred and fifty years before. In 1886 Mitchell went out to map wind-swept Tory Island, and Cruise went out to map the gneiss forming those two northernmost outposts of Ireland, Inishtrahull and The Tor Rocks.

In 1886 the Survey mapped 751 square kilometres of ground, traced 1,790 kilometres of geological boundary, and examined 304 kilometres of coastline, all of this territory being in County Donegal. It was the Survey's last full field-season. That year Cruise, Egan, Kilroe, McHenry, Mitchell, Nolan, and Symes all completed the mapping of their assigned ground; only 94 square kilometres of Ireland remained to be surveyed, 78 square kilometres in Kinahan's ground around Rathmelton and Church Hill, and 16 square kilometres in that of Wilkinson out upon the Rosguil Peninsula. The following summer – the summer of 1887 – Kinahan and Wilkinson took to the field to complete the Survey's task. It was a good year in which to finish; it was Queen Victoria's golden jubilee year. In London on 21 June 1887 there processed in triumph through thronged streets that Queen who, with the much lamented Albert at her side, had long ago paused to admire Griffith's map on display at the Great Dublin Industrial Exhibition. At just about the same time Kinahan and Wilkinson in far off County Donegal were wending their solitary ways across barren hillsides as they closed off their geological boundaries, were pushing aside the brambles as they struggled to explore their final stream sections, and were making their last collections of samples from some wave-washed cliff. It was Wilkinson who completed his ground first; Kinahan decided that he would have to defer the examination of some of his ground until access became easier after the harvest. The harvest was early – it had been an unusually hot, dry summer – but Kinahan stayed on at his field-station at Rathmelton until the end of October. It must have been one evening towards the close of that month when Kinahan plied his hammer to the very last rock exposure to be examined during the primary geological survey of Ireland. Having peered at the resultant sliver of rock through his lens, he must have entered some comment in his notebook, recorded the site upon his field-sheet, closed his map-case, and turned his steps homeward. It was all a routine that he knew so well; thousands of other days in the field had ended for him in precisely the same way as the evening shadows lengthened. But this evening was really different. The geological campaign which had opened in County Wexford in the summer of 1845 was now at long last concluded. And how appropriate that it should have been Kinahan who made those final observations which brought the primary survey of Ireland to its close. He was a stormy character certainly, but in terms of service within Ireland he was the Survey's senior officer. For well over thirty years he had carried his hammer and his maps through some of Ireland's most rugged terrain; he surely deserved to strike the final hammer blow on behalf of the Survey. As he trudged home that last evening did his memories perhaps return to that now distant occasion in August 1854 when, fresh out of Trinity College, he had reported for training at Bantry, there to find himself in the inspiring company of Jukes, Ramsay, and Sir Henry De La Beche himself. In those days a one-inch geological map of Ireland was still merely a project under discussion, but now in Donegal he and Wilkinson had just fashioned those final pieces necessary to complete the great one-inch geological mosaic.[82]

Although the primary survey of Ireland was now complete, the Survey had before it three further years of work connected with the mapping. That work was of five types. Firstly, pattern copies of the final one-inch sheets had to be prepared and the publication of the sheets arranged. Three new one-inch sheets were published in 1887, four in 1888, five in 1889, and a final batch of seven in 1890. Secondly, the accompanying sheet memoirs had to be completed, and the last of them – the memoir to sheets 22, 23, 30, and part of 31 – appeared in 1891. Thirdly, the regular preparation of duplicate copies of the six-inch field-sheets had ceased in 1881 in order to expedite the progress of the Survey and this deficiency now had to be remedied. In October 1888 Hull reported that some two hundred six-inch field-sheets still had to be copied. Fourthly, eleven sheets of longitudinal sections had to be prepared for northern Ireland, and the last of these (Sheet 35) was not published until December 1894. Finally, as a result of the Science and Art Department's opposition to six-inch field-mapping, certain areas of County Donegal had been mapped only upon the one-inch scale. One such area was the Slieve League Peninsula, and it was now decided to re-map the peninsula on the six-inch scale. The task was assigned to Kilroe, and he completed his re-investigation during the summer of 1889. Other areas of the county were not accorded similar treatment, and to this day there remain parts of Donegal which have never received a Survey examination at anything larger than the one-inch scale.[83]

While the final Irish one-inch sheets and memoirs were being prepared for publication,

Geikie decided that it was desirable for some of the Irish staff to visit Scotland in order to be shown something of the exciting discoveries that the Scottish Survey had been making in the Northwest Highlands. Following upon Lapworth's seminal work in Sutherland in 1882, Geikie had immediately set some of his most experienced staff to work in the region, and their revelation of the presence of a series of folds and over-thrusts of awe-inspiring complexity had begun to excite the attention of the entire geological world. Geikie himself had badly burned his fingers in the region; for far too long he had clung to his own and Murchison's false interpretation of the stratigraphy of the Northwest Highlands as a normal sedimentary sequence. Now he had made ample acknowledgement both of his own error and of the brilliance of Lapworth's tectonic interpretation,[84] but he was determined that he and the Survey would not be caught out for a second time. If the rocks of the Northwest Highlands enshrined some lesson that the Irish Survey should learn, then that lesson must be administered before the Irish Survey finally committed itself in front of the geological world through the publication of the remaining one-inch sheets and memoirs for County Donegal. The party selected for the Scottish excursion consisted of Kilroe, Kinahan, McHenry, Nolan, and Wilkinson, and they crossed from Belfast to Greenock on the night of 4-5 June 1888. In the Northwest Highlands they visited much of the same ground around Inchnadamff as Hull had seen in 1880, and their guide was the leader of Geikie's team in the area, the redoubtable and much loved Benjamin Neeve Peach (1842-1926). The Irish party was back in Dublin by 24 June and each member of the group was then required to write for Geikie a report relating what they had seen in Scotland to their experience of Irish geology. Those reports were not very enlightening; the Irishmen understandably found it difficult to relate the geology of Donegal to that of the Northwest Highlands. This, for instance, is an extract from Nolan's equivocal report dated 10 July 1888:

> I do not think we shall find evidence for anything like the extensive displacements that have occurred in Sutherlandshire, yet it seems not improbable that some such movements have played an important part in shaping the geology of Donegal and may serve to explain much that is now obscure.

Geikie himself remained persistent. If the ancient rocks of Ireland contained secrets comparable to those which the Scottish rocks had so long concealed from him, then Geikie was determined to unmask those secrets. In the spring of 1889 he therefore brought Peach over to Ireland, and he and Peach scoured counties Mayo and Galway in the company of McHenry, Geikie's favourite Irish travelling companion, and of Hyland, the newly recruited petrologist. From Benwee Head, County Mayo, they travelled to Belmullet, where W.B. Leonard had been drowned in 1876, to Achill Island, and thence via Westport and the southern shores of Clew Bay to Killary Harbour, a feature which Murchison had long ago described as 'perhaps the finest thing in Ireland'. There they spent a few days examining rocks that had been first mapped upon the six-inch scale by Ganly in 1840, and then they moved on to Clifden, where Jukes had lain in a coma in 1868, along the margin of the quartzites forming The Twelve Pins, past the old constabulary barracks at Recess which for many years had been a home for the Kinahan family, and so on to Galway city. But Geikie had to admit that the excursion had been disappointing.[85] Even the discerning eye of Peach had seen little more in the rocks than had the men of the Irish Survey some twenty years before. Late in life Geikie remembered that excursion of 1889 and he regretted 'that the tracts of Mayo and Galway which we examined had already been mapped by the Survey, and the maps of the ground had been published, while the key to the structure of the country was still undiscovered'.[86]

As Geikie travelled through Ireland in that spring of 1889 he must have carried with him one problem which was administrative rather than geological. What should happen to the Geological Survey of Ireland now that it had accomplished its appointed task of mapping the country's geology? Back in London there were politicians and civil servants who had always regarded the Geological Survey of Great Britain and Ireland as engaged upon a task of finite duration; once the mapping of the British Isles was completed, then the Survey could be disbanded. Now the Geological Survey of Ireland found itself a potential sacrificial victim to Whitehall's economy axe. Geikie saw no hope of maintaining the Irish Survey at anything like its 1889 levels of staffing and finance. On 24 October 1888 even Hull had written to Geikie saying:

> ... unless for a few isolated districts, I regard the field work of the Survey as so accurate and complete in all its details that I cannot regard a general revision of the Geological Survey of Ireland as either necessary or desirable.

But on the other hand Geikie was convinced — and rightly so — that Ireland had a continuing need for a Geological Survey presence. Some compromise was called for. Early in March 1889 Geikie recommended to his superiors that the Geological Survey of Ireland should be allowed to survive in an abbreviated form.[87] A small staff should be left in Dublin and charged with the task of handling routine geological enquiries from the public, with the responsibility of keeping the one-inch sheets available and up-to-date, and with the duty of undertaking occasional field-revision as the need arose.

Negotiations around this proposal evidently took place throughout much of the remainder of 1889, and not until 2 December was Geikie in a position to communicate to Hull the official news of the final verdict. Hull himself, together with Kinahan, were both to retire on pension, each having served with the Survey for close upon forty years. Symes and Wilkinson were to be transferred to the Geological Survey of Scotland. Cruise, Mitchell, and Wynne were all to be pensioned off, Cruise and Mitchell because Geikie regarded them as being geologists of inferior quality (Mitchell had developed a drink problem while living in the hut at Fintown in 1885), and Wynne because he was now aged fifty-four. The Irish Survey was to be left in the charge of Nolan, who was given the new title of Senior Geologist, and his staff was to consist of Egan, Hyland, Kilroe, and McHenry, together with Clark, the Fossil Collector.

Over the next few months 14 Hume Street must have seen many a farewell as Hull's men began to go their several ways. Symes and Wilkinson became members of the Geological Survey of Scotland as from 1 April 1890. Cruise, Mitchell, and Wynne retired on 31 August, and on that same day there ended Kinahan's long and turbulent association with the Survey. One can but wonder whether, on his final day in Hume Street, Kinahan chose to come down the creaking stairway from the room on the top floor which had been his since the close of the mapping programme — he had complained of his banishment to what he termed a garret — down to the first floor, past the entrance to the office inhabited by the despised Wynne, and to the door of the Director's room, there to knock as a preparatory to taking of Hull a polite adieu. Perhaps he did make the descent only to discover that Hull had seen fit to leave the building 'on urgent private business'. Hull certainly was seated in his room a few weeks later on 30 September; it was his own final day in Hume Street before retirement. The Director General was over from London. As he sat awaiting Geikie's arrival, Hull could look through the windows and across the street to the red-brick house that was Griffith's birthplace. Between those windows there stood the two bookcases containing Portlock's library. The walls of the room were decorated with a few geological scenes bearing testimony to the artistic talents of Du Noyer. The room thus incapsulated something of the spirit of those men who for a hundred years had striven to complete the geological map of Ireland. Upon Hull's desk at that very moment there lay a document recording the virtual consummation of that prodigious task. It was a document entitled 'Agenda for Dr Geikie' and it listed those publications arising from the geological survey of Ireland which still remained to be completed. Under the heading 'One-inch maps' was a single curt entry: 'All published except Sheet 10, copies of which are now in Colourer's hands'. That final sheet — Sheet 10 — was published on 21 November 1890, and a very attractive and striking sheet it turned out to be, depicting, as it does, a region of most complex geological structure lying in northern County Donegal between Creeslough and Letterkenny. Originally

Sheet 10 was priced at 3 shillings, as by then were all the full sheets in the one-inch series but, on the eve of World War I, when all the sheets were re-priced in accordance with the complexity of their geological colouring, sheet 10 became at 14 shillings and 3 pence the second most expensive of the Irish one-inch sheets — second only to Sheet 94 which covers the Recess district of County Galway and sold at 15 shillings and 9 pence. The intricate pattern of the various rock types present had indeed rendered Sheet 10 a veritable sheet of many colours (Plate 4).

This book opened with a reference to the artistry of Miss Susanna Drury, but from that point onwards the story of the geological map of Ireland has been unfolded without our having had to make so much as a single allusion to a female contribution towards the completion of that great enterprise. For the sake of structural symmetry, however, perhaps we should conclude by making reference to another lady — to Mrs E. Williams of 4 Stratford Villas, Camden Square, Camden Town, London. During the closing decades of the nineteenth century she was employed to work at home colouring those maps of the Geological Survey of Great Britain and Ireland which were destined for distribution to the public. As we contemplate one of the early copies of that colourful final Irish one-inch sheet — Sheet 10 — it is Mrs Williams's handiwork that excites our admiration. Toiling in the grimy air of Victorian London, close to the Metropolitan Cattle Market and hard by the Midland Railway's line out of St Pancras, hers was a landscape far removed from the Irish landscape of purple mountains, tranquil rivers, sandy eskers, brown peat-bogs, and storm-beaten cliffs — an Irish landscape which had become so familiar to generations of geologists as they painstakingly traced out those geological lines which Mrs Williams's brush now had to follow with such care and precision.

REFERENCES AND NOTES FOR CHAPTER SEVEN

1. Some sources for the life of Hull: SIN; Edward Hull, *Reminiscences of a strenuous life*, London, 1910, pp. 120 + iii; *Q. Jl. Geol. Soc. Lond.* 74, 1918, proceedings liv; *Proc. R. Soc.* 90B, 1919, xxviii-xxxi; *Geol. Mag.* N.S. dec. 6, 4, 1917, 528, 553-555; *Ir. Nat.* 27, 1918, 17-19.
2. RP. Notes in Ramsay's diary for 1858.
3. In this chapter all documents not given a specific location will be found in the archives of the Geological Survey of Ireland.
4. Hull also received his salary as Professor of Geology in the Royal College of Science for Ireland.
5. *The coal-fields of Great Britain: their history, structure, and resources*, London, five editions 1861-1905; *A treatise on the building and ornamental stones of Great Britain and foreign countries*, London, 1872, pp. xxiv + 333; *Mount Seir, Sinai and western Palestine*, London, 1885, xvi + 227; *Memoir on the geology and geography of Arabia Petraea, Palestine, and adjoining districts*, London, 1886, pp. x + 145.
6. The society's MS minute books are in the Library of Trinity College Dublin.
7. In the 1950s Portlock's library was handed over to the Royal Dublin Society.
8. Archibald Geikie, *Memoir of Sir Andrew Crombie Ramsay*, London, 1895, 319-321.
9. *Ibid.*, 321.
10. *Ibid.*, 319.
11. Archibald Geikie, *A long life's work: an autobiography*, London, 1924, 319.
12. ADB; *Geol. Mag.* N.S. dec. 3, 4, 1887, 334-336; *Nature, Lond.* 36, 1887, 62.
13. SIN.
14. *A description of the soil-geology of Ireland, based upon Geological Survey maps and records, with notes on climate*, Dublin, 1907, pp. 12 + 300. The volume contains a fine colour-printed map at a scale of 1:633,600 showing the surface geology of Ireland.
15. A. Geikie, *Confidential report on the Irish Branch of the Geological Survey*, PROL, DSIR 9/71.
16. See reference 42, p.
17. PROL, DSIR 9/71, (ref. 15).
18. Geikie, *op. cit.*, 1924, 211, (ref. 11).
19. *Ir. Nat.* 7, 1898, 153.
20. PROL, DSIR 9/71, (ref. 15).
21. Hull, *op. cit.*, 1910, 37-38, (ref. 1).
22. Royal Irish Academy, 24 o 35/36/37.
23. Geikie, *op. cit.*, 1924, 211, (ref. 11).
24. *Instructions for the guidance of the officers of the Irish Branch of Her Majesty's Geological Survey of the United Kingdom*, Dublin, 1876, pp. 23.
25. Sadly, all these notebooks were destroyed about 1950.
26. E. Hull, On the progress of the Geological Survey of Ireland, *Geographical Magazine* 1, 1874, 309, and *Rep. Br. Ass. Advmt. Sci.*, Belfast, 1874, part 2, 83.
27. J.E. Richey, The Tertiary ring complex of Slieve

Gullion (Ireland), *Q. Jl. Geol. Soc. Lond.* 88, 1932, 776-849.
28. J. Nolan, On a remarkable volcanic agglomerate near Dundalk, *J.R. Geol. Soc. Irel.* 4 (4), 1876-77, 233-239. See also J. Nolan, On the ancient volcanic district of Slieve Gullion, *Rep. Br. Ass. Advmt. Sci.*, Dublin, 1878, 527-528, and *Geol. Mag.* N.S. dec. 2, 5, 1878, 445-449.
29. Edward Battersby Bailey *et al., Tertiary and Post-Tertiary geology of Mull, Loch Aline, and Oban*, Memoirs of the Geological Survey of Scotland, Edinburgh, 1924, 7.
30. Edward Hull, *The physical geology & geography of Ireland*, London, 1878, pp. xvi + 291; George Henry Kinahan, *Manual of the geology of Ireland*, London, 1878, pp. xx + 444.
31. *Geological map of Ireland founded on the maps of the Geological Survey, of Sir Richard Griffith and of Proff. J. Beete Jukes*, Edward Stanford, London, 30 March 1878. See also E. Hull, Note on a new geological map of Ireland, *J.R. Geol. Soc. Irel.* 5(2), 1878-79, 104-105.
32. E. Hull, On the upper limit of the essentially marine beds of the Carboniferous group of the British Isles and adjoining continental districts, *Q. Jl. Geol. Soc. Lond.* 33, 1877, 613-651.
33. R. Griffith, Of the order of succession of the strata of the south of Ireland, with a particular reference to the Killarney district of the county of Kerry, *J. Geol. Soc. Dubl.* 3 (2), 1845, 150-160.
34. On the geological age of the rocks forming the southern highlands of Ireland, generally known as "The Dingle Beds" and "Glengariff Grits and Slates" (Jukes), *Q. Jl. Geol. Soc. Lond.* 35, 1879, 699-723; On the geological relations of the rocks of the south of Ireland to those of north Devon and other British and continental districts, *ibid.*, 36, 1880, 255-276. See also On the relations of the Carboniferous, Devonian, and Upper Silurian rocks of the south of Ireland to those of North Devon *Scient. Trans. R. Dubl. Soc.* N.S. 1(11), 1880, 135-150.
35. Hull, *op. cit.*, 1879, (ref. 34).
36. *Hansard's Parliamentary Debates*, series 3, 260, 3 May 1881, col. 1655.
37. A.B. Wynne, On some points in the physical geology of the Dingle and Iveragh promontories, *J.R. Geol. Soc. Irel.* 6 (1), 1881, 1-7 and *Scient. Proc. R. Dubl. Soc.* N.S. 2, 1878-80, 590-596.
38. J.B. Archer, Richard Griffith and the first published geological maps of Ireland, pp. 143-171 in Gordon Leslie Herries Davies and Robert Charles Mollan (editors), *Richard Griffith 1784-1878*, Royal Dublin Society, 1980.
39. Hull, *op. cit.*, 1879, (ref. 34).
40. Hull, *op. cit.*, 1880, 137-138, (ref. 34).
41. *Nature, Lond.* 18, 10 October 1878, 627-628.
42. But see also R. Griffith, Notes on the stratigraphical relations of the sedimentary rocks of the south of Ireland, *J. Geol. Soc. Dubl.*, 8 (1), 1857-58, 2-15.
43. Hull, *op. cit.*, 1880, 138f, (ref. 34).
44. G.H. Kinahan, On the hiatus said to have been found in the rocks of West Cork, *Rep. Br. Ass. Advmt. Sci.* Swansea, 1880, 574-575.
45. George William Lamplugh *et al., The geology of the country around Cork and Cork Harbour*, Memoirs of the Geological Survey, Ireland, Dublin, 1905, 10.
46. Kinahan, *op. cit.*, 1880, (ref. 44).
47. Hull, *op. cit.*, 1910, 89-94, (ref. 1).
48. J.G. Capewell, The stratigraphy and structure of the country around Sneem, Co. Kerry, *Proc. R. Ir. Acad.* 58B (8), 1957, 167-183.
49. Lamplugh *et al., op. cit.*, 1905, (ref. 45).
50. William Bourke Wright *et al., The geology of Killarney & Kenmare*, Memoirs of the Geological Survey of Ireland, Dublin, 1927, pp. viii + 111.
51. E. Hull, On a proposed Devono-Silurian formation, *Q. Jl. Geol. Soc. Lond.* 38, 1882, 200-209.
52. Geikie, *op. cit.*, 1924, 205, (ref. 11).
53. Maxwell Henry Close and George Henry Kinahan, *The general glaciation of Iar-Connaught and its neighbourhood, in the counties of Galway and Mayo*, Dublin, 1872, pp. 20.
54. George Henry Kinahan, *Economic geology of Ireland*, published between 1887 and 1889 in three parts constituting volume 8 of the *Journal of the Royal Geological Society of Ireland*.
55. Edward Greenly, *A hand through time*, London, 1938, vol. 1, 129.
56. G.H. Kinahan, Suggestions about denudation, *Geol. Mag.* 6, 1869, 109-115.
57. George Henry Kinahan, *Valleys and their relation to fissures, fractures, and faults*, London, 1875, chap. 9.
58. Kinahan, *op. cit.*, 1878, 314, (ref. 30).
59. G.H. Kinahan, The Scalp, County Dublin, *Ir. Nat.* 2 (12), 1893, 241-245.
60. Geikie, *op. cit.*, 1924, 205, (ref. 11).
61. IGS 1/17.
62. *Geol. Mag.* N.S. dec 2, 9, 1882, 131-133, 190.
63. RP 1/50 and 1/51.
64. E. Hull, On the geological age of the Ballycastle coal-field, *J.R. Geol. Soc. Irel.* 2 (3), 1869-1870, 260-275; G.H. Kinahan, *Geol. Mag.* 8, 1871, 573-574.
65. E. Hull, Observations on the general relations of the drift deposits of Ireland to those of Great Britain, *Geol. Mag.* 8, 1871, 294-299; G.H. Kinahan, Middle Gravels (?), Ireland, *ibid.*, 9, 1872, 265-268.
66. G.H. Kinahan, Diagram of the Irish Palaeozoic rocks, *Trans. Manchr. Geol. Soc.* 15, 1878-1880, 176-183.
67. *Proc. R. Ir. Acad.* series 2, 2, 1875-77, *passim*.
68. Royal Irish Academy MS minutes of the Com-

mittee of Science for 23 June 1879 and 22 December 1879. See also *Nature, Lond.* 20, 30 October 1879, 641.
69. G.H. Kinahan, Report on the rocks of the Fintona and Curlew Mountain districts, *Proc. R. Ir. Acad.* series 2, 3 (5), 1880, 475-500.
70. His name is also on five revised sheets for southern County Wexford issued in the years after 1878, but four of these sheets are largely occupied by the sea.
71. *28th report of the Science and Art Department*, H.C. 1881 (2970), XXXVII, 55, 56, 58, 69, 70.
72. R. Harkness, On the rocks of portions of the Highlands of Scotland south of the Caledonian Canal; and on their equivalents in the north of Ireland, *Q. Jl. Geol. Soc. Lond.* 17, 1861, 256-271. See also his On the metamorphic fossiliferous rocks of the County of Galway, *ibid.*, 22, 1866, 506-513.
73. R.I. Murchison and A. Geikie, On the altered rocks of the Western Islands of Scotland, and the north-western and central Highlands, *Q. Jl. Geol. Soc. Lond.* 17, 1861, 171-232.
74. C. Lapworth, The secret of the Highlands *Geol. Mag.* N.S. dec. 2, 10, 1883, 120-128, 193-199, 337-344; On the close of the Highland Controversy, *ibid.* N.S. dec. 3, 2, 1885, 97-106.
75. E. Hull, On the geological structure of the northern Highlands of Scotland, *J.R. Geol. Soc. Irel.* 6 (1), 1881, 56-68; *Scient. Proc R. Dubl. Soc.* N.S. 3, 1880-82, 34-46.
76. In 1879 Lapworth (*Geol. Mag.* N.S. dec. 2, 6, 1879, 1-15) suggested that the term 'Ordovician' be employed for the rocks lying between the base of the Lower Llandovery and the base of the Lower Arenig. The Survey — and especially Geikie — disliked the term and it was 1913 before Ordovician rocks featured upon an Irish Geological Survey map.
77. *Nature, Lond.* 24, 1881, 81-82; E. Hull, On the Laurentian beds of Donegal and other parts of Ireland, *Rep. Br. Ass. Advmt. Sci.*, York, 1881, 609; *J.R. Geol. Soc. Irel.* 6(2), 1882, 115-116; *Scient. Trans. R. Dubl. Soc.* N.S. 1 (18), 1882, 243-256.
78. There is a long memoir by Hull upon this subject and dated 10 November 1887 in GSI Letter Book 9, 1887. See also the 1891 memoir devoted to northwestern Donegal (sheets 3, 4, 5, 9, 10, 11, 15, 16).
79. E. Hull, On the occurrence of an outlying mass of supposed Lower Old Red Sandstone and conglomerate in the promontory of Fanad, County Donegal, *J.R. Geol. Soc. Irel.* 7 (1), 1886, 74; *Rep. Br. Ass. Advmt. Sci.*, Aberdeen, 1885, 1016.
80. J.R. Kilroe, The discovery of two Carboniferous outliers on Slieve League, Co. Donegal, *Scient. Proc. R. Dubl. Soc.* N.S. 6, 1888-1890, 63-66.
81. Archibald Geikie, *Landscape in history*, London, 1905, 62; Geikie, *op. cit.*, 1924, 220-221 (ref. 11); Geikie's preface to the 1891 memoir on sheets 22, 23, 30, 31.
82. The one-inch geological map of Ireland is so large that it can rarely have been displayed as a single unit. From 1890, however, there was in the Science and Art Museum in Dublin (later the National Museum of Ireland) a relief model of Ireland made by the pupils of the Model School, Marlborough Street, Dublin, and to this model geological colours were affixed under McHenry's direction. The model was destroyed when the geological galleries of the museum were cleared to make way for civil servants following the establishment of the Dail in nearby Leinster House in the 1920s. See Grenville Arthur James Cole, *Description of the raised map of Ireland*, National Museum of Science and Art, Dublin, 1909, pp. 16 + 10 plates.
83. Sidney Berdoe Neal Wilkinson *et al.*, *The geology of the country around Londonderry*, Memoirs of the Geological Society of Ireland, Dublin, 1908, iii, 2.
84. A. Geikie, B.N. Peach, and J. Horne, The crystalline rocks of the Scottish Highlands, *Nature, Lond.* 31, 13 November 1884, 29-35.
85. A. Geikie, Recent researches into the origin and age of the Highlands of Scotland and the west of Ireland, *Proc. R. Instn Gt. Br.* 12, 1889, 528-546. See also *Nature, Lond.* 40, 1889, 299-302, 320-324.
86. Geikie, *op. cit.*, 1924, 229-230, (ref. 11).
87. PROL, DSIR 9/71, (ref. 15).

EPILOGUE

Although the primary mapping of Ireland conducted by the Geological Survey was completed in 1887, that year by no means saw an end to the story of Irish geological cartography. There remained then — and there still remains today — a great deal to be learned about the rocks of Ireland. Over the last hundred years, from Malin Head in the north to Cape Clear in the south, from the Blasket Islands in the west to the Ards Peninsula in the east, geologists have been out in the field mapping, re-mapping, and re-mapping yet again. Those geologists have possessed skills and knowledge immensely superior to the attainments of their nineteenth-century precursors, and as a result they have, through their mapping, achieved insights far beyond those vouchsafed to the men of an earlier age. Some of those who have mapped Ireland's rocks during the years since 1887 have been officers of the Geological Survey of Ireland or, since its formation in 1947, of the Geological Survey of Northern Ireland; some of them have come from universities at home and overseas; and some of them have been men from mining companies engaged upon a search for Ireland's hidden mineral resources. The work of all these geologists has brought into being — continues to bring into being — a multitude of geological maps possessed of an accuracy and sophistication far beyond that possible in the last century. But the story of the geological mapping of Ireland over the years since 1887 is a story that must be left to be told by some future scribe.

INDEX

Académie Royale des Sciences, 4
Accommodation in the field, problems of, 148, 153, 164, 171, 185, 204
Achill Island, 227
Act of Union, 10, 35
Adare, Lord, 108-109
Adelaide Hospital, Dublin, 205
Agassiz, Louis, 61, 150
Aghanloo Parish, Co. Londonderry, 13, 15, 91, 92
Agrological maps, 150
Aher, David, 20
Albert, Prince, 57, 75, 81, 226
Allen, Bog of, 20
Anascaul, Co. Kerry, 74
Andrews, John H., 87
Antrim basalts, early maps of, 1, 4-5
Antrim Town, 179, 196
Aqaba, Gulf of, 195
Archer, Jean B., 74, 79, 123
Ards Peninsula, 232
Arigna, Co. Roscommon, 119
Arklow, Co. Wicklow, 129, 219
Armagh City, 168
Arrowsmith, Aaron, 39, 40, 45, 60, 61, 66, 79, 81
Arthurstown, Co. Wexford, 134
Ashford, Co. Wicklow, 142
Athboy, Co. Meath, 10
Athlone, Co. Westmeath, 108, 113
Athy, Co. Kildare, 25
Audley, Lord, 44
Aughnacloy, Co. Tyrone, 106
Aughrim, Co. Wicklow, 219
Australia, 133, 142, 156, 157, 183, 225
Auvergne, The, 4, 46, 184, 213
Aveline, William T., 122, 157
Avoca, Co. Wicklow, 8, 132, 147, 151, 217, 218, 219, 220, 222, 223
Avoca Lodge, Co. Wicklow, 217

Bailey, Edward B., 109, 217
Bailieborough, Co. Cavan, 160
Baily, William H., 173, 186, 187-188, 190, 202, 203, 205, 212, 216, 221, 225
Bakewell, Robert, 56
Ballina, Co. Mayo, 204
Ballingarry Hills, Co. Limerick, 176
Ballycastle, Co. Mayo, 190, 202, 204, 205
Ballycastle coalfield, Co. Antrim, 221
Ballyellis House, Co. Cork, 46
Ballyhale, Co. Kilkenny, 186
Ballymacadam, Co. Tipperary, 44, 177
Ballymena, Co. Antrim, 223
Ballyshannon, Co. Donegal, 224
Balteagh Parish, Co. Londonderry, 91
Baltinglass, Co. Wicklow, 8
Banbridge, Co. Down, 205
Bank of Ireland, Dublin, 2
Banks, Joseph, 7
Banteer, Co. Cork, 47
Bantry, Co. Cork, 164, 226
Bantry Bay, 164, 210, 211
Barlow, Peter, 62
Barrington, Richard M., 217
Barton, Richard, 2
Base-map problem, 19, 26, 41, 43, 44-46, 47, 49, 61, 62, 101-102, 136-137, 149, 151, 154, 160-162
Beagle, H.M.S., 157
Beaufort, Daniel A., 45
Belderg, Co. Mayo, 21
Belfast, 78, 100, 102, 103, 106, 107, 129, 141, 142, 160, 179, 227
Belfast Naturalists' Field Club, 160
Belmullet, Co. Mayo, 190, 205, 224, 227
Benwee Head, Co. Mayo, 227
Berehaven, Co. Cork, 171
Berger, Jean F., 27-28, 29, 34, 61, 77
Bermuda, 113
Bessborough, Earl of, 147, 150
Best, Edward, 220
Big Wind, 65
Blackrock, Co. Dublin, 106
Blackwood, Francis P., 156
Blackwood, Henry, 156
Blasket Islands, Co. Kerry, 232
Blessington, Co. Wicklow, 8
Blumenbach, Johann F., 38
Board of Agriculture, 7, 8, 9, 14
Board of Works, 62, 79, 137, 185
Bog Commissioners, 20-21, 24, 37, 39, 41, 45, 53, 54
Boldero, Henry G., 108, 109
Bone, Charles, 174
Booterstown, Co. Dublin, 150
Bordes, Lt., 93, 94
Borris, Co. Carlow, 142
Borrisokane, Co. Tipperary, 41, 168
Boulton, Matthew, 35
Boundary Survey, 47, 49, 58, 65, 79, 85, 88, 116, 118
Boyle, Co. Roscommon, 200
Boys, John, 13
Bray Head, Co. Wicklow, 8
Bristow, Henry W., 122, 215
British Association
York	1831	58
Edinburgh	1834	58, 60
Dublin	1835	60-61, 66, 70, 77, 81, 85, 99, 113, 168
Liverpool	1837	66
Newcastle-upon-Tyne	1838	65, 66, 67
Glasgow	1840	75, 148
Cork	1843	73, 113, 168, 169
York	1844	118, 120
Swansea	1848	149
Belfast	1852	78, 106, 168, 172
Liverpool	1854	172
Dublin	1857	170
Manchester	1861	188
Edinburgh	1871	58
Dublin	1878	208, 209
Swansea	1880	212

See also 97, 113, 216
Broad Haven, Co. Mayo, 204
Brown, Henry, 119
Browne, Cara A., 157, 185, 186, 191
Bryce, James, 78
Buckland, William, 28, 29, 61, 77, 91, 118, 120
Bunclody, Co. Wexford, 140, 142, 178
Bunmahon, Co. Waterford, 136
Burgoyne, John F., 62

233

Burke, Thomas H., 204
Burren, Co. Clare, 185, 206
Bushmills, Co. Antrim, 4
Butcher, Norman, 123
Butler, Isaac, 2
Butler's Bridge, Co. Cavan, 66

Cahore Point, Co. Wexford, 167
Calp limestone, 17, 43
Cambria, Paddle Steamer, 170
Camden, William, 2
Camolin, Co. Wexford, 147
Cape Clear, Co. Cork, 232
Cape Colony, 167, 186
Cappagh, Co. Cork, 44
Cappoquin, Co. Waterford, 178, 182
Carboniferous rocks, classification of Irish, 16, 17, 27, 43, 66, 69, 75-76, 175, 182-184, 208
Carlingford, Co. Louth, 200
Carlo, dog, 164
Carlow, Co., geological map of, 160
Carlow Town, 134, 158
Carrickart, Co. Donegal, 72
Carrickfergus, Co. Antrim, 104
Carrol, James, 44
Castlebar, Co. Mayo, 185
Castlecomer, Co. Kilkenny, 20, 23, 185-186
Castlecomer coalfield, *see* Leinster Coal District
Castlemaine Harbour, Co. Kerry, 167
Castle Saunderson, Co. Cavan, 43
Cavendish, Lord Frederick, 204
Chair of Kildare, 182
Charleville, Co. Cork, 44
Christ's College, Cambridge, 156
Church Hill, Co. Donegal, 226
Cirques, 218
Civil Service Commissioners, 188, 202
Clane, Co. Kildare, 35
Clara, Vale of, Co. Wicklow, 163
Clare, Earl of, 108
Clare, Co., early map of, 8
Clark, Richard, 212, 228
Clew Bay, Co. Mayo, 206, 227
Clifden, Co. Galway, 72, 190, 205, 227
Clogher, Co. Louth, 191
Clogher, Co. Tyrone, 49
Clondavaddog, Co. Donegal, 5
Clonmel, Co. Tipperary, 129
Close, Maxwell H., 58, 217
Clough, Charles T., 148
Coalbrookdale Company, 57
Cobh, Co. Cork, 75
Colby, Thomas:
    early years 85
    as a surveyor 85
    in Ireland 85-122
    and six-inch maps 86
    and historico-statistical inquiry 86-87
    at Geological Society of London 87
    at Geological Society of Dublin 87
    plans geological survey 87-88, 90
    and J.W. Pringle 88-95
    and Carmichael Smyth 93-95
    geology stopped 94-95
    and J.E. Portlock 95-106
    the memoir project 97-104, 108-109, 110
    geology in Britain 98-99, 100, 101
    the Templemore memoir 99-100, 103
    the 'geological office' 100-103
    the office closed 102-104
    and engineering geology 103-104
    the Londonderry report 104-106
    and Henry James 106-122
    and struggle over Irish geology 107-122
    and De La Beche 110-115, 119-122, 126
    and Richard Griffith 115-118
    see also 129, 130, 136, 138, 162, 168
Colby, Mrs T., 86
Cole, Grenville A.J., 215
Coleraine, Co. Londonderry, 60, 223
Colthurst Family, 38
Connaught Coal District, 41-43, 45, 72, 118, 207
Connemara, Co. Galway, 46, 60, 132, 184, 190, 200, 218, 224
Conybeare, William D., 28, 29, 61, 77, 91
Cook, James, 95
Copley Medal, 17
Cork Beds, 182, 183-184
Cork City, 47, 73, 168, 212, 215
Cork, Co., Smith's map of, 3, 4
Cork, Co., Townsend's map of, 15-17, 28, 70
Corps Royal des Mines, 95
Courtown, Co. Wexford, 142
Creeslough, Co. Donegal, 224, 228
Crimean War, 109, 172
Croll, James, 189
Cronbane, Co. Wicklow, 24
Crossfarnoge Point, Co. Wexford, 165
Crotta House, Co. Kerry, 195
Cruise, Richard J., 167, 189, 203, 205, 212, 214, 218, 220, 221, 223, 225, 228
Cruise's Hotel, Limerick, 168
Crystal Palace Exhibition, 57, 167
Cuilcagh Plateau Country, 65, 66, 204
Custom House, Dublin, 114, 129, 133, 143, 146, 159

Dalkey, Co. Dublin, 150
Darwin, Charles, 156, 157, 181-182
Davis, William M., 178
Davy, Edmund, 57
Davy, Humphry, 35
Dead Sea, 195
De Dunstanville, Lord, 35, 36
De La Beche:
    his character 111
    plans an Irish survey 107-109
    struggle over Irish geology 107-122, 126
    and Thomas Colby 110-115, 126
    use of John Phillips 113-115
    and Sir Robert Peel 120-121
    defeats Colby 119-122
    and Henry James 107, 138
    work in Ireland 131, 132, 134, 150, 175, 183
    his *Instructions* 127, 142-143
    the base-map problem 136-137, 149-150
    and Thomas Oldham 138-141, 152
    and railway geology 148-149
    appoints J.B. Jukes to Survey 157
    sends Jukes to Ireland 158-159
    dispute with Jukes 161-162
    and Survey of Scotland 163-164
    his death 168

see also 49, 60, 81, 98, 100, 101, 102, 104, 105, 107, 127, 129, 130, 131, 133, 136, 137, 138, 142, 143, 144, 148, 149, 151, 152, 170, 171, 172, 191, 196, 197, 200, 207, 222, 226
Department of Science and Art, 160, 167, 222, 223, 226, 227-228
Desmarest, Nicolas, 4, 30
Devonian System, 183-184
Dingle Beds, *see* Dingle Group
Dingle Group, 183, 208-216
Dingle Peninsula, 70, 74, 80, 130, 164, 169-170, 182-183, 209, 210, 211, 212
Diocesan School, Londonderry, 13
Dollands, opticians, 112
Donegal, Co., mapping of, 222, 223, 224, 225
Donnybrook, Co. Dublin, 11
Doran, Patrick, 100
Down, Co., early map of, 8
Downshire, Marquis of, 108
Doyle, John, 172-173
Drayton Manor, Staffordshire, 120
Drift deposits, representation of, 8-9, 11, 14-15, 20-21, 53, 69, 99, 134, 142, 150, 151, 172-174
Dripsey Castle, Co. Cork, 38
Drogheda Grammar School, 202
Dromahair, Co. Leitrim, 72
Drumachose Parish, Co. Londonderry, 91
Drumkeeran, Co. Leitrim, 65
Drummond, Thomas, 62
Drury, Susanna, 1, 4, 30, 225, 229
Dublin Castle, 147
Dublin, Co., geological map of, 160, 161
Dublin Society, *see* Royal Dublin Society
Dubourdieu, John, 8
Duffin, William E.L'E., 202
Duncannon, Co. Wexford, 134
Dundalk, Co. Louth, 28, 206
Dungannon, Co. Tyrone, 72, 106
Dungarvan, Co. Waterford, 3, 159, 178
Dunlavin, Co. Wicklow, 147, 148
Dunlewy, Co. Donegal, 224
Dunloe, Gap of, Co. Kerry, 170
Dunmanway, Co. Cork, 65
Du Noyer, George V., 100, 101, 102, 130, 141, 148-149, 152, 160, 164, 165-167, 168, 169, 173, 177, 178, 179, 181, 183, 184, 185, 187, 188, 189, 190, 216, 228
Dunquin, Co. Kerry, 70, 169, 183, 212
Dutton, Hely, 8

East India Company, 152
East Indies, 156
Eccles, Thomas, 164, 212
Edgeworth, Maria, 20
Edgeworth, Richard L., 20
Edgeworth, William, 41
Edgesworthstown, Co. Longford, 196
Edinburgh, University of, 36, 53
Edwards, Lewis, 129
Egan, Frederick W., 189, 190, 200, 203, 205, 207, 223, 225, 228
Engineering geology map, 103-4
Enniscorthy, Co. Wexford, 134, 136, 218
Enniskillen, Co. Fermanagh, 49
Enniskillen, Earl of, 81
Errigal Parish, Co. Londonderry, 13
Eton College, 10
Examinations, 6, 188, 202

Exposition Universelle, Paris, 74, 171

Fair Head, Co. Antrim, 4
Fanad Peninsula, Co. Donegal, 220-221, 225
Fenian troubles, 185-186, 204
Fenwick, Lt., 90-91, 94, 98
Fermoy, Co. Cork, 213, 214
Ferriters Cove, Co. Kerry, 70, 130, 164, 169
Fethard, Co. Wexford, 134, 135
Fintona, Co. Tyrone, 221
Fintown, Co. Donegal, 204, 224, 228
Fitton, William H., 19, 39, 44, 90
Fitzgerald, Thomas J., 7
Fitzwilliam Place, Dublin, 57, 60, 81, 82, 212
Flanagan, James, 100, 129, 130, 133, 134, 147, 148, 152, 160, 163, 164, 167, 168, 186, 202
Fluvialism, 178, 218
Fly, H.M.S., 156, 157, 183
Foot, Frederick J., 165, 168, 173, 184, 185, 186, 189, 190, 195, 205
Forbes, Edward, 78, 114, 115, 119, 122, 131, 133-134, 137, 138, 140, 147, 163, 169, 183, 187
Forsters, printers, 59
Foynes, Co. Limerick, 66
Fraser, Robert, 8-9, 11, 13, 14, 15, 19, 21, 28, 39, 101
Freiberg School of Mines, 18, 24, 36, 51, 88, 89

Galvan, Charles, 185-186
Galway Bay, 205, 217
Galway City, 227
Gandon, James, 129
Ganly, Patrick:
   as a field-geologist   65-67, 71, 79-80
   revision of Griffith's map   72-76, 134, 136, 161, 182-183, 210-211, 212, 213
   and cross-bedding   59
   relationship with Richard Griffith 72, 77, 78-80
   his work used by Griffith   79-81
   his degree   77
   the *Ganly Papers*   65, 71, 81
   see also 59, 69, 74, 115, 132, 134, 143, 175, 184, 209, 210, 211, 212, 213, 214, 217, 227
Gardner, James, 63, 67, 69
Garinish Island, Co. Cork, 164
Geikie, Archibald:
   becomes Director General   200, 215
   visits to Ireland   200, 205, 227
   studies in Irish geology   200, 202, 227
   and the Munster revision   213-215
   and G.H. Kinahan   217-220, 222
   on Slieve League   225
   reduces Irish Survey   227-228
   his confidential report   202, 203, 222, 228
   last visit to Edward Hull   228
   see also 58, 59, 78, 79, 113, 188, 189, 190, 197, 200, 201, 215, 217, 218, 219, 220, 222, 223-224, 225, 227
Geikie, James, 189, 197
Genesis, Book of, 59, 196
Geneva Colliery, Co. Kilkenny, 182
Geognostical maps, 28-29, 43, 51, 63
Geological Maps,
   character of   1, 3-4, 12, 28, 30
   methods of constructing   11, 44

Index

need for 5-6, 6-7, 18, 21-22, 23, 24, 28, 39, 40
temporal dimension of, 11, 27, 28, 43
see also Geognostical maps and Petrographic maps
Geological map of Ireland, early schemes for, 4, 5-6, 18, 21, 22
Geological maps of Irish counties:
    Carlow 151, 160
    Cork 15-17, 28, 70
    Dublin 160, 161
    Kildare 151, 160
    Kilkenny 10-13, 27, 28
    Londonderry 13-15, 105-106
    Waterford 161
    Wexford 160, 161
    Wicklow 8-9, 149-150, 151, 160
Geological sections, 23, 24, 25, 41, 43, 50, 61, 65, 66, 70, 75, 90-91, 101, 131, 147, 148-149, 150, 179, 207, 226
Geological Society of Dublin, *see* Royal Geological Society of Ireland
Geological Society of London, 13, 27-29, 36, 38, 39, 40, 43, 51, 60, 64, 67, 77, 78, 87, 88, 90, 91, 96, 97, 98, 118, 120, 178, 184, 209-210, 212, 215-216, 223
Geological Survey of Ireland:
    founded 121-122
    under Henry James 126-138
    recruits staff 127-131
    uniforms 112, 131-132
    and Richard Griffith 133, 134, 136
    James's programme 132-137
    field-work starts 133-137
    financial problems 137
    James resigns 137-138
    under Thomas Oldham 138-153
    quality of early work 142-143, 160
    move to 'The Green' 143
    collects museum specimens 143, 146-147, 150, 151-152, 162-163
    railway sections 148-149
    first publications 149-151
    sales of early maps 151, 160
    base-map problems 136-137
    Oldham resigns 152-153
    under J.B. Jukes 156-191
    more county maps 160-161
    further base-map problems 160-161
    one-inch sheets 161-162, 170-177, 205, 206, 207, 222, 226, 228
    drift problems 172-173
    six-inch field-sheets 132, 161, 162, 165-167
    Jukes's staff 160, 165, 167-168, 173
    rules 165
    expansion of 1850s 167-168
    sheet memoirs 177-179, 191, 198, 207, 226
    sections 147, 150, 207, 226
    expansion of 1867 188-190, 198-199, 202
    Dingle rocks 183
    Cork Beds 183
    Fenian troubles 185-186
    Jukes's illness 190-191
    under Edward Hull 195-229
    microscope petrology 198, 221
    move to Hume Street 198-200
    Hull's staff 200, 202-203
    buys tent and hut 204
    methods 205-206
    Land War 204-205
    six-inch sheets published 161, 207, 223
    at Slieve Gullion 207
    Munster revision 208-216, 218, 221
    Irish Laurentian rocks 224-225
    in Donegal 222-225
    visits to Scotland 223-224, 226-227
    mapping completed 222-229
    retirements 228-229
    salaries 127, 158, 185, 192, 197-198, 203
    promotion 203
    ill-health 185, 205
    disputes within 131, 141, 142, 143, 151-152, 186-190, 212-213, 216-222
    problems of 137, 146, 147, 148, 163, 164, 165, 181-184, 184-186, 203-205
    progress of 136, 147, 148, 151, 163, 170-171, 174, 179, 180, 205, 207-208, 222, 225-226
Geological Survey of Northern Ireland, 232
Geological Survey of Scotland, 122, 163-164, 189, 197, 200, 217, 219, 223-224, 227, 228
Geology, Irish loss of interest in, 46, 48, 51, 54, 57
Geology, popularity of, 1, 5, 100
Geology, practical and applied, 5, 18, 47, 49
Geology, submarine, 53
Geology as a 'military' campaign, 109-110, 134
Geomorphology, American School of, 178
Giant's Causeway, Co. Antrim, 1, 4, 225
Giant's Causeway Portrush and Bush Valley Railway and Tramway, 202
Giesecke, Charles L., 49, 51, 56
Gilbert, Grove K., 178
Gilligan's Cove, Co. Wexford, 167
Glacial Erosion Theory, 200
Glacial Submergence Theory, 181-182
Gladstone, William E., 188
Glasgow, University of, 8
Glasnevin, Dublin, 191
Glen, Co. Donegal, 224
Glenamoy, Co. Mayo, 204, 205
Glenarm, Co. Antrim, 91
Glenbride, Co. Wicklow, 148
Glengarriff, Co. Cork, 164, 209, 212
'Glengarriff Grits', 209-216
Glenmalur, Co. Wicklow, 150
Goggin, Cornelius, 57
Gold Mines Valley, Co. Wicklow, 5, 17
Gorey, Co. Wexford, 217
Goulburn, Edward, 115
Grafton Street, Dublin, 151, 172
Graisberry and Campbell, printers, 25
Grand Canal, 7, 23, 138, 148
Great Barrier Reef, 157
Great Dublin Industrial Exhibition, 57, 75, 226
Great Famine, the, 126, 129, 147, 148
Great Southern Hotel, Killarney, 204
Greenly, Edward, 148

Greenough, George B., 23, 38-39, 40, 45, 46, 60, 61, 65, 66, 71, 117
Griffith, Richard:
early years	22, 35-37
joins Dublin Society	22, 37
and Bog Commission	20-21, 23-24, 25, 37, 39, 54
and Leinster Coal District	22-27, 28, 37, 39, 48
as Mining Engineer	24, 35, 37, 38, 44, 51
idea of a map	38-40
reasons for making map	23, 40, 48, 49, 59
first displays his map	35, 39-40
other mining district surveys	40-44, 47, 50, 175
and the base-map problem	41, 43, 44-46, 47, 49, 60, 61, 62
triangulation of Ireland	45-46, 47, 81
section of northern Ireland	43, 50
in private practice	44, 47
Engineer of Public Works	46-47, 57, 58
Director of Boundary Survey	47, 49, 58, 65, 79, 85, 116
criticism at Royal Dublin Society	48-51
and the Ordnance Survey	48-50, 91-92, 102, 115-118
resigns as Mining Engineer	35, 50-51, 57, 58
Commissioner of Valuation	50-51, 57-61, 79, 80, 116
and Geological Society of Dublin	57-58, 80, 85,
and British Association	58, 60-61, 65, 66, 67, 70, 73, 75, 77, 78, 80, 81
map displayed 1835	61, 66, 70
his unofficial survey	58-81
and Railway Commission	61-71, 133
and palaeontology	43, 66, 116
progress of the map	44, 45, 46, 47-48, 50-51, 54, 58-59, 60-61, 62, 64-81, 84, 133
problems in Munster geology	64, 69-70, 210-211, 212, 213
relationship with Patrick Ganly	65-67, 72-77, 79-80, 81, 136
authorship of his map	77-81
tries to become the official survey	82, 115-118
and the Geological Survey	133, 134, 136
and Wollaston Medal	78
death	81
auction of property	81-82
see also	88, 89, 91, 95, 100, 101, 102, 127, 132, 133, 134, 139, 143, 149, 150, 199, 200, 208, 209, 210, 217, 224, 228

Grubb, Thomas, 57

Hachured one-inch maps, 206-207
Hacket, Mr, 2
Hall, Sir James, 36
Hamilton, Charles W., 67, 80, 117, 118, 119, 182, 183
Hamilton, William, 4-5, 18
Hamilton, William R., 57
Hampshire, H.M.S., 195
Hampstead House, Glasnevin, 191, 197
Harkness, Robert, 159, 160, 223
Harding, Thomas, 18
Hardinge, Henry, 87
Hardman, Edward T., 202, 205, 208, 223, 225
Harris, Walter, 2, 3
Harz Mountains, 6
Haughton, Samuel, 77, 158, 159, 179, 181, 192
Hawaii, 95
Hawkins, William, 35
Hawkins Street, Dublin, 20, 35, 39
Hendersyde Park, Kelso, 37
Henfrey, George, 137
Henry Mountains, 178
Higgins, William, 6
Hodges and Smith, Dublin, 67, 69, 151, 172, 174, 190
Holywood, Co. Down, 191
Hook Head, Co. Wexford, 134
Horne, John, 189
Hoskins, Pierce, 168, 185
Houghton, Edward, 37
Hull, Edward:
early years	196-197
joins Geological Survey	196
his poor work	197
joins Scottish Survey	197
takes over Irish Survey	197-198
as an author	198, 208
his geological style	198, 208-209
and microscope petrology	198, 221
in Middle East	195, 197, 218, 222
his innovations	198, 206-207
his map of Ireland	208, 209
his character	195-198
retires	228-229
Munster revision	208-216, 218, 221
mapping of Donegal	222, 223, 224, 225, 226
Laurentian rocks	224-225
conflict with G.H. Kinahan	197, 212-213, 216-222

Hume Street, Dublin, 199, 200, 203, 209, 214, 215, 216, 217, 218, 220, 223, 228
Huron, Lake, 95
Hutton, James, 5, 17, 36
Huxley, Thomas H., 157
Hyland, John S., 203, 227, 228

Iar-Connaught, Co, Galway, 224
Ill-health of surveyors, 205
India, 100, 139, 152-153, 158, 167, 168, 186, 187, 203, 218
Inishowen Peninsula, Co. Donegal, 221
Inishtrahull, Co. Donegal, 225
Inistioge, Co. Kilkenny, 10
Irish Civil War, 10
Irish Fishery Board, 53
Irish Insurrection of 1798, 19
Irish Mining Board, 17-18
Island and Coast Society of Ireland, 198
Island Magee, Co. Antrim, 94
Iveragh Peninsula, Co. Kerry, 209, 213

James, Henry:
early years	107
joins J.E. Portlock	102, 104, 107
joins H.T. De La Beche	104, 107, 143
takes over geology	106-122

struggle over geology	107-122
a geological dilettante	112, 119-129, 137
as Irish Local Director	126-138
his duties	127
mapping programme	132-137
financial problems	137
resignation	137-138

see also 74, 140, 141, 142, 143, 146, 153, 171, 177, 207, 216, 222

James, Trevor E., 130-131, 134, 216
Jameson, Robert, 36, 64, 131, 139
Java, 157
Jenkins, Lionel, 2-3
Jones, John F., 81
Joyce, Patrick W., 202
Jukes, Joseph B.:

early years	156-158
joins Survey	157
appointed to Ireland	158
dislike of Ireland	158-159, 185, 186
his staff	160, 167-168
base-map problems	160-162
belief in six-inch mapping	158, 161-162, 164, 165-167
his field programme	162-165, 170-171
and one-inch maps	171-177, 179, 180, 191
and sheet memoirs	177-180, 191
and rivers of southern Ireland	178
and classification of the Carboniferous	182
and the Dingle rocks	182-183
and the Cork Beds	183
and the Devonian System	183-184
relations with staff	186-190, 218
and expansion of 1867	188-190
his geological map	190
his illness	190-191
his death	191
his character	157-158
his conception of the Survey	179-180

see also 111, 195, 196, 197, 198, 199, 200, 202, 203, 204, 205, 206, 207, 208, 209, 210, 215, 216, 217, 218, 221, 222, 223, 226, 227

Jukes, Mrs J.B., 158, 164, 184, 186, 190, 191
Jukes-Browne, Alfred J., 157

Kane, Robert, 114, 120, 121, 129, 138, 140, 143, 146, 150, 151-152, 163
Kanturk, Co. Cork, 44, 175
Kells, Book of, 195
Kelly, John, 26, 41, 43, 46, 50, 59, 61, 65, 69, 72, 78, 79, 80, 81, 168, 175, 186, 187
Kenmare, Co. Kerry, 157, 171, 190, 209, 215
Kenmare River, 210, 211
Kennedy, John S., 168
Kilbrew, Co. Meath, 24
Kilflynn, Co. Kerry, 195
Kilkenny City, 168
Kilkenny, Co., Tighe's map of, 10-13, 27, 28
Killala Bay, 204
Killaloe, Co. Clare, 178, 187
Killarney, Co. Kerry, 44, 117, 170, 186, 204, 209, 214, 215
Killarney, Lakes of, 167, 170
Killary Harbour, 72, 168, 209, 227
Killiney, Co. Dublin, 221
Kilmore Quay, Co. Wexford, 165

Kilroe, James, 202, 203, 225, 226, 227, 228
Kiltorcan, Co. Kilkenny, 163, 186
Kimberley Goldfield, 225
Kinahan, George H.:

joins Survey	164
conflict with J.B. Jukes	189, 190
passed over for Directorship	197, 218-219
and microscope petrology	221
conflict with Edward Hull	197, 212-213, 216-222
sent to Donegal	223
completes survey in Donegal	225-226
retires	228-229
his character	197, 202, 203, 217-218
his geology	181-182, 217, 218, 219-220
and guns	204-205

see also 27, 167, 168, 176, 185, 186, 187, 188, 192, 202, 203, 204, 205, 208, 223, 225, 227

Kinahan, Mrs G.H., 216, 219
Kinahan, Gerrard, 220
Kinahan, Katherine, 204
King, William, 159
King's College London, 113
Kingscourt, Co. Cavan, 205
Kinsale, Co. Cork, 168
Kirwan, Richard, 5, 6, 7, 17-18
Kirwanian Chemical and Natural Historical Society of Dublin, 18
Kitchener, Earl, 195, 198, 204
Knight, Patrick, 59, 77, 80
Knockmahon, Co. Waterford, 70, 79, 136, 147

Lambay Island, Co. Dublin, 198
Lancey, Lt., 90-91, 94
Land War, 204-205
Lansdowne Arms, Kenmare, 190
Lapworth, Charles, 224, 227
Larcom, Thomas A., 62, 63, 64, 66, 67, 97, 98, 99, 102, 103, 104, 105, 108, 109, 113, 114, 115, 117, 120, 132
Larne, Co. Antrim, 207
Laurentian rocks in Ireland, 224-225
Leane, Lough, Co. Kerry, 167, 170
Leinster, Central, Stephens's map of, 18-20
Leinster Coal District, 11, 20, 21-27, 37, 39, 40, 41, 42, 48, 81, 129, 131, 132, 134, 149, 175, 179, 182, 208, 223
Leinster House, Dublin, 61, 70, 231
Leinster Lawn, 57
Leinster Mountains, Griffith's bog map of, 21, 25
Leonard, Hugh, 189, 190, 202, 204, 205, 218
Leonard, William B., 189, 190, 202, 204, 205, 227
Leske, Nathanael G., 6
Leskean Mineral Collection, 6, 17, 35
Letterfrack, Co. Galway, 205
Letterkenny, Co. Donegal, 220, 222, 223, 228
Lifford, Co. Donegal, 49, 119
Limerick City, 168, 169
Lincoln, Lord, 121
Lisburn, Co. Antrim, 129
Lismore, Co. Waterford, 3, 213, 214
Lister, Martin, 1
Lloyd, Humphrey, 57, 81
Londonderry, City of, 4, 5, 13, 28, 99-100, 102, 119
Londonderry, Co., Berger in, 27-28
Londonderry, Co., Ordnance Survey work in, 89-

Sheets of Many Colours

92, 93, 97, 102, 104-106, 107, 129
Londonderry, Co., geological report on, 104-106
Londonderry, Co., Sampson's maps of, 13-15, 28
Londonderry Grand Jury, 15
Listowel-Killorglin Lowland, 195
Longland, Joseph, 119
Lonsdale, William, 70
Loughbrickland, Co. Down, 112
Loughrea, Co. Galway, 184
Loughs, Irish:
    Allen   41, 119
    Carlingford   107
    Conn   21
    Corrib   191
    Derg   202, 224
    Foyle   13, 89
    Key   190, 195
    Leane   167
    Mask   205
    Shin   223
    Swilly   5, 200, 223
Loveday, Richard J., 119
Lowry, Wilson, 25
Lucan, Co. Dublin, 196
Lyell, Charles, 58, 61, 81
Lyons, James A., 119

MacAdam, James, 78
McCarthy, Bucknal, 20
McCoy, Frederick, 66, 116, 127-128, 133, 134, 136, 139, 141, 142, 144, 147, 149, 151, 216
Macculloch, John, 44-45, 46, 71, 95
McHenry, Alexander, 167, 190, 202, 203, 205, 209, 210, 211, 212, 213, 214, 215, 218, 220, 223, 225, 227, 228, 231
McInnes, Daniel, 119
McLauchlan, Henry, 98
MacNeill, John, 140
Macroom, Co. Cork, 67
Maghera, Co. Down, 202
Magilligan's Strand, Co. Londonderry, 89, 91
Malin Head, Co. Donegal, 175, 232
Malin More, Co. Donegal, 204
Mallet, Robert, 57
Mallow, Co. Cork, 47, 57, 64, 171
Manchester Geological Society, 221
Manorcunningham, Co. Donegal, 200
Manorhamilton, Co. Leitrim, 65
Marochetti, Baron, 57
Mathew, Father, 126
Maton, William G., 13
Mayo, Co., Griffith's bog map of, 20-21
Medlicott, Henry H., 168, 187
Medlicott, Joseph G., 152, 153, 160, 167, 168
Medlicott, Samuel, 168
Melbourne, 142
Meyler, Dr, 48-49
Microscope petrology, 198, 221
Middle East, 195, 197, 198, 218, 222
Military General Service Medal, 169
Millar, James, 56
Millicent, Co. Kildare, 35
Millstreet, Co. Cork, 171
Milners' Patent Fire-Resisting Safes, 191
Mining Records Office, London, 124
Mr Buff, dog, 167
Mitchell, William F., 202, 203, 225, 228
Mohs, Friedrich, 88

Monkstown, Co. Cork, 131, 159
Mooney, Denis, 173
Mountains, Irish:
    Benbo   42
    Benbulbin   43, 66
    Binevenagh   13
    Boggeragh   47, 70, 73
    Carrauntoohil   181
    Comeragh   52, 165, 166
    Croghan   8, 17, 53, 150
    Curlew   42, 221
    Derrynasaggart 70, 73
    Galty   15, 52
    Kilworth   15, 213
    Knockmealdown 15, 52, 213
    Monavullagh   52, 166
    Mount Eagle   169
    Mourne   200, 205, 207
    Nagles   213
    Ox   42, 224
    Sheehy   70
    Slieve Aughty   8
    Slieve Bernagh   8
    Slieve Bloom   52
    Slieve Gallion   13, 49
    Slieve Gullion   207
    Slieve League   96, 225
    Slievenamon   11
    Slieve Russell   43
    Stack's   195
    Sugar Loafs   8
    Twelve Pins   227
Mountjoy House, Dublin, 88, 91, 93, 94, 95, 97, 101, 103, 107, 108, 121
Muckross, Co. Kerry, 170
Mudge, William, 85
Mullingar, Co. Westmeath, 148
Munchausen, Baron, 170
Munster Coal District, 43-44, 45, 47, 48, 175, 182, 208
Munster geology, problems in, 64, 69-70, 73, 183, 208-216, 218, 221
Munster Plateau, 15, 16, 17, 47
Murchison, Roderick I., 58, 61, 63, 64, 70, 73, 74, 77, 80, 81, 98, 111, 118, 120, 130, 158, 168-170, 172, 173, 177, 178, 179, 181, 182, 183, 184, 185, 186, 187, 188, 189, 190, 196, 197, 198, 200, 207, 212, 218, 219, 224, 227
Murphy, Thomas, 129, 134, 137
Museum of Economic Geology, Craig's Court, London, 101, 109, 110, 121, 124, 143, 146
Museum of Economic Geology, Dublin, *see* Museum of Irish Industry
Museum of Irish Industry, 121, 143, 146, 159, 167, 168, 179, 181, 188, 192, 197, 202

National Library of Ireland, 52, 53
National Museum of Ireland, 160, 163
National Museum of Wales, 113
Nature, 224
Necker, Louis A., 36
Nelson, Lord, 156
Neptunian Theory, 4, 5, 17, 27, 36
Neville, Arthur R., 19
Newcastle, Co. Down, 43, 205
Newcastle West, Co. Limerick, 47
Newfoundland, 156
New Guinea, 157, 183

New Ross, Co. Wexford, 136
Nicholson, Samuel, 59, 78, 80
Nicholson, William, 35
Niger, River, 220
Nimmo, Alexander, 20, 52-53, 69
Nolan, Joseph, 167, 189, 203, 205, 207, 225, 227, 228
Norfolk Island, 157
Norreys, Denham, 104
North Cork Syncline, 213, 214
Northwest Highlands of Scotland, 223-224, 226-227

O'Connor Don, The, 108
O'Grady, Murtaugh, 28
O'Hara, Charles, 5
O'Kelly, Joseph, 168, 173, 176, 185, 188, 202, 203, 205, 209, 211, 212, 225
Oldham, Thomas:
early years	138-139
discovery of *Oldhamia*	140
appointed Local Director	140
conflict with Frederick McCoy	141, 142
as Local Director	141-153
relations with Robert Kane	140, 143, 146, 150, 151-152
collection of soils	146-147, 150, 151-152
completes sections	147, 150
railway sections	148-149
first publications	149-151
and base-map problem	151
resignation	152-153
later years	153

see also 100, 101, 102, 105, 113, 156, 158, 167, 168, 171, 177, 179, 185, 196, 207, 216, 222

Oldham, Mrs T., 152-153
Omdurman, battle of, 195
One-inch field-mapping, 102, 158, 163, 223, 226
One-inch geological maps, 99-100, 162, 170-177, 179-180, 191, 198, 205-208, 222, 226, 228
Ordnance Survey:
in Ireland	47, 85
six-inch mapping	86, 132
Richard Griffith's relations with	48-49
and J.W. Pringle	88-95
in Co. Londonderry	89-92, 93
Carmichael Smyth enquiry	93-95
and J.E. Portlock	95-106
the memoir project	97-104, 108-109
geology in England & Wales	98-99, 100, 101, 107
Templemore memoir	99-100
the 'geological office'	100-103
uniforms	112
the office closed	102-104
the Londonderry Report	104-106
and Henry James	106-122
struggle over Irish geology	107-122

see also 19, 45, 48, 58, 59, 61, 62, 63, 64, 65, 66, 67, 70, 85, 128, 129, 131, 133, 136-137, 138, 139, 143, 149, 150, 151, 160, 162, 163, 165, 172, 173, 174, 175, 179, 206, 207, 214

Owen, Richard, 120
One-inch relief model, 231

Paardeberg, battle of, 195
Palestine Exploration Fund, 195
Parliament, British, 10, 210

Parliament, Irish, 5, 7, 10, 18
Parliament House, Dublin, 2
Patrickson, Major, 107
Peach, Benjamin N., 189, 227
Peel, Robert, 108, 110, 120-122, 126
Penny, James, 129, 136, 152
Perceval, Robert, 36
Petrographic maps, 13, 15, 88, 97, 119
Pettigo, Co. Donegal, 49, 202
Phillips, John, 61, 64, 73, 74, 99, 107, 109, 110, 113-115, 117-118, 122, 133, 134, 136, 140, 182
Phlogiston Theory, 17
Phoenix Park, Dublin, 62, 88, 156, 204
Phoenix Park Murders, 204
Physico-Historical Society of Ireland, 1-4, 7
Pickersgill, Henry W., 169
Playfair, John, 36
Plot, Robert, 1
Plutonic Theory, 5, 6, 36
Pomeroy, Co. Tyrone, 70
Portarlington, Co. Leix, 35
Porter, John S., 129
Portlock, Joseph E.:
early years	95-96
as a geologist	96-97
the memoir project	97-104
Templemore memoir	99-100, 103
his 'geological office'	100-103
and the base-map problem	101-102
compared to H.T. De La Beche	102
his office closed	102-104
relations with Thomas Colby	103-106
the Londonderry Report	104-106
sent to Corfu	105
his later years	106
his library	198, 228, 229

see also 70, 94, 107, 108, 109, 112, 114, 118, 122, 129, 130, 132, 136, 139, 140, 141, 143, 147, 148, 149, 179, 225

Portlock Harbour, 95
Portlock Point, 95
Portrush, Co. Antrim, 4
Potato blight, 126
Preston, William, 5
Pringle, John W., 54, 88-95, 101, 109, 132
Prior, Thomas, 2

Queen's College Belfast, 142, 146, 147, 159
Queen's College Cork, 146, 147, 159, 223
Queen's College Galway, 127, 146, 147, 159
Queenstown, Co. Cork, 75
Questionnaire surveys, 2, 6-7

Raglan Road, Dublin, 209
Railway Commission, The, 61-71, 133
Railways, Irish:
Dublin and Belfast Junction	149
Dublin and Drogheda	148
Dublin and Kingstown	126
Great Southern and Western	148-9
Midland Great Western	148

Ramsay, Andrew C., 111, 113, 121, 122, 127, 133, 134, 137, 138, 140, 141, 157, 158, 159, 160, 161, 163, 164, 170, 184, 187, 188, 189, 197, 200, 201, 210, 213, 215, 217, 220, 222, 223, 226
Raspe, Rudolf E., 170
Rathangan, Co. Kildare, 35
Rathdrum, Co. Wicklow, 163

Rathlin Island, Co. Antrim, 225
Rathmelton, Co. Donegal, 223, 226
Rathmines, Dublin, 57, 188
Rathnew, Co. Wicklow, 10
Rattlesnake, H.M.S., 157
Recess, Co. Galway, 205, 218, 227, 229
Reeks, Trenham, 187, 189, 200
Richardson, William, 5
Richmond National Institution, Dublin, 85
Rimington, Lt., 95
Rivers, Irish:

Avonmore	163
Bandon	178
Bann	13
Barrow	25, 27
Blackwater	178, 182
Boyne	149
Gweebarra	224
Kings	148
Lagan	28
Liffey	35, 129
Nore	10, 25
Roe	13
Shannon	178
Slaney	178
Suck	20
Suir	12, 44

Robe, Lt., 94
Rocque, John, 19
Rosanna, Co. Wicklow, 10
Roscommon Grand Jury, 41, 45
Rosguil Peninsula, Co. Donegal, 226
Ross Bay, Co. Kerry, 167
Rosscarbery, Co. Cork, 15
Roundstone, Co. Galway, 205
Roundwood, Co. Wicklow, 147
Royal College of Science for Ireland, 197, 198, 199, 202, 229, *see also* Museum of Irish Industry
Royal Dublin Society:

and Susanna Drury	1
and applied geology	6-17
and mineralogy	6, 7
the Leskean Collection	6, 35
and Donald Stewart	7
and county statistical surveys	7-17, 22
and early geological maps	7-17, 28, 30
Fraser's Wicklow	8-9
Tighe's Kilkenny	10-13
Sampson's Londonderry	13-15
Townsend's Cork	15-17
and Kirwan	17-18
and a Mining Board	17-18
and the Bog Commissioners	20-21, 39
and Leinster Coal District	21-27, 28, 37, 39
proposed geological map of Dublin	22
Griffith as Mining Engineer	24-27, 28, 35, 37, 38, 51, 57
and Griffith's map	38-40, 44, 47-51, 50-51
other mining district surveys	40-44, 47, 50
Griffith's northern Ireland section	43, 50
Griffith's triangulation	45-46
criticism of Griffith	48-51
and Giesecke	49, 51
Griffith's resignation	35, 50-51, 58
Great Dublin Industrial Exhibition	57, 226

see also 2, 35, 53, 54, 57, 60, 61, 81, 83, 91-92, 127, 148, 149, 168
Royal Engineers, 95, 107, 109, 121, 126, 204
Royal Geological Society of Ireland, 44, 57-58, 59, 60, 62, 66, 67, 72, 73, 75, 77, 79, 80, 85, 87, 97, 100, 104, 112, 117, 118, 119, 127, 129, 138, 140, 141, 142, 147, 148, 150, 151, 153, 159, 161, 162, 172, 174, 178, 184, 198, 216, 221, 224
'Royal Hammerers', 132
Royal Irish Academy, 5, 6, 17, 53, 65, 77, 108, 140, 150, 151, 174, 204, 216, 221
Royal Irish Constabulary, 185, 186, 204
Royal Irish Regiment of Artillery, 35
Royal Military Academy, Woolwich, 85, 107
Royal Military College, Marlow, 110
Royal Observatory, Greenwich, 180
Royal Regiment of Artillery, 35, 89
Royal Sappers and Miners, 112, 114, 119
Royal School of Mines, London, 124
Royal Society of Arts, 21-22, 27, 40
Royal Society of Edinburgh, 36
Royal Society of London, 1, 17, 81, 86, 120, 138, 140, 170, 221
Rutland Square, Dublin, 114
Rutty, John, 2

St Andrews, University of, 53
St Columba, College of, 148
St Columb's Cathedral, Londonderry, 5
St John's College, Cambridge, 10, 156
St Stephen's Green, number 51, 143, 146, 152, 153, 158, 172, 187, 191, 196, 198, 199
Sallins, Co. Kildare, 35
Salter, John W., 169, 183, 187
Sampson, George V., 8, 13-15, 19, 28, 39
Science and Art Museum, Dublin, 231
Scientific Establishment for Pupils, London, 35
Scotland, Parochial Survey of, 7
Scouler, John, 83, 192
Sedgwick, Adam, 58, 61, 70, 74, 81, 111, 115, 117, 118, 142, 156, 158
Selwyn, Alfred, 158
Seymour, Lord Webb, 36
Shannon Improvement Commissioners, 79
Shaw, Mrs G.B., 17
Shearer, John, 119
Shepheard's Hotel, Cairo, 195
Silurian rocks in Ireland, 70, 73, 74
Sinclair, John, 7
Singapore, 157
Six-inch geological mapping, 65, 66, 72, 81, 101, 102, 115, 117, 119, 132, 133, 135, 143, 144, 145, 149, 158, 161, 162, 163-164, 165, 171, 178, 179, 181, 191, 205-207, 223, 224, 226, 227
Six-mile Cross, Co. Tyrone, 106
Slane, Co. Meath, 24
Slieveardagh Coalfield, *see* Leinster Coal District
Slieveardagh Plateau, 11, 52
Slieve Gullion Ring Complex, Co. Armagh, 207
Slieve League Peninsula, Co. Donegal, 204, 226
Slievenamuck Fault, Co. Tipperary, 179
Sligo Town, 200, 223
Smith, Charles, 2, 3, 4, 5, 16
Smith, James, of Jordanhill, 61
Smith, William, 13, 21, 40, 61, 67, 77, 80, 113
Smyth, James Carmichael, 93-95, 100, 102

Smyth, Warington W., 131, 134, 137, 138, 147, 150, 151, 175
Sneem, Co. Kerry, 46, 209, 211, 213
Society for the Encouragement of Arts, Manufactures and Commerce, *see* Royal Society of Arts
Soil samples, collection of, 146-147, 150, 151-152, 162-163
South Cork Regiment of Militia, 168
Southwell, Lord, 2
Sowerby, James de C., 70
Spencer, Earl, 215
Stackallen House, Co. Meath, 148
Stanford, Edward, 190, 208
Stephens, Walter, 18-19, 22, 39, 44
Stewart, Donald, 7, 8, 148
Stewart, Robert, 57
Stokes, Whitley, 18
Stotherd, Lt., 98
Strabane, Co. Tyrone, 119
Strickland, Hugh E., 148
Sullivan, William K., 163, 192
Swinford, Co. Mayo, 205
Symes, Richard G., 184, 188, 202, 205, 221, 223, 224, 225, 228

Taghmon, Co. Wexford, 144, 145
Tarbert, Co. Kerry, 169
Tate, Ralph, 160
Taylor, Alexander, 19, 39, 40, 45, 52, 53, 79
Taylor, J., 11, 42
Telford, Thomas, 53
Templemore, Co. Tipperary, 186
Templemore Parish, Co. Londonderry, 13, 99-100, 102, 103, 105, 108
Thackeray, William M., 168
The Devil's Glen, Co. Wicklow, 218
The Geological Magazine, 187, 198, 219, 221
The Scalp, Co. Wicklow, 218
The Tor Rocks, Co. Donegal, 225
Thomastown, Co. Kilkenny, 11
Thomson, Wyville, 160
Thurles, Co. Tipperary, 186
Tighe, Mary, 10
Tighe, William, 8, 10-13, 15, 27, 28-29, 39
Tigroney, Co. Wicklow, 24
Toe Head, Co. Cork, 175
Tommy, dog, 164
Toormakeady, Co. Mayo, 205
Tory Island, Co. Donegal, 225
Tower of London, 89
Townsend, Horatio, 8, 15-17, 39, 70
Traill, William A., 189, 200, 202, 203, 205, 207
Tralee, Co. Kerry, 169, 186, 195
Trinity College Dublin, 2, 4, 5, 13, 15, 18, 19, 35, 36, 57, 61, 77, 113, 114, 115, 138, 140, 158, 159, 164, 168, 181, 189, 195, 196, 202, 226
Troughton, Edward, 43, 46

Ulster Coal Districts, 43, 44, 50, 207, 221, 223
University College Galway, 65, 82
University College Liverpool, 203
University College London, 156
Upper Leeson Street, Dublin, 191
Upper Merrion Street, Dublin, 199

Upper Sackville Street, Dublin, 85
Utah, 178

Vallancey, Charles, 6, 7, 10, 13, 20, 37
Valuation Survey, 50, 57-61, 65, 67, 72, 77, 79, 80, 82, 116, 117, 168, 185, 186
Vesuvius, 6
Victoria, Queen, 57, 75, 81, 126, 148, 169, 226
Vivarès, François, 1
Vivian, Richard H., 99
Volcanic Theory, 4, 5
Von Giesecke, Karl L.M., 33

Wadi Araba, 195, 204
Waldie, Maria Jane, 37
Warburton, Henry, 118
Warren, James L., 189, 190, 202, 205
Warrenpoint, Co. Down, 200
Waterford city, 3, 129, 136, 160
Waterford Harbour, 135
Waterloo, battle of, 46, 54, 88, 109, 212
Watson, White, 6, 25
Weather, problems of Irish, 47, 65, 163, 164-165, 169-170, 185, 203-204, 225
Weaver, Thomas, 24, 51-52, 56, 61, 64, 77, 182, 183
Webster, Letitia, 99
Webster, Thomas, 192
Weld, Isaac, 6, 10
Wellington, Duke of, 85, 199, 200
Werner, Abraham G., 51
West Africa, 220
Western Australia, 225
Westport, Co. Mayo, 205, 227
Wexford Town, 133, 142, 222
Whewell, William, 61, 86
White Ball Head, Co. Cork, 182
Whyte, Letitia, 112
Wicklow, Co., Fraser's map of, 8-9, 28
Wicklow Glens, 200
Wicklow Town, 8, 10
Wilkinson, Berdoe, 204
Wilkinson, Sidney B.N., 189, 204, 205, 224, 226, 227, 228
Williams, Mrs E., 229
Willson, Walter L., 129, 133, 134, 147, 148, 152, 160, 164, 165, 167, 168, 177, 186, 187
Wollaston Medal, 78
Woodenbridge, Co. Wicklow, 219, 220, 225
Woodstock House, Co. Kilkenny, 10
Woodward, Henry, 219
Woodwardian Museum, Cambridge, 142
Wright, Captain, 89, 94
Wright, John R., 98
Wright, Romley, 98
Wright, William B., 207, 215
Wyley, Andrew, 129, 137, 141, 142, 147, 152, 160, 164, 165, 167, 171, 177, 186, 187, 208
Wynne, Arthur B., 167, 173, 176, 177, 178, 183, 186, 203, 218, 220, 228

Young, John, 108, 109
Youghal, Co. Cork, 178, 212, 213

Zirkel, Ferdinand, 203
Zoological Gardens, Dublin, 61, 156